Entropy

PRINCETON SERIES IN APPLIED MATHEMATICS

EDITORS

Ingrid Daubechies, *Princeton University*
Weinan E, *Princeton University*
Jan Karel Lenstra, *Georgia Institute of Technology*
Endre Süli, *University of Oxford*

TITLES IN THE SERIES

Chaotic Transitions in Deterministic and Stochastic Dynamical Systems: Applications of Melnikov Processes in Engineering, Physics, and Neuroscience
Emil Simiu

Selfsimilar Processes
Paul Embrechts and Makoto Maejima

Self-Regularity: A New Paradigm for Primal-Dual Interior Point Algorithms
Jiming Peng, Cornelis Roos and Tamás Terlaky

Analytic Theory of Global Bifurcation: An Introduction
Boris Buffoni and John Toland

Entropy
Andreas Greven, Gerhard Keller and Gerald Warnecke, Editors

The Princeton Series in Applied Mathematics publishes high quality advanced texts and monographs in all areas of applied mathematics. Books include those of a theoretical and general nature as well as those dealing with the mathematics of specific applications areas and real-world situations.

Entropy

Andreas Greven

Gerhard Keller

Gerald Warnecke

Editors

PRINCETON UNIVERSITY PRESS

PRINCETON AND OXFORD

Copyright © 2003 by Princeton University Press
Published by Princeton University Press,
41 William Street, Princeton, New Jersey 08540

In the United Kingdom: Princeton University Press,
3 Market Place, Woodstock, Oxfordshire OX20 1SY

All rights reserved

Library of Congress Cataloguing-in-Publication Data

Entropy / edited by Andreas Greven, Gerhard Keller, and Gerald Warnecke.
  p. cm. – (Princeton series in applied mathematics)
  Includes bibliographical references and indexes.
  ISBN 0-691-11338-6 (acid-free paper)
  1. Entropy. I. Greven, Andreas, 1953– II. Keller, Gerhard, 1954– III. Warnecke, Gerald, 1956–
    IV. Series.

QC318.E57E55 2003
536′.73—dc21   2003051744

British Library Cataloguing-in-Publication Data is available

This book has been composed in Times and LucidaSans
Typeset by T&T Productions Ltd, London
Printed on acid-free paper ⊛

www.pupress.princeton.edu

Printed in the United States of America

10 9 8 7 6 5 4 3 2 1

Contents

Preface   xi

List of Contributors   xiii

Chapter 1. Introduction
A. Greven, G. Keller, G. Warnecke   1

1.1   Outline of the Book   4
1.2   Notations   14

PART 1. FUNDAMENTAL CONCEPTS   17

Chapter 2. Entropy: a Subtle Concept in Thermodynamics
I. Müller   19

2.1   Origin of Entropy in Thermodynamics   19
2.2   Mechanical Interpretation of Entropy in the Kinetic Theory of Gases   23
    2.2.1   Configurational Entropy   25
2.3   Entropy and Potential Energy of Gravitation   28
    2.3.1   Planetary Atmospheres   28
    2.3.2   Pfeffer Tube   29
2.4   Entropy and Intermolecular Energies   30
2.5   Entropy and Chemical Energies   32
2.6   Omissions   34
    References   35

Chapter 3. Probabilistic Aspects of Entropy
H.-O. Georgii   37

3.1   Entropy as a Measure of Uncertainty   37
3.2   Entropy as a Measure of Information   39
3.3   Relative Entropy as a Measure of Discrimination   40
3.4   Entropy Maximization under Constraints   43
3.5   Asymptotics Governed by Entropy   45
3.6   Entropy Density of Stationary Processes and Fields   48
    References   52

PART 2. ENTROPY IN THERMODYNAMICS — 55

Chapter 4. Phenomenological Thermodynamics and Entropy Principles
K. Hutter and Y. Wang — 57

- 4.1 Introduction — 57
- 4.2 A Simple Classification of Theories of Continuum Thermodynamics — 58
- 4.3 Comparison of Two Entropy Principles — 63
  - 4.3.1 Basic Equations — 63
  - 4.3.2 Generalized Coleman–Noll Evaluation of the Clausius–Duhem Inequality — 66
  - 4.3.3 Müller–Liu's Entropy Principle — 71
- 4.4 Concluding Remarks — 74
- References — 75

Chapter 5. Entropy in Nonequilibrium
I. Müller — 79

- 5.1 Thermodynamics of Irreversible Processes and Rational Thermodynamics for Viscous, Heat-Conducting Fluids — 79
- 5.2 Kinetic Theory of Gases, the Motivation for Extended Thermodynamics — 82
  - 5.2.1 A Remark on Temperature — 82
  - 5.2.2 Entropy Density and Entropy Flux — 83
  - 5.2.3 13-Moment Distribution. Maximization of Nonequilibrium Entropy — 83
  - 5.2.4 Balance Equations for Moments — 84
  - 5.2.5 Moment Equations for 13 Moments. Stationary Heat Conduction — 85
  - 5.2.6 Kinetic and Thermodynamic Temperatures — 87
  - 5.2.7 Moment Equations for 14 Moments. Minimum Entropy Production — 89
- 5.3 Extended Thermodynamics — 93
  - 5.3.1 Paradoxes — 93
  - 5.3.2 Formal Structure — 95
  - 5.3.3 Pulse Speeds — 98
  - 5.3.4 Light Scattering — 101
- 5.4 A Remark on Alternatives — 103
- References — 104

Chapter 6. Entropy for Hyperbolic Conservation Laws
C. M. Dafermos — 107

- 6.1 Introduction — 107
- 6.2 Isothermal Thermoelasticity — 108
- 6.3 Hyperbolic Systems of Conservation Laws — 110
- 6.4 Entropy — 113
- 6.5 Quenching of Oscillations — 117
- References — 119

Chapter 7. Irreversibility and the Second Law of Thermodynamics
J. Uffink   121

7.1	Three Concepts of (Ir)reversibility	121
7.2	Early Formulations of the Second Law	124
7.3	Planck	129
7.4	Gibbs	132
7.5	Carathéodory	133
7.6	Lieb and Yngvason	140
7.7	Discussion	143
	References	145

Chapter 8. The Entropy of Classical Thermodynamics
E. H. Lieb, J. Yngvason   147

8.1	A Guide to Entropy and the Second Law of Thermodynamics	148
8.2	Some Speculations and Open Problems	190
8.3	Some Remarks about Statistical Mechanics	192
	References	193

PART 3. ENTROPY IN STOCHASTIC PROCESSES   197

Chapter 9. Large Deviations and Entropy
S. R. S. Varadhan   199

9.1	Where Does Entropy Come From?	199
9.2	Sanov's Theorem	201
9.3	What about Markov Chains?	202
9.4	Gibbs Measures and Large Deviations	203
9.5	Ventcel–Freidlin Theory	205
9.6	Entropy and Large Deviations	206
9.7	Entropy and Analysis	209
9.8	Hydrodynamic Scaling: an Example	211
	References	214

Chapter 10. Relative Entropy for Random Motion in a Random Medium
F. den Hollander   215

10.1	Introduction		215
	10.1.1	Motivation	215
	10.1.2	A Branching Random Walk in a Random Environment	217
	10.1.3	Particle Densities and Growth Rates	217
	10.1.4	Interpretation of the Main Theorems	219
	10.1.5	Solution of the Variational Problems	220
	10.1.6	Phase Transitions	223
	10.1.7	Outline	224
10.2	Two Extensions		224
10.3	Conclusion		225

10.4	Appendix: Sketch of the Derivation of the Main Theorems	226
	10.4.1 Local Times of Random Walk	226
	10.4.2 Large Deviations and Growth Rates	228
	10.4.3 Relation between the Global and the Local Growth Rate	230
	References	231

Chapter 11. Metastability and Entropy
E. Olivieri   233

11.1	Introduction	233
11.2	van der Waals Theory	235
11.3	Curie–Weiss Theory	237
11.4	Comparison between Mean-Field and Short-Range Models	237
11.5	The 'Restricted Ensemble'	239
11.6	The Pathwise Approach	241
11.7	Stochastic Ising Model. Metastability and Nucleation	241
11.8	First-Exit Problem for General Markov Chains	244
11.9	The First Descent Tube of Trajectories	246
11.10	Concluding Remarks	248
	References	249

Chapter 12. Entropy Production in Driven Spatially Extended Systems
C. Maes   251

12.1	Introduction	251
12.2	Approach to Equilibrium	252
	12.2.1 Boltzmann Entropy	253
	12.2.2 Initial Conditions	254
12.3	Phenomenology of Steady-State Entropy Production	254
12.4	Multiplicity under Constraints	255
12.5	Gibbs Measures with an Involution	258
12.6	The Gibbs Hypothesis	261
	12.6.1 Pathspace Measure Construction	262
	12.6.2 Space-Time Equilibrium	262
12.7	Asymmetric Exclusion Processes	263
	12.7.1 MEP for ASEP	263
	12.7.2 LFT for ASEP	264
	References	266

Chapter 13. Entropy: a Dialogue
J. L. Lebowitz, C. Maes   269

References	275

PART 4. ENTROPY AND INFORMATION — 277

Chapter 14. Classical and Quantum Entropies: Dynamics and Information
F. Benatti — 279

- 14.1 Introduction — 279
- 14.2 Shannon and von Neumann Entropy — 280
  - 14.2.1 Coding for Classical Memoryless Sources — 281
  - 14.2.2 Coding for Quantum Memoryless Sources — 282
- 14.3 Kolmogorov–Sinai Entropy — 283
  - 14.3.1 KS Entropy and Classical Chaos — 285
  - 14.3.2 KS Entropy and Classical Coding — 285
  - 14.3.3 KS Entropy and Algorithmic Complexity — 286
- 14.4 Quantum Dynamical Entropies — 287
  - 14.4.1 Partitions of Unit and Decompositions of States — 290
  - 14.4.2 CNT Entropy: Decompositions of States — 290
  - 14.4.3 AF Entropy: Partitions of Unit — 292
- 14.5 Quantum Dynamical Entropies: Perspectives — 293
  - 14.5.1 Quantum Dynamical Entropies and Quantum Chaos — 295
  - 14.5.2 Dynamical Entropies and Quantum Information — 296
  - 14.5.3 Dynamical Entropies and Quantum Randomness — 296
- References — 296

Chapter 15. Complexity and Information in Data
J. Rissanen — 299

- 15.1 Introduction — 299
- 15.2 Basics of Coding — 301
- 15.3 Kolmogorov Sufficient Statistics — 303
- 15.4 Complexity — 306
- 15.5 Information — 308
- 15.6 Denoising with Wavelets — 311
- References — 312

Chapter 16. Entropy in Dynamical Systems
L.-S. Young — 313

- 16.1 Background — 313
  - 16.1.1 Dynamical Systems — 313
  - 16.1.2 Topological and Metric Entropies — 314
- 16.2 Summary — 316
- 16.3 Entropy, Lyapunov Exponents, and Dimension — 317
  - 16.3.1 Random Dynamical Systems — 321
- 16.4 Other Interpretations of Entropy — 322
  - 16.4.1 Entropy and Volume Growth — 322
  - 16.4.2 Growth of Periodic Points and Horseshoes — 323
  - 16.4.3 Large Deviations and Rates of Escape — 325
- References — 327

Chapter 17. Entropy in Ergodic Theory
M. Keane 329
    References 335

Combined References 337

Index 351

Preface

Imagine the following scene. A probabilist working in stochastic processes, a physicist working in statistical mechanics, an information theorist and a statistician meet and exchange their views about entropy. They are talking about the same notion, and, although they may be using distinct technical variants of it, they will probably understand each other. Someone working in the ergodic theory of dynamical systems may join those four. He will probably have to explain to the others precisely what he means by entropy, but the language he uses is still that of probability (except if he is talking about topological entropy, etc.). In any case, the five of them all see the historical roots of their notion of entropy in the work of Boltzmann and Gibbs. In another group, including people working in thermodynamics, on conservation laws, in rational mechanics and in fluid dynamics, the talk is about entropy from their perspective. However, because they probably see the common root of their concept of entropy in the work of Carnot, Clausius and later Boltzmann, the views they exchange are rather different.

Now imagine that both groups come together and each picks up parts of the discussion of the others: they will realize that entropy appears to be an important link between their respective fields of research. But very soon they realize that they have major difficulties in understanding each other—the gap between their notions of entropy seems too large, despite the closely related historical roots and their common interest in gaining a better understanding of complex systems.

This little scene is basically what happened during the meetings of the following three priority research programs sponsored by the German Science Foundation (DFG) since 1995:

Ergodic Theory, Analysis and Efficient Simulation of Dynamical Systems (DANSE);

Analysis and Numerics of Conservation Laws (ANumE);

Interacting Stochastic Systems of High Complexity.

A related program was launched at the same time by the Netherlands Organisation for Scientific Research (NWO):

European Institute for the Study of Randomness (EURANDOM).

Although all these programs were designed independently and each of them involves researchers from rather distinct fields, they have as a common theme the mathematical study of complex nonlinear systems and their applications. Therefore, it is hardly surprising that *entropy*, one of the key concepts in 20th-century science, devised in order to quantify various aspects of the complexity of systems, shows up at prominent places in the scientific work of all four programs. What did surprise us, however, was the large number of different meanings attributed to this word within the community of mathematicians and scientists from fields close to mathematics. And it was at a meeting of one of these programs that the idea for a symposium on entropy was born, a symposium which would present the concept and use of entropy in a variety of fields in mathematics and physics. Giving room to invited talks by leading specialists and leaving much space for discussions, the hope was that the meeting would help to clarify differences as well as common aspects of entropy as used in these various fields.

The symposium took place at the Max Planck Institute for the Physics of Complex Systems, Dresden (Germany), from 25 to 28 June 2000, with a program of invited talks and discussions. This ranged from expository introductions of the basic concepts to various surveys of current research in the areas of thermodynamics, continuum mechanics, stochastic processes, statistical physics, dynamical systems, ergodic theory and coding. During the meeting a general consensus emerged that it would be useful for the community of mathematicians and physicists to have the essence of the talks and discussions available in a book that conveys the thought processes and discussions at the conference.

The present volume is the result of the joint efforts of the symposium's invited speakers and a number of referees. They have all worked hard to shape the individual contributions in order to create a book that can serve as an advanced introduction to the many facets of entropy in mathematics and physics. As organizers of the symposium and editors of this book, we have added an introduction in which we describe how the whole material is organized and why we organized it in this way, and also a common alphabetical index for all contributions.

Our thanks go, first of all, to the speakers of the symposium, who not only provided written versions of their talks, but who also took part in the dynamic (and sometimes rather complex and nonlinear) process of turning a collection of individual contributions into a (hopefully) coherent volume on entropy. All this, however, would not have been possible without the very efficient support from the directors and the staff of the Max Planck Institute for the Physics of Complex Systems at Dresden (Germany). They, along with DFG and NWO, provided the necessary financial resources. We are heavily indebted to all of them.

List of Contributors

F. BENATTI
*Dipartimento di Fisica Teorica,
Università di Trieste and Istituto
Nazionale di Fisica Nucleare Sezione
di Trieste, Strada Costiera 11,
I-34014 Trieste, Italy
(fabio.benatti@trieste.infn.it)*

C. M. DAFERMOS
*Division of Applied Mathematics,
Brown University, Providence,
RI 02912, USA
(dafermos@cfm.brown.edu)*

F. DEN HOLLANDER
*EURANDOM, PO Box 513,
5600 MB Eindhoven,
The Netherlands
(denhollander@eurandom.tue.nl)*

H.-O. GEORGII
*Mathematisches Institut der
Universität München,
Theresienstr. 39,
D-80333 München,
Germany
(georgii@imu.de)*

A. GREVEN
*Universität Erlangen,
Mathematisches Institut,
Bismarckstrasse 1 ½,
D-91054 Erlangen, Germany
(greven@mi.uni-erlangen.de)*

K. HUTTER
*Institute of Mechanics,
Darmstadt University of Technology,
Hochschulstr. 1, 64289 Darmstadt,
Germany
(hutter@mechanik.tu-darmstadt.de)*

M. KEANE
*Department of Mathematics and
Computer Science,
Wesleyan University, Middletown,
CT 06459, USA
(mkeane@wesleyan.edu)*

G. KELLER
*Universität Erlangen,
Mathematisches Institut,
Bismarckstrasse 1 ½,
D-91054 Erlangen, Germany
(keller@mi.uni-erlangen.de)*

J. L. LEBOWITZ
*Center for Mathematical Sciences
Research, Rutgers, The State
University of New Jersey,
110 Frelinghuysen Road, Piscataway,
NJ 08854-8019, USA
(lebowitz@math.rutgers.edu)*

E. H. LIEB
*Princeton University, Jadwin Hall,
PO Box 708, Princeton,
NJ 08544-708, USA
(lieb@math.princeton.edu)*

C. MAES
*Instituut voor Theoretische Fysica,
Katholieke Universiteit Leuven,
B-3001 Leuven, Belgium
(christian.maes@fys.kuleuven.ac.be)*

I. MÜLLER
*Technische Universität Berlin,
Strasse des 17. Juni 135,
10623 Berlin, Germany
(im@thermodynamik.tu-berlin.de)*

E. OLIVIERI
*Dipartimento di Matematica,
II Università di Roma, Tor Vergata,
Via della Ricerca Scientifica,
00133 Roma, Italy
(olivieri@mat.uniroma2.it)*

J. RISSANEN
*Helsinki Institute for Information
Technology, Technical University of
Tampere, Finland and University of
London, Royal Holloway, Egham,
Surrey TW20 0EX, UK
(jorma.rissanen@mdl-research.org)*

J. UFFINK
*Institute for History and Foundations
of Science, Utrecht University,
PO Box 80.000, 3508 TA Utrecht,
The Netherlands
(uffink@phys.uu.nl)*

S. R. S. VARADHAN
*Courant Institute of Mathematical
Sciences, 251 Mercer St., New York,
NY 10012, USA
(varadhan@cims.nyu.edu)*

Y. WANG
*Institute of Mechanics, Darmstadt
University of Technology,
Hochschulstr. 1, 64289 Darmstadt,
Germany
(wang@mechanik.tu-darmstadt.de)*

G. WARNECKE
*Institute of Analysis and Numerical
Mathematics, Otto-von-Guericke
University Magdeburg,
Universitätsplatz 2,
D-39106 Magdeburg, Germany
(gerald.warnecke@mathematik.uni-
magdeburg.de)*

J. YNGVASON
*Institut für Theoretische Physik,
Universität Wien, Boltzmanngasse 5,
A 1090 Vienna, Austria
(yngvason@thor.thp.univie.ac.at)*

L.-S. YOUNG
*Courant Institute of Mathematical
Sciences, 251 Mercer St., New York,
NY 10012, USA
(lsy@cims.nyu.edu)*

Entropy

Chapter One

Introduction

Andreas Greven, Gerhard Keller
Universität Erlangen

Gerald Warnecke
Otto-von-Guericke University Magdeburg

The concept of entropy arose in the physical sciences during the 19th century, in particular in thermodynamics and in the development of statistical physics. From the very beginning the objective was to describe the equilibria and the evolution of thermodynamic systems. The development of the theory followed two conceptually rather different lines of thought. Nevertheless, they are symbiotically related, in particular through the work of Boltzmann.

The historically older line adopted a macroscopic point of view on the systems of interest. They are described in terms of a relatively small number of real variables, namely, temperature, pressure, specific volume, mass density, etc., whose values determine the macroscopic properties of a system in thermodynamic equilibrium. Clausius, building on the previous intuition of Carnot, introduced for the first time in 1867 a mathematical quantity, S, which he called *entropy*. It describes the heat exchanges that occur in thermal processes via the relation

$$dS = \frac{dQ}{T}.$$

Here Q denotes the amount of heat and T is the absolute temperature at which the exchange takes place. Among the scientists who developed this idea further are some of the most influential physicists of that time, most notably Gibbs and Planck, as well as the mathematician Carathéodory.

The other conceptual line started from a microscopic point of view on nature, where macroscopic phenomena are derived from microscopic dynamics: any macrostate is represented by many different microstates, i.e. different configurations of molecular motion. Also, different macrostates can be realized by largely differing numbers of corresponding microstates. Equilibria are those macrostates which are most likely to appear, i.e. they have the largest number of corresponding microstates (if one talks about the simplest case of unconstrained equilibria). In particular, two names are associated with this idea: Boltzmann and Maxwell. Boltzmann argued that the Clausius entropy S associated with a system in equilibrium is proportional to the logarithm of the number W of microstates which form the macrostate of this equilibrium,

$$S = k \ln W.$$

The formula is engraved on his tombstone in Vienna. Here the idea in the background, even though not explicitly formulated, is a relation between the encoded information specifying a microstate and the complexity of the system.

Since then both of these approaches to entropy have led to deep insights into the nature of thermodynamic and other microscopically unpredictable processes. In the 20th century these lines of thought had a considerable impact on a number of fields of mathematics and this development still continues. Among the topics where these ideas were most fruitful are stochastic processes and random fields, information and coding, data analysis and statistical inference, dynamical systems and ergodic theory, as well as partial differential equations and rational mechanics. However, the mathematical tools employed were very diverse. They also developed somewhat independently from their initial physical background. After decades of specialization and diversification, researchers involved in one of the areas of research mentioned above are often unprepared to understand the work of the others. Nowadays, it takes a considerable effort to learn and appreciate the techniques of a complementary approach, and the structure of this book reflects this and the state of the whole field. On the other hand, it was our intention with the symposium and this book to help bridge the gaps between the various specialized fields of research dealing with the concept of entropy. It was our goal to identify the unifying threads by inviting outstanding representatives of these research areas to give surveys for a general audience. The reader will find examples of most of the topics mentioned above in this book. To a large extent, the idea of entropy has developed in these areas in the context of quite different mathematical techniques, as the reader will see, but on the other hand there are common themes which we hope to exhibit in this book and which surfaced in the discussions taking place during the meeting.

We found two major methods in which entropy plays a crucial role. These are *variational principles* and *Lyapunov functionals*. They provide common tools to the different research areas.

Variational principles come in many guises. But they always take the following form, which we discuss for the case of the Gibbs variational principle. There, the sum of an energy functional and the entropy functional are minimized or maximized, depending on a choice of sign (see, for example, Section 2.3.1 on p. 29 and Section 3.6 on p. 50). Along the macroscopic line of thought such a variational problem appears as a first principle. It is clear that such a principle is crucial in thermodynamics and statistical physics. However, the reader will learn that it also plays a key role in the description of dynamical systems generated by differentiable maps, for populations evolving in a random medium, for the transport of particles and even in data analysis. As we proceed in the description of the various contributions we will show in more detail how this variational principle plays its role as an important tool.

For the microscopic line of thought, take the situation of a system where a macrostate corresponds to many microstates. In the Gibbs variational principle, one term reflects the number of possibilities to realize a state or path of a dynamical system or of a stochastic process under a macroscopic constraint such as, for example, a fixed energy. Then for a gain, e.g. in energy, one may have to 'pay' by a reduction in the number of ways in which microstates can achieve this gain. This results in a competition between energy and entropy that is reflected in the variational principle. Bearing this in mind it is not surprising that in many contributions to this book, a major role is played by information or information gain. There are two sides to the coin: one side is the description of the *complexity* of the problem to specify a microstate satisfying a macroscopic constraint; the other is the problem of *coding* this information. It is therefore also plausible that there are deeper connections between problems which at the surface look quite different, e.g. the description of orbits of a map on the one hand and data analysis on the other. We hope that these seemingly mysterious relations become clearer by reading through this book.

Another such common theme in many areas touched on in this book is the close connection of entropy to stability properties of dynamical processes. The entropy may, for instance, be interpreted as a Lyapunov function. Originally introduced for systems of ordinary differential equations, Lyapunov functions are an important tool in proving asymptotic stability and other stability properties for dynamical systems. These are functions that are increasing or decreasing with the dynamics of a system, i.e. the trajectories of solutions to the system cut the level sets of the function. This is exactly the type of behavior that the entropy exhibits due to the second law of thermodynamics. It is also exhibited in the axiomatic framework of Lieb–Yngvason when considering sequences of adiabatically accessible states, especially if the sequence of transitions is irreversible. For time-dependent partial differential equations in the framework of evolution equations, i.e. considering them as ordinary differential equations on infinite-dimensional spaces, such as Sobolev spaces, one has the analogous tool of Lyapunov functionals. Basically, in all these cases asymptotic stability tells us in which state (or states), in the long run, a dynamical system ends up.

There are other theoretical implications concerning the stability of a system as well, which are coming up in the chapters by Dafermos and Young. For instance, on the one hand, Sections 6.3 and 6.4 on pp. 110 and 113 of the chapter by Dafermos are devoted to illuminating the implications of entropy inequalities for stability properties of weak, i.e. discontinuous, solutions to hyperbolic conservation laws. Similarly, entropy inequalities are also used as a tool for systems of parabolic differential equations. On the other hand this might, at first sight, contrast with the role played by entropy in the chapter by Young, where it is connected to the instability of dynamical systems with chaotic behavior. However, instability of individual trajectories is a feature of the microscopic level which is responsible for the macroscopic stability of the actually observed invariant measures. Keeping in mind that the ergodic theory of dynamical systems is an equilibrium theory that cannot describe the dynamics of transitions between invariant measures, the role of entropy in variational principles, which single out the actually observed invariant measure among all other ones, comes as close as possible to the role of a Lyapunov functional.

1.1 Outline of the Book

Now we outline and comment on the contents of this book. In the first part, basic concepts, terminology and examples from both the macroscopic and the microscopic line of thought are introduced in a way that should provide just the right amount of detail to prepare the interested reader for further reading. The second part comprises five contributions tracing various lines of evolution that have emerged from the macroscopic point of view. The third part collects five contributions from the 'probabilistic branch,' and in a final part we offer the reader four contributions that illustrate how entropic ideas have penetrated coding theory and the theory of dynamical systems.

Due to limited time at the symposium, a number of further important topics related to entropy could not be dealt with appropriately, and as a consequence there are a number of deplorable omissions in this book, in particular regarding the role of entropy in statistics, in information theory, in the ergodic theory of amenable group actions, and in numerical methods for fluid dynamics.

Part 1 The concept of entropy emerging from the macroscopic point of view is emphasized by Ingo Müller's first contribution (p. 19). Starting from Clausius' point of view, Boltzmann's statistical ideas are added into the mix and the role of entropy as a governing quantity behind a number of simple real world phenomena is discussed: the elastic properties of rubber bands as an illustration for configurational entropy, the different atmospheres of the planets in our Solar System as a result of the 'competition between entropy and energy,' and the role of entropy in the synthesis of ammonia are just a few examples.

The approach to entropy in statistical physics developed its full strength in the framework of probability theory. It had a large impact on information theory, stochastic processes, statistical physics and dynamical systems (both classical and quantum). Concepts, terminology and fundamental results which are basic for all these branches are provided by Hans-Otto Georgii (p. 37), the second introductory chapter of this book.

Part 2 In the second part we have collected expositions on the macroscopic approach. There are two chapters on thermodynamics in the context of continuum mechanics. The first one, by Hutter and Wang, reviews alternative approaches to nonequilibrium thermodynamics, while in the second Müller surveys one of these approaches more extensively—the theory of extended thermodynamics—which he developed with various collaborators. Next, in the chapter by Dafermos on conservation laws, we pursue the bridge from continuum mechanics to differential equations further. Finally, we have two chapters that reflect the views of physicists on the fundamental issues concerning the understanding of entropy and irreversibility. Surprisingly, it turns out that there has not been until now one unified, universally accepted theory on the *thermodynamics of irreversible processes*. There are competing theories and concepts with a certain overlap and some fundamental differences. This diversity is a bit confusing, but bringing more clarity and unity to this field is also a challenging research goal. The contributions we have collected here are attempting to expose this diversity to a certain extent and clearly point out some of the conceptual differences that need to be resolved.

Interestingly, the uncertainty about some of the basic physical concepts involved is reflected by the fact that the state of the mathematical theory of the equations that follow from these theories is also very inadequate. Numerical methods play an increasingly important role in the effort to sort this out. This is because numerical computations allow us to explore more deeply consequences of mathematically complicated theories, i.e. the large systems of nonlinear differential equations they lead to. They may allow us to make quantitative predictions and comparisons with physical reality. Entropy also plays a key role in numerical analysis and numerical computations, an important topic that is missing here. However, the analytical tools are actually discussed in the chapter by Dafermos. The link between analytical theory and numerical approximations is particularly close in this field since key existence results were actually obtained via convergence proofs for numerical schemes. In numerical computations it is important to have verified that the method being used is consistent with an entropy condition, otherwise unphysical shock discontinuities may be approximated. Also, it is useful to monitor the entropy production in computations, because false entropy production is an indicator of some error in the scheme or the implementation.

Let us now turn to the contributions. The first chapter in this part is a survey by Kolumban Hutter and Yongqi Wang (p. 57) on some theories of irreversible thermodynamics. It highlights two different forms of the second law of thermodynamics, namely, the Coleman–Noll approach to the Clausius–Duhem inequality and the Müller–Liu entropy principle in a detailed comparison. The structure of these descriptions of thermodynamics is exposed in very concise form. The interesting lesson to the nonexpert is that there are different theories available that have much common ground but marked differences as well, a theme we will encounter again in Uffink's chapter. The second approach that Hutter and Wang discuss is then exposed much more extensively in the next chapter, by Ingo Müller (p. 79). The reader is given an introduction to the theory of *extended thermodynamics*, which provides a macroscopic theory for irreversible thermodynamic processes. It builds a vital link to the microscopic approach of kinetic gas theory formulated in terms of the Boltzmann equation and the density function it has as solution. From these one can rigorously derive an infinite sequence of balance equations involving higher and higher moments of the density function. In order to have a finite number of equations, these are truncated and certain closure laws have to be derived. This approach leads to symmetric hyperbolic systems of partial differential equations that are automatically endowed with an entropy function. The entropy production plays an important role in deriving closure laws for these systems. The reader who wishes to learn more should consult the book on extended thermodynamics that Müller together with Ruggeri recently published in updated form.

Shock discontinuities, i.e. an instantaneous transition from one physical state to a quite different state, are a good approximation to physical reality when we are modelling at the scale of continuum physics. This means that the free mean path between collisions of neighboring particles in a fluid is orders of magnitude smaller than the scale in which we are making observations. In a close-up view of shock waves we see that they do have a finite width. Since at least the work of Becker in the 1930s, it has been known that this width is only of the order of a few free mean path lengths, i.e. in the realm of microscopic kinetic gas theory. It is one of the important achievements of extended thermodynamics that its moment theories provide a precise description of the transition between the physical states across these shocks.

The chapter by Constantine Dafermos (p. 107) is intended to be a teaser to the book he recently wrote, where the ideas and concepts as well as the methods touched on in the chapter are expanded in much more detail. In the theory of partial differential equations a very significant open problem is the lack of an existence and uniqueness theory for nonlinear hyperbolic systems in more than one space dimension. Though such a theory may not be available in full generality such a theory should be found at least in the class of symmetric systems of conservation laws. Such systems appear naturally in various physical contexts, such as the moment systems derived from kinetic theory and discussed by Müller in his chapter. Dafermos gives a further example from isothermal thermoelasticity.

The peculiarity of these nonlinear differential equations is that they generically tend not to have smooth differentiable solutions. One has to use the concept of distributions which provides a framework in which functions may be differentiated in a generalized sense infinitely often. In this context one has a well-defined concept of solutions to differential equations that have jump discontinuities. This is not just a mathematical trick, namely, when you are looking for a solution, just make the solution space in which you are searching bigger and you may find something, possibly useless. However, making the solution space larger has its price. One tends to obtain too many solutions, i.e. one actually does get an abundance of unphysical solutions along with the reasonable one. If we allow jumps in a solution, then the solution may jump up and down almost arbitrarily in an oscillatory manner. This behaviour and certain types of mathematically feasible jump discontinuities are not observed in nature and must be excluded. Next to providing initial conditions in time and appropriate boundary conditions in the case of bounded solution domains, an additional mathematical condition is needed to specify the physically relevant solutions. Solutions of this nature are a very good description of shock waves on a macroscopic scale. These waves are well known and audible in connection with detonations or supersonic flight.

It has become a generally accepted jargon in the field to call such mathematical selection principles *entropy conditions*, or *E-conditions*, as in Chapter 6, even if they do not involve the concept of entropy explicitly. This is due to the fact that in all physically relevant hyperbolic systems of differential equations this selection is due to the second law of thermodynamics in one form or another. A microscopic approach to this question is presented for a special case in the contribution by Varadhan (Chapter 9).

The kinetic theory of gases has as its foundation the classical mechanics of gas molecules. Its mathematical formulation contains no irreversibility of any kind. The sheer enormity of the number of particles involved makes classical mechanics useless for quantitative description and prediction. Going to a macroscopic formulation by averaging procedures, the reversibility of classical mechanics is to a certain extent lost. This is a quite remarkable fact and the cause of some uneasiness with the notion of irreversibility, as discussed by Uffink and by Lieb and Yngvason in their respective chapters. Mathematically, irreversibility clearly shows up in the structure of shock wave solutions of conservation laws. In one-dimensional systems the characteristic curves of one family merge in the shock and most of the information they carry is lost. Already, for the simple Burgers equation, equation (6.9) in the chapter by Dafermos, one can consider for time $t \geqslant T > 0$ an admissible solution with only one shock. It may result from infinitely many different monotonely nonincreasing initial data at time $t = 0$. They have all merged into the same solution by the time $t = T$. This loss of information on the initial data is exactly linked to the property that the mathematical entropy has a jump discontinuity at the shock. But, as long as the entropy and other physical states remain smooth we may go back in time along characteristics to reconstruct the initial data.

A problem of the macroscopic theory of thermodynamics is that there are two levels of theory involved. One level, closer to continuum mechanics, is usually formulated in precise mathematical language of differential forms and differential equations. It involves the energy balance in the first law of thermodynamics, the Gibbs relation stating that the reciprocal of temperature is an integrating factor for the change in heat allowing the entropy to be defined as a total differential, and the second law, e.g. in the form of Clausius as a differential inequality. Such equations and the differential equations incorporating irreversible thermodynamics into continuum mechanics are discussed in the chapters by Müller, Hutter and Wang, and Dafermos.

A second, more fundamental level of macroscopic theory lies at the basis of these analytic formulations, and this is addressed in the final two chapters of this part. As pointed out by Hutter and Wang, there is some controversy about the foundations of irreversible thermodynamics leading to a certain extent to differing formulations of the second law of thermodynamics. But this is not surprising since some of the basic concepts, even the definition of fundamental quantities such as temperature and entropy, have not yet found a universally accepted form. The chapter by Jos Uffink (p. 121) summarizes a more detailed paper he recently published on the foundations of thermodynamics. He discusses historically as well as conceptually the basic notions of time-reversal invariance, irreversibility, irrecoverability, quasistatic processes, adiabatic processes, the *perpetuum mobile* of the second kind, and entropy. Clearly, there is much confusion on the precise meaning of some of them. Mathematical formalism and physical interpretation are not as clear as in other areas of physics, making thermodynamics a subject hard to grasp for nonspecialists and, as all authors of this part seem to agree, for the specialists as well. The distinctions, comparisons of theories and criticisms pointed out by Uffink are a helpful guide to understanding the development of the field and discussions of theoretical implications. He covers some of the same historical ground as Müller's survey in Part 1, but with a different emphasis. In passing he sheds a critical light on the controversial and more philosophical issues of the arrow of time and the heat death of the Universe, the statements in connection with entropy that have stirred the most interest among the general public.

The final contribution of Part 2 is a survey by Elliott Lieb and Jakob Yngvason (p. 147) of their very recent work attempting to clarify the issues of introducing entropy, adiabatic accessibility and irreversibility by using an axiomatic framework and a more precise mathematical formulation. This is a clear indication that the foundations of thermodynamics remain an active area of research. Such a more rigorous attempt was first made almost a century ago by Carathéodory. Uffink discusses these two more rigorous approaches of Carathéodory and Lieb and Yngvason in his chapter. Here we have a more detailed introduction to the latter approach by its proponents. To a mathematician this is clearly the type of formalization to be hoped for, because it allows precise formulations and the proof of theorems. They provide a clear and precise set of axioms. From these they prove theorems such as the existence and, up to an affine transformation, uniqueness of the nonincreasing entropy function on states as well as statements on its scaling properties. Another theorem states that temperature is a function derived from entropy. The fact that energy flows from the system with higher temperature to the system with lower temperature is a consequence of the nondecrease of total entropy. An interesting fact, also pointed out by Uffink,

INTRODUCTION

is that the Lieb–Yngvason theory derives this entropy from axioms that are time-reversal invariant, another variant of the issue already mentioned above. The goal of this theory is to have very precise mathematical formulations that find universal acceptance in the scientific communities involved. As Lieb and Yngvason point out at the end of Section 8.2, discovering the rules of the game is still part of the problem. This intellectually challenging field is therefore open to further exploration.

Part 3 The third part of this book deals with the role of entropy as a concept in probability theory, namely, in the analysis of the large-time behavior of stochastic processes and in the study of qualitative properties of models of statistical physics. Naturally, this requires us to think along the lines of the microscopic approach to entropy. We will see, however, that there are strong connections to Part 2 via hyperbolic systems and via Lyapunov functions (compare the sections on the Burgers equation in the contributions of Dafermos and Varadhan), and to Part 4, in particular to dynamical systems, due to similar phenomena observed in the large-time behavior of dynamical systems and stochastic systems. The first two contributions of Part 3 deal with the role of entropy for stochastic processes, the last three deal with entropy in models of statistical physics. All these contributions show that variational principles play a crucial role. They always contain a term reflecting constraints of some sort imposed on the system and an entropy term representing the number of possibilities to realize the state of the system under these constraints.

The fundamental role that entropy plays in the theory of stochastic processes shows up most clearly in the theory of *large deviations*, which tries to measure probabilities of rare (untypical) events on an exponential scale. Historically, this started by asking the question how the probability that a sum of n independent random variables exceeds a level $a \cdot n$, when a is bigger than the mean of a single variable, decays as n tends to infinity. More precisely, can we find the *exponential rate* of decay of this event as n tends to infinity? The role of entropy in the theory of large deviations is the topic treated in the contribution by Shrinivasa R. S. Varadhan (p. 199). He starts by explaining how entropy enters in calculating the probability of seeing untypical frequencies of heads in n-fold and independent coin tossing. If one observes the whole empirical distribution of this experiment, one is led to Sanov's Theorem now involving relative entropies. Along this line of thought one passes next from independent trials to Markov chains and spatial random fields with a dependence structure which is given by Gibbs measures, the key object of statistical physics. In all these situations one is led to *variational problems* involving entropy on the one hand, since one has to count possible realizations, and a term playing the role of an energy arising from a potential defined as the logarithm of relative probabilities. This competition between entropy and energy is explained in various contexts ranging from coin tossing to occupation numbers of Markov chains. At this point it is useful to remark that in Part 4 very similar structures play a role in the contribution by Young. However, these questions cannot only be asked in spatial situations given, for example, by the realization of a random field at a given time, but one can take the whole space-time picture of the process into view. This allows us, for example, to study small random perturbations of dynamical systems and to determine asymptotically as the perturbation gets small the probabilities of paths of the stochastic process leaving small tubes around the path of the dynamical system in an exponential scale. The contribution concludes with a discussion of the hydrodynamic rescaling of stochastic particle systems and their relation to Burgers equations. Of particular interest in this model is the problem of selecting the proper solution in the presence of shocks and here again entropy is the key. This connects directly with the contribution by Dafermos in Part 2.

An important application of large-deviation theory is the study of stochastic processes evolving in a random medium. This means that the parameters of a stochastic evolution are themselves produced by a random mechanism. The contribution of Frank den Hollander (p. 215) discusses an interesting and typical example for this situation in which the relative entropy is the key quantity. It is an example from the area of spatial population models. Here particles live on the integer lattice, they migrate to the left and to the right with certain probabilities and split into several new particles (or become extinct) with a probability distribution of the offspring size which depends on the location. The large-time behavior exhibits phase transitions from extinction to explosion of the population. Here the *variational problem* energy versus entropy of statistical physics appears in a different guise. The energy is replaced by the local growth parameter given by the random medium. The entropy enters since it regulates asymptotically the numbers of paths which can be used by individuals migrating from a point A to a point B, when A and B become very distant. However, since the mathematical structure is exactly the one occurring in statistical physics, the system exhibits various phase transitions in the parameter regulating the drift of motion. Depending on that drift we see populations growing or becoming extinct and here one can distinguish both local and global extinction. Even though the model is rather special, the phenomena and structures exhibited are quite representative for systems in random media.

The two contributions by Enzo Olivieri and Christian Maes use stochastic evolutions to discuss the phenomenon of *metastability* and *entropy production*, two key topics for the broader theme of entropy in statistical physics. Metastable behavior of a physical system is a phenomenon which should be understood by means of probabilistic microscopic models of the system. To make a step in this direction is the goal of the contribution by Olivieri (p. 233). He discusses two examples of metastable behavior, supersaturated vapor as an example of a conservative system (the number of particles is conserved), and ferromagnetic systems below the Curie temperature for a nonconservative context. He discusses these two systems in the framework, respectively, of van der Waals and Curie–Weiss theory in great detail. He continues the discussion with a critical review of the shortcomings of this approach and passes on to the description of the 'restricted ensemble approach' and the 'pathwise approach' to this problem of describing metastable behavior. This pathwise approach is then formulated precisely in the context of the stochastic Ising model for ferromagnets and finally a description of the general probabilistic mechanism behind metastable behavior via first-exit problems of Markov chains is provided. The concept of the first descent tube of trajectories is used to relate metastable behavior to appropriate concepts of time entropy. Again, in deciding which route a system takes to pass from the metastable to the stable state, a key role is played by the competition between the number of paths producing particular features of the route and the probability of such paths. Hence again a *variational problem* describing this competition is in the background, and again it has the form of the sum of an entropy and an energy term.

The concept of entropy in nonequilibrium is treated in the contribution by Maes (p. 251). The key quantity in this approach is the entropy production rate. One proves that it is always nonnegative and shows that the system is time reversible if and only if there is no entropy production. The second point is to study the local fluctuations of the entropy production rate. Here these ideas are developed in the framework of spatially extended systems with many interacting components and a simple stochastic evolution mechanism, the so-called asymmetric exclusion process. In this system particles move according to independent random walks with a drift and the additional rule is imposed that every site can only be occupied by at most one particle. Due to this term the system exhibits interaction between the components. (Nevertheless, the equilibria of the system are product measures.) Here the entropy production is defined by considering the system and the time-reversed system and forming the logarithmic ratio of the number of those microstates with which the original macrostate can be realized.

The third part of the book concludes with a contribution by Joel Lebowitz and Christian Maes (p. 269) of a quite different form, namely, a dialogue on entropy between a scientist and an angel. Their conversation touches on such topics as the Boltzmann equation, the second law of thermodynamics and the von Neumann entropy in quantum systems. The reader will see many of the topics in this part in a different light and feel stimulated to have another look at the previous contributions.

Part 4 The final part of this book is devoted to several offspring of the probabilistic concept of entropy discussed already in Part 3, namely, to its use in information compression and dynamical systems. Although these two directions deal with nearly complementary aspects of entropy, they both use Shannon's concept of entropy as a measure of information (see also Section 3.2 on p. 39).

For a thorough understanding of why information compression and uncertainty produced by dynamical systems are two sides of the same coin, *coding* is the basic ingredient. Information compression is the art of describing data given as a (long) string of symbols from some alphabet in a 'more efficient' way by a (hopefully shorter) string from a possibly different alphabet, here from the set {0, 1}. The classical result in this situation is Shannon's source-coding theorem for memoryless sources saying roughly that for randomly generated data there exists a simple coding procedure such that the average code length for data of length n is nearly n times the entropy of the source. In dynamical systems one wants to measure the 'diversity' of the set of all orbits of length n when such orbits are observed with 'finite precision,' i.e. one splits the state space into a finite number of sets and observes only which sets are visited in the successive time steps. This led Kolmogorov and Sinai to code individual orbits by recording the outcomes of such coarse-grained observations at times 1 to n and to apply ideas from information theory to these code words and their probability distributions.

This circle of ideas is the starting point of Fabio Benatti's contribution (p. 279). Beginning with a condensed description of the basic notions and results leading to the concept of *Kolmogorov–Sinai entropy* for the dynamics of a classical measure-theoretic dynamical system, Benatti gradually develops the appropriate concepts for quantum dynamical systems which generalize the classical ones. In this way discrete probability distributions are replaced by density matrices, products of probability distributions by tensor products of density matrices, the Shannon entropy of a discrete probability distribution by the von Neumann entropy of a density matrix and the coding theorem for classical memoryless sources by an analogous result for quantum memoryless sources. Finally, the generalization from discrete probability distributions to general probability spaces translates into a passage from density matrices to positive operators on Hilbert spaces. Whereas these transfers from the classical to the quantum world are relatively straightforward, the situation changes drastically when one tries to pass from the classical Kolmogorov–Sinai entropy to a quantum dynamical entropy. The idea to model a coarse-grained observation device for a classical system as a finite measurable partition of the classical phase space has no obvious quantum counterpart. Two approaches (which both coincide with Kolmogorov–Sinai entropy on classical dynamical systems) are discussed: the *Connes–Narnhofer–Thirring entropy*, where partitions of the state are used, and the *Alicki–Fannes entropy*, which is based on partitions of unity.

It is worthwhile noting that the above-mentioned problem to find a shortest description of a given finite string of data is not inherently probabilistic, but that Kolmogorov and, independently, Chaitin offer a purely deterministic approach using the concept of complexity of such a string. It is reassuring, though, that both theories meet—at least in the classical context as pointed out in Section 14.3 of Benatti's contribution. A much more prominent role is played by complexity in Jorma Rissanen's contribution (p. 299). Again the coding of strings of data is the starting point, but now the optimal code word is the shortest 'program' for a universal Turing machine from which the machine can reproduce the encoded data. Following some unpublished work from Kolmogorov's legacy, we look at (essentially) the shortest descriptions which consist of two parts: one that describes 'summarizing properties' of the data and another one that just specifies the given data string as a more or less random string within the class of all strings sharing the same summarizing properties. Then the summarizing properties represent only the 'useful' information contained in the data. Unfortunately, it is impossible to implement a data analysis procedure based on Kolmogorov's complexity because the complexity of a string is uncomputable. A more modest goal can be achieved, however. Given a suitably parametrized class of models for the observed data, one can define the complexity of a string of data relative to the model class and decompose the shortest description of the data relative to the model class into some 'useful' information and a noninformative part. Now the close relation between effective codes and probability distributions expressed by Shannon's source-coding theorem allows us to formalize the notion of a 'parametrized class of models' by a parametrized class of probability distributions so that it is not too surprising that actual calculations of complexity and information of data involve not only Shannon entropy but also classical statistical notions like the Fisher information matrix.

There is one more notion from dynamical systems which Benatti connects to entropy: *chaos* in classical and also in quantum systems. The basic idea, briefly indicated here, is the following. Unpredictability of a deterministic dynamical system, which is caused by an extreme sensitivity to initial conditions, is noticed by an external observer because typical coded trajectories of the system are highly complex and resemble those produced by a random source with positive entropy. This is the theme of Lai-Sang Young's contribution (p. 313), where entropy means dynamical entropy, i.e. the logarithmic growth rate of the number of essentially different orbits as mentioned before. If not really all orbits are taken into account but only most of them (in the sense of an underlying invariant probability measure), this growth rate coincides with the Kolmogorov–Sinai entropy introduced above. If all orbits are counted, the corresponding growth rate is called topological entropy. Both notions are connected through a variational principle saying that the topological entropy is the supremum of all Kolmogorov–Sinai entropies with respect to invariant measures. For differentiable dynamical systems with a given invariant measure the Kolmogorov–Sinai entropy, which is a global dynamical characteristic, can be linked closely to more geometric local quantities, namely, to the Lyapunov exponents which describe how strongly the map expands or contracts in the directions of certain subspaces, and to certain partial dimensions of the invariant measure in these directions. In the simplest case, when the invariant measure is equivalent to the

Riemannian volume on the manifold and has at least one positive Lyapunov exponent, this very general relationship specializes to *Pesin's Formula*. In fact, the validity of Pesin's Formula characterizes exactly those invariant measures which are physically observable and known as the *Sinai–Ruelle–Bowen* measures. This characterization can be cast in the form of a *variational principle* involving the Kolmogorov–Sinai entropy and, as 'energy' terms, the average Lyapunov exponents.

There are more interpretations of entropy in dynamical systems. Young describes how the counting of periodic orbits and how the volume growth of cells of various dimension is linked to topological entropy, and how large-deviation techniques, omnipresent in Part 3 of this book, can be successfully applied to dynamical systems with positive entropy. Michael Keane (p. 329) describes the breakthrough that Kolmogorov–Sinai entropy, called mean entropy in this contribution, brought for the classification of measure-theoretic dynamical systems. It culminated in *Ornstein's* proof that two Bernoulli systems are isomorphic if and only if they have the same Kolmogorov–Sinai entropy. As, in turn, many systems occurring 'in nature' are isomorphic to Bernoulli systems, this result clarifies the measure-theoretic structure of a great number of dynamical systems.

All the results above, however, are rather meaningless for systems of zero entropy. The structure theory of such systems is therefore a challenging field of research, and Keane presents an example—the so-called binomial transformation—that conveys the flavour of these kinds of system, which in many respects are rather different from positive entropy systems.

1.2 Notations

In view of the broad spectrum of contributions from rather different fields—all with their own history, mathematical techniques and their time-honored notations—we feel that a few additional remarks might be useful to help readers to interpret the various notations correctly.

For the benefit of readers not familiar with continuum mechanics we would like to point out a subtle point concerning time derivatives that may be somewhat confusing at first. The fundamental laws of mechanics, such as conservation of mass, i.e. no mass is created or lost, or Newton's law, namely, that the time change in momentum is equal to the acting forces, are most simply and familiarly formulated in the so-called *Lagrangian framework*, where one is following trajectories of massive particles, single-point particles or (in continuum mechanics) an arbitrary volume of them. The rate of change in time, i.e. the time derivative, in this framework is written as the total derivative d/dt or frequently by a dot over the physical state variable that is changing in time, e.g. $m\dot{u} = f$ for Newton's law with mass m, velocity u and force field f. This derivative is called the *material derivative* and must be distinguished from the partial derivative in time arising in the context of the *Eulerian framework*. In the latter all physical state variables are simply described as a field, not in the sense of algebra, but rather meaning that they are given as functions of time and space, maybe vector-valued or higher-order tensor-valued. Particle trajectories are not given explicitly but may be obtained by integrating the velocity field. The rate of change in time in this framework is given by the partial derivatives $\partial/\partial t$, the rate of change in space by the spatial gradient ∇ consisting of the partial spatial derivatives of the physical variables. The connection between the two types of time derivatives d/dt and $\partial/\partial t$ is given by the chain rule. Following a particle trajectory of an Eulerian variable such as the mass density ρ gives the derivatives in the following form

$$\frac{d\rho}{dt}(t, x(t)) = \frac{\partial \rho}{\partial t} + \nabla \rho \cdot \dot{x} = \frac{\partial \rho}{\partial t} + \boldsymbol{u} \cdot \nabla \rho,$$

where $\boldsymbol{u}$ denotes the velocity field in the Eulerian framework. This point is discussed in more detail in any good introductory textbook on continuum mechanics.

The standard set-up used in probability theory is a basic probability space $(\Omega, \mathcal{A}, P)$ with a set Ω describing possible outcomes of random experiments, a σ-algebra $\mathcal{A}$ describing possible events, and a basic measure P representing the probabilities of these events. On that space random variables are defined, often called X, which are maps from Ω into some set E, the possible values which the random variable can attain. This might be real numbers, vectors, or even measures. In the latter case such random variables are called random measures. One very important example is the empirical distribution arising from repeating a random experiment n times. The law specifying the probability that this observed distribution is close to a given one is then in fact a probability law on the set of probability laws. Random variables are in principle maps from one measure space into another, i.e. $X : \Omega \to E$, and one would expect to write $\{\omega \in \Omega \mid X(\omega) \in A\}$ for the event that X takes values in a set A, but in fact one typically writes $\{X \in A\}$. Hence you might encounter symbols like L_n, Y_t, etc., which are random objects but no ω floats around to remind you, and the state spaces of the random variables might vary from $\mathbb{R}$ over spaces of continuous functions to spaces of measures. Furthermore, for the expectation $\int_\Omega X(\omega) P(d\omega)$ of the random variable X one writes $\int X \, dP$ and even shorter $E[X]$, and here neither ω nor Ω or P appears explicitly. There is a good reason for this since the interest is usually not so much in Ω and P but in the values a certain random variable takes and in its distribution. This distribution is a probability measure on the space E where the random variable takes its values. Analysing properties of this random variable the basic space is then very often of little relevance.

Another point to be careful about is to distinguish in convergence relations whether random variables, i.e. measurable functions, actually converge in some sense (almost surely, in mean or in probability), or whether only their laws converge, that is, expectations of continuous bounded functions of those variables converge to a limit thus determining a limiting probability distribution.

In ergodic theory and the theory of dynamical systems (with discrete time), the basic object is a measurable transformation, call it f, acting on a more or less abstract state space, often called X. Typical examples of state spaces are manifolds or sequence spaces like $A^{\mathbb{Z}}$, where A is a finite symbol set. Properties of trajectories x, $f(x)$, $f^2(x)$, ... obtained from repeated application of the transformation are studied relative to an invariant measure, say μ, where invariance means that $\mu(f^{-1}A) = \mu(A)$ for each measurable subset A of X. In this book only the most common setting is treated where μ is a probability measure. Typically, a given map f has many invariant measures, and one problem is to single out (often using a variational principle) those which are dynamically most relevant. In particular, for each such measure (X, μ) is a probability space and the scene is set for carrying over concepts from the world of probability to dynamical systems. Note, however, that in contrast to the probabilistic approach, where properties of the underlying probability space play a minor role, the specific structure of X and μ can be of central interest in the context of dynamical systems.

PART 1
Fundamental Concepts

Chapter Two

Entropy: a Subtle Concept in Thermodynamics

Ingo Müller
Technical University of Berlin

Thermodynamics started from a concern for more efficient heat engines, but that motive was soon left behind when Clausius discovered entropy and was thus able to find a universal relation between thermal and caloric properties of matter. From a molecular point of view entropy is related to the tendency of matter to spread out in a random walk maintained by thermal motion: entropy grows as the random walk proceeds. The tendency of entropy to grow is often in competition with the opposing tendency of energy to decrease. As a result of the competition, the free energy of matter tends to a minimum. The free energy combines energy and entropy in a manner that favours energy at low temperatures and entropy at high temperatures. There are numerous examples: planetary atmospheres, osmosis, phase transitions, chemical reactions. Thermodynamic equilibrium results as a compromise between the opposing energetic and entropic trends.

2.1 Origin of Entropy in Thermodynamics

The concept and name of entropy originated in the early 1850s in the work of Rudolf Julius Emmanuel Clausius (1822–1888) (see Clausius 1887), and that work was at first primarily concerned with the question of which cycle is best suited for the conversion of heat into work and which substance is best used for the conversion. Clausius based his argument on the plausible axiom that

> heat cannot pass by itself from a cold to a hot body.

In order to exploit that axiom Clausius considered two competing Carnot cycles[1] (see Figure 2.1) working in the same temperature range, one as a heat engine and one as a refrigerator; the refrigerator consumes the work the engine provides. And, by comparing the amounts of heat Q passed from top to bottom and vice versa, he came to the conclusion that among all efficiencies

> the efficiency of a Carnot cycle is maximal and universal.

---

[1] An engine performs a Carnot cycle if it exchanges heat with a heat bath at two temperatures only.

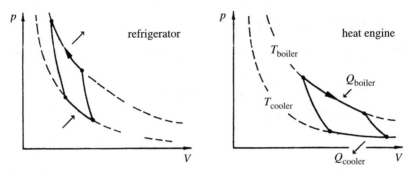

Figure 2.1. Two competing Carnot engines (pressure p, volume V)-diagrams.

'Maximal' means that no cycle in the same range of temperature has a bigger efficiency than a Carnot cycle, and 'universal' means that all working substances provide the same efficiency in a Carnot cycle.

It is easy to calculate the efficiency of the Carnot engine of an ideal gas:

$$e = \frac{\text{Work}}{Q_{\text{boiler}}} = \frac{Q_{\text{boiler}} - Q_{\text{cooler}}}{Q_{\text{boiler}}} = 1 - \frac{Q_{\text{cooler}}}{Q_{\text{boiler}}} = 1 - \frac{T_{\text{cooler}}}{T_{\text{boiler}}}, \quad (2.1)$$

where T is the Kelvin temperature. And, since by Clausius' result this efficiency is universal, it holds not only for ideal gases but for all substances, be they water, mercury, sulphur, or steel.

It is only fair to say that Nicolas Léonard Sadi Carnot (1796–1832) (see Carnot 1824) anticipated Clausius by 30 years, but no one could understand Carnot's reasoning. Carnot believed in the caloric theory of heat, by which the heat passing through the cycle from boiler to cooler is unchanged in amount. This is quite wrong and it is a kind of miracle that Carnot, despite his erroneous concepts, obtained a correct result. Carnot's trouble was that he did not know the balance of energy, or the *first law of thermodynamics*,

$$\boxed{\text{1st law}} \quad \frac{d(U+K)}{dt} = \dot{Q} + \dot{A}, \quad (2.2)$$

by which the rate of change of energy, internal and kinetic, is equal to the sum of the heating and working.[2]

Clausius, working 30 years later, knew this law and he also knew that

$$\dot{A} = -p\frac{dV}{dt} \quad (2.3)$$

holds, if the working is expended reversibly for a volume change. With that superior knowledge Clausius showed that it is not the heat that passes through

---

[2] Heating $\dot{Q}$ and working $\dot{A}$ are not time derivatives but rates of exchange of heat and work through the surface of a system which has an internal energy U and a kinetic energy K. The quantity Q, when it occurs, is the time integral of $\dot{Q}$ over the duration of a heat exchange.

ENTROPY: A SUBTLE CONCEPT IN THERMODYNAMICS

the cycle from boiler to cooler unchanged in amount. He proves that (see equation (2.1))

$$\left.\frac{Q}{T}\right|_{\text{boiler}} = \left.\frac{Q}{T}\right|_{\text{cooler}} \tag{2.4}$$

holds, so that it is the quantity Q/T that passes unchanged through the cycle. And that quantity is the *entropy*. Clausius said: I have proposed to call this quantity entropy from the Greek word $\tau\rho o\pi\eta$ for change.

Seeing that Clausius was anticipated by Carnot, albeit in a mysterious manner, we cannot give full credit to Clausius for the above result on the efficiency of a Carnot cycle. However, Clausius must be credited with recognizing the concept of entropy and, above all, with liberating that concept from its origins in cycles and, in particular, Carnot cycles. He shows that in an arbitrary process the rate of change of entropy, called S, satisfies the inequality

$$\boxed{\text{2nd law}} \quad \frac{dS}{dt} \geq \frac{\dot{Q}}{T} \quad \text{for} \begin{cases} \text{an irreversible process,} \\ \text{a reversible process,} \end{cases} \tag{2.5}$$

which important relation became known as the *second law of thermodynamics*. We now discuss the significance of that law.

First, we stick to reversible processes, where the equality holds in (2.5). We may then eliminate $\dot{Q}$ between the first and second laws and, neglecting the kinetic energy, we obtain the *Gibbs equation* (Josiah Willard Gibbs (1839–1903) (see, for example, Gibbs 1960)):

$$\frac{dS}{dt} = \frac{1}{T}\left(\frac{dU}{dt} + p\frac{dV}{dt}\right). \tag{2.6}$$

The importance of this equation can hardly be overestimated; it saves time and money and it is literally worth billions to the chemical industry. The reason is the integrability condition implied by the Gibbs equations, namely,

$$\left(\frac{\partial U}{\partial V}\right)_T = -p + T\left(\frac{\partial p}{\partial T}\right)_V. \tag{2.7}$$

This equation relates the thermal equation of state

$$p = p(V, T)$$

and the caloric equation of state

$$U = U(V, T).$$

Both equations depend on the material and they must be known for the calculation of nearly all thermodynamic processes in the material. The thermal equation of state is easy to measure, since p, V, and T are all three of them measurable quantities and need only be related. But the caloric equation of state is difficult, because we cannot measure the internal energy U. Therefore, the determination of the caloric equation of state proceeds in a roundabout manner by the measurement of specific heats which—like all caloric measurements—are difficult, time consuming, expensive, and unreliable to boot. And that is where the integrability condition (2.7) helps: it relates the V-dependence of U to the thermal equation of state and therefore greatly reduces the number of laborious caloric measurements.

From the few remaining specific heat measurements and the integrability condition (2.7), we may determine the caloric equation of state $U = U(V, T)$ by integration to within an additive constant. And, once the thermal and caloric equations of state are known, we may use the Gibbs equation to calculate $S(T, V)$ or $S(T, p)$ by integration to within another additive constant. The values of these constants are unimportant except when it comes to chemical reactions (see Section 2.5 below).

In an irreversible process the second law (2.5) implies that the entropy cannot decrease in an adiabatic system, i.e. a system without heating. Thus it appears that the second law reveals a teleological tendency of nature. Both Clausius and Gibbs would succinctly express this tendency—and the first and second laws of thermodynamics—with the words:

> Die Energie der Welt ist konstant.
> Die Entropie der Welt strebt einem Maximum zu.

'Welt' means the Universe, that being the only system which one could reasonably be sure suffers no heating. The philosophical implications of these statements were and are eagerly discussed by philosophers, historians and sociologists, and both Nietzsche (1872) and Spengler (1919) had something to say about them.

Now, looking again at the Gibbs equation we recognize that $1/T$ is an integrating factor for the heating

$$\dot{Q} = \frac{dU}{dt} + p\frac{dV}{dt}.$$

Constantin Carathéodory (1873–1950) (see Carathéodory 1909) emphasized this mathematical aspect of the Gibbs equation and, in doing so, provided an important alternative formulation of Clausius' axiom, which is called the *inaccessibility* axiom:

> An adiabatic process with
>
> $$\frac{d(U + K)}{dt} = \dot{A}$$
>
> cannot reach all states U, V in the neighborhood of an initial state U_i, V_i.

In common parlance, one may express this by saying that friction heats, never cools. (In a more extensive account of thermodynamics, the Carathéodory axiom would deserve more than just this short note; the axiom puts thermodynamics on a firm foundation for bodies in which reversible working may amount to more than just a change of volume. Moreover, the Carathéodory approach to the second law may be used more easily than the Clausius approach for an extrapolation of the entropy concept to thermodynamic field equations. Recently, Alt (1998) proposed such an extrapolation.)

2.2 Mechanical Interpretation of Entropy in the Kinetic Theory of Gases

Despite the efforts of Clausius and Carathéodory, the entropy remained a somewhat implausible concept of little suggestive meaning. Clearly, what was needed was an interpretation in terms of molecules or atoms. And that is why everyone in the last decades of the 19th century worked on what was then called the mechanical theory of heat. Clausius, James Clerk Maxwell (1831–1879) (see Maxwell 1866) and Ludwig Edward Boltzmann (1844–1906) (see Boltzmann 1910) were all considering the distribution function $f(x, p, t)$, where

$f \, dx \, dp$ is the number of atoms in the volume element dx
with momenta between p and $p + dp$.

Boltzmann succeeded in writing a differential equation for that function, an equation which is now called the Boltzmann equation. It has the form[3]

$$\frac{\partial f}{\partial t} + p_i \frac{\partial f}{\partial x_i} = \int (f' f_1' - f f_1) \sigma g \sin \vartheta \, d\vartheta \, d\phi \, dp_1 \qquad (2.8)$$

and we see that it is essentially a continuity equation in the phase space spanned by x and p. However, it has a complicated production term on the right-hand side which accounts for collisions between the atoms. The derivation of that production term was Boltzmann's finest achievement.

From this equation, by integration, Boltzmann derived conservation laws for the following macroscopic quantities:

$$\left. \begin{array}{rl} \text{mass:} & M = \iint mf \, dx \, dp, \\ \text{momentum:} & P_i = \iint p_i f \, dx \, dp, \\ \text{energy:} & U + K = \iint \frac{p^2}{2m} f \, dx \, dp. \end{array} \right\} \qquad (2.9)$$

---

[3] For the derivation of the Boltzmann equation and for the interpretation of its symbols, we refer the reader to any textbook on the kinetic theory of gases (see, for example, Müller 1985, pp. 114–148).

He also derived an inequality, a balance law with a positive production for a quantity which—because of the positive production—could only be the entropy

$$S = \iint (-k \ln f) f \, d\mathbf{x} \, d\mathbf{p}, \qquad (2.10)$$

where k is a universal constant, the Boltzmann constant.

Inspection of (2.9) shows that the mass is just the sum, or integral, of the masses m of all atoms, while the momentum is the integral over all atomic momenta p_i and the energy is the integral over all atomic kinetic energies $p^2/2m$. What about entropy? The entropy is the integral over the logarithm of the distribution function itself; this is hardly a suggestive quantity with an intuitive meaning and, furthermore, it is restricted to monatomic gases.

However, in the long and fierce debate that followed Boltzmann's interpretation of entropy, this nonsuggestive expression was eventually hammered into something suggestive and, moreover, something amenable to extrapolation away from gases.

Actually, 'hammered' is the right word. The main proponents went into a prolonged and acrimonious debate. Polemic was rife and good sense in abeyance. After 50 years, when the dust settled and the fog lifted, one had to realize that quantum mechanics could have started 30 years earlier, if only the founders of thermodynamics—not exempting Boltzmann, Maxwell and Gibbs—had put the facts into the right perspective.

What had to be done was this: $f \, d\mathbf{x} \, d\mathbf{p}$, the number of atoms in the phase space element $d\mathbf{x} \, d\mathbf{p}$, had to be replaced by N_{xp}, the number of atoms occupying a single point and having a single momentum. Those two quantities, $f \, d\mathbf{x} \, d\mathbf{p}$ and N_{xp}, are related by the number of points in $d\mathbf{x} \, d\mathbf{p}$. That number is proportional to $d\mathbf{x} \, d\mathbf{p}$ or equal to $Y \, d\mathbf{x} \, d\mathbf{p}$ (say) with Y as a factor of proportionality. Thus we have

$$N_{xp} = \frac{f \, d\mathbf{x} \, d\mathbf{p}}{Y \, d\mathbf{x} \, d\mathbf{p}}. \qquad (2.11)$$

If this is used to eliminate f from the Boltzmann expression for entropy, and if the integral is converted into a sum, and if the Stirling formula is used, one can, after some calculation, come up with the expression

$$S = k \ln \frac{N!}{\prod_{xp} N_{xp}!} - kN \ln(YN) \qquad (2.12)$$

of which the additive constant was conveniently dropped.

The argument of the logarithm can be interpreted as the number of possibilities to realize a distribution $\{N_{xp}\}$ of atoms in a gas. And from there it was an easy step of extrapolation that led to the famous formula

$$S = k \ln W, \qquad (2.13)$$

where W is now the number of possibilities to realize any distribution appropriate to any body of many individual elements. And that is the mechanical interpretation of entropy, embodied in equation (2.13), which Boltzmann, incidentally, never wrote down, but which is engraved on the headstone at his grave.

Having mentioned quantum mechanics, I should explain that setting the number of points in $d\mathbf{x}\,d\mathbf{p}$ equal to $Y\,d\mathbf{x}\,d\mathbf{p}$ is tantamount to the quantization of phase space. Indeed, $1/Y$ is thus identified as the smallest cell in phase space that can accommodate a point which might, or might not, be occupied by atoms. Max Karl Ernst Ludwig Planck (1858–1947) (see Planck 1912) explained the relation between entropy and the smallest cell of phase space when he attempted to introduce chemists to the idea of quantization. He maintained that they have implicitly used quantization all along; indeed, without quantization, i.e. with $1/Y = 0$, the entropy would have come out as infinite (see (2.12)). Planck identified $1/Y$ as h^3, where h is the universal Planck constant. Thus entropy is the only classical quantity that depends on the Planck constant.

The Kelvin temperature also has a mechanical interpretation; it measures the kinetic energy of the thermal motion of atoms and molecules. We know that atoms are constantly in motion; the atoms of a gas fly straight until they collide, and the atoms in a solid oscillate about their equilibrium position. The mean kinetic energy of that *thermal motion* is given by

$$\frac{p^2}{2m} = \tfrac{3}{2}kT. \tag{2.14}$$

For room temperature the mean speed is of the order of several hundred meters per second. Of course, heavy particles are slower and light particles are faster than the average.

We proceed to discuss the entropy as given by W, the number of ways of realizing a distribution, and—in order to emphasize the generic character of that expression—we shall treat two cases in juxtaposition: a monatomic gas and a rubber molecule (see Figure 2.2). For simplicity of argument we stick to spatial (so-called configurational) distributions as opposed to distributions in phase space.

2.2.1 Configurational Entropy

A rubber molecule may be seen as a long chain of N links with independent orientations $\mathbf{n}^{(l)}$ and an end-to-end distance r (see Figure 2.2). A typical state, or *microstate*, is also given in the figure. The distribution of the molecules is given by $\{N_{n^1}, \ldots, N_{n^P}\}$, the number of links having a particular orientation, and the entropy is given by $k \ln W$ with

$$W = N! \Big/ \prod_{i=1}^{P} N_{n^i}!,$$

since that is the number of microstates that can realize the distribution.

Much the same holds for the gas. The distribution is $\{N_{x^1}, \ldots, N_{x^P}\}$, the number of atoms sitting in a particular position, and the configurational entropy is

$$S = k \ln N! \Big/ \prod_{i=1}^{P} N_{x^i}!.$$

Let us familiarize ourselves with these entropies. Do they have the growth property that we expect from entropy? Yes, they do. In order to see that most easily we focus on a rubber molecule and we assume that *each microstate occurs just as frequently as any other one in the course of the thermal motion.* (This *a priori assumption of equal probability* of each microstate is the only reasonable unbiased assumption, given the random character of the thermal motion that kicks one microstate into another every split second.) For the rubber molecule this means that the fully stretched microstate occurs just as frequently as the kinky microstate of Figure 2.2.

This means also, however, that the kinky *distribution* with a small end-to-end-distance r occurs more frequently than the fully stretched one, because it can be realized by more microstates; for an illustration see Figure 2.3. Most common is the distribution with the most realizations, i.e. with the biggest entropy. For the rubber molecule the distribution with the most microstates is the isotropic one for which an equal number of links point into equal solid angles; rather obviously, the end-to-end distance is zero when they do. For many links, i.e. big N, the number of microstates of that distribution is bigger, or much bigger, than the number of microstates of all other distributions together.

Therefore, if the molecule starts out straight—with $W = 1$, i.e. $S = 0$— the thermal motion will very quickly mess up this straight distribution and kick the molecule into a kinky distribution and, eventually with overwhelming probability, into the distribution of maximum W, i.e. maximum S, which we call the *equilibrium distribution*. That is the nature of the growth of entropy as the molecule tends to equilibrium.

Can we prevent this growth of entropy? Yes, we can. If we wish to keep the molecule straight, we need only give it a proper pull at the ends each time the thermal motion kicks it. And, if the thermal motion kicks the molecule 10^{12} times per second, we may apply a constant force at the ends. That is the nature of entropic forces and entropic elasticity.

The force needed to maintain the end-to-end distance r of the chain is proportional to r; we therefore speak of an *entropic elastic spring*. The elastic constant is proportional to the temperature, so that the spring is stiff for a high intensity of the thermal motion and soft for a low intensity.

Much the same can be said about the atoms in a gas enclosed in a volume V (see Figure 2.2). Here the distribution with the highest number of microstates is the homogeneous distribution, where all of the YV points within V are occupied

ENTROPY: A SUBTLE CONCEPT IN THERMODYNAMICS

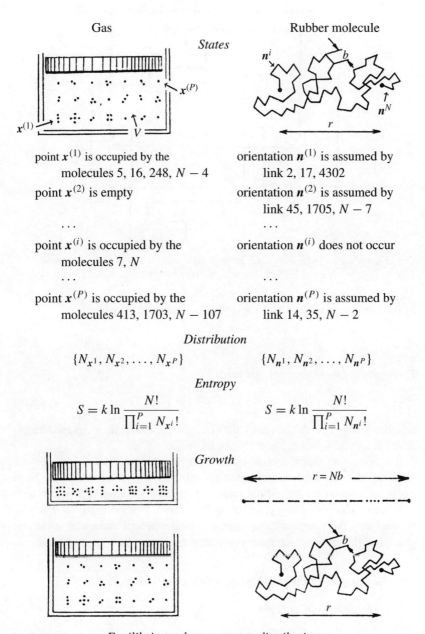

Figure 2.2. Entropies of a gas and a rubber molecule in juxtaposition.

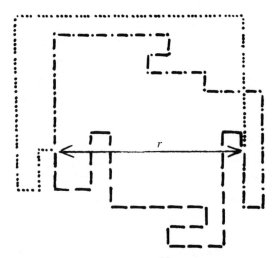

Figure 2.3. Three out of the 2.086×10^{15} microstates of the distribution $\{N_N, N_S, N_E, N_W\} = \{9, 9, 2, 12\}$ with the four orientations N, S, E, W ($P = 4$).

by the same number of atoms. Obviously, we must then have

$$N_{x^i} = \frac{N}{YV}. \tag{2.15}$$

Therefore, the maximum entropy reads

$$S = k \ln V^N + k \ln Y^N \tag{2.16}$$

and that means that the tendency for entropic growth tends to spread the gas molecules evenly over as big a volume as possible.

It is often said that the value of the entropy of a distribution is a measure of the disorder in the arrangement of its particles. This interpretation is most easily understood for the rubber molecule. Indeed, the stretched-out, orderly distribution has zero entropy, while the disordered, kinky distribution has positive entropy. Also, in a gas, the order is perfect when all molecules are in one place, and the gas is all the more disorderly the more points are occupied.

2.3 Entropy and Potential Energy of Gravitation

2.3.1 Planetary Atmospheres

We recall that the entropy of a gas is maximal for a homogeneous distribution of the atoms. At least, that is what we have concluded. But such a distribution does not prevail in planetary atmospheres—such as the atmosphere of the Earth—where the lower layers are denser than the upper ones. We know why this is so: it

ENTROPY: A SUBTLE CONCEPT IN THERMODYNAMICS

is the effect of the gravitational field. Indeed, in this case we have a competition of two conflicting tendencies:[4]

(i) the potential energy which tends to become minimal by assembling all molecules on the planetary surface; and

(ii) the entropy which tends to become maximal by distributing the molecules over as large a volume as possible, i.e. over all space.

When gravitation and thermal motion compete in this way, it is neither the potential energy E_{pot} that reaches a minimum, nor the entropy S that reaches a maximum. The prize goes to the free energy

$$F = E_{pot} - TS, \qquad (2.17)$$

which becomes minimal; and temperature determines which dominates, energy or entropy.

For a *high* temperature the energetic term in (2.17) may be ignored and the free energy becomes minimal, because the entropy becomes maximal. Therefore, the atmosphere of a hot planet scatters over the whole space and, indeed, mercury, the hottest planet, has none left.

For a *low* temperature the entropic term in the free energy may be ignored and the free energy becomes minimal, because the energy does. The cold planets, far from the Sun, have therefore retained all their gases, even the light hydrogen which was overabundant when the planetary system was formed. That is why those planets—Jupiter, Uranus, Neptune—are so big, and relatively light.

Our Earth stands in the middle—its temperature is too high to have kept its hydrogen, but too low to lose the heavier gases like nitrogen or oxygen. Their molecules are heavier than those of hydrogen so that their thermal motion is slower. As a result the molecules of oxygen and nitrogen do not reach the escape velocity as readily as hydrogen molecules.

2.3.2 Pfeffer Tube

A tube closed by a water-permeable membrane is dipped into a reservoir filled with water (see Figure 2.4). The water level is the same in both until we throw a little salt into the tube. The salt cannot pass through the membrane and therefore it will dissolve homogeneously in its little bit of water inside the tube, and that might be considered the end of it. However, nature is cleverer than that! The entropy of the salt could grow, if it had more volume. Therefore, the salt pulls water into the tube so that it can expand; or else, the water pushes itself into the tube so that the salt can expand. Of course, 'pulling' and 'pushing' are anthropomorphic terms; in reality, it is blind chance that is at work here. The system moves to a distribution of salt and water with more microstates.

---

[4] See Müller (2001) for an explicit analysis of the competition in some cases: Pfeffer tube, ammonia synthesis, droplets in vapor and shape-memory alloys.

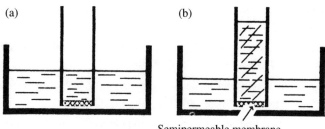

Semipermeable membrane

Figure 2.4. Pfeffer tube (a) without and (b) with salt.

Note, however, the *system* moves to a better distribution: the salt profits in the process by increasing its entropy, while the water has to pay by increasing its potential energy. Profit and payment eventually balance and the tube may be filled as high as 10 m for reasonable data, e.g. 2 l of water, 1 g of salt, and a tube diameter of 1 cm^2.

2.4 Entropy and Intermolecular Energies

If entropy and potential energy compete in the manner described above, the energy need not be the potential energy of gravitation. It may be the potential energy of the interaction between atoms and molecules. When the molecules of a gas are far apart, they feel no forces and the potential energy between them is zero. But when two or more molecules are close, there is mutual attraction, the potential energy is then negative. It has a minimum when the molecules are in contact. The intermolecular forces are called van der Waals forces after the physicist Johannes Diderik van der Waals (1837–1923), who conceived them.

Depending on the type of atom or molecule the attraction is strong or weak. If it is strong—as it is for water molecules—the molecules are all close together under normal conditions, i.e. the water is condensed, it forms a liquid. The potential energy is then small, because all molecules are trapped in the potential minima which they offer to each other. And the entropy is small, because the volume of the liquid is small. When the temperature is raised, the intensity of the thermal motion of the molecules grows and this enables them to jump out of the potential wells and to form a gas, or a vapor. In that case the entropy dominates in its tendency to distribute the molecules over a large volume.

Condensation and evaporation are not something that can only be experienced by the trillions of molecules in a steam cylinder (say). We may simulate them with only seven molecules on the computer screen. Figure 2.5 shows what we see upon 'cooling.' In (a) we see molecules flying about on the screen with great speeds over the whole area.[5] They represent a vapor and the potential energy

---

[5] Figure 2.5 shows stills of a movie. Therefore the speeds of the molecules cannot be appreciated.

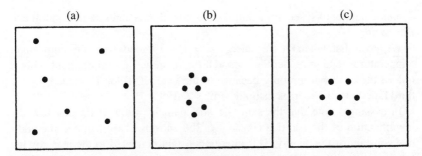

Figure 2.5. Seven atoms simulating (a) a gas, (b) a liquid, and (c) a solid.

is negligibly small compared to the kinetic energy. The cooling is simulated by instructing the computer to withdraw half of the kinetic energy from every molecule that hits the wall. Consequently, the molecules become slower, their kinetic energy drops and eventually it becomes smaller than the depth of the potential well that the molecules offer to each other. In that case the molecules will cling together and form a droplet of liquid (see Figure 2.5(b)). This is not the end, however. It so happens that a regular arrangement of molecules creates tiny potential wells which are too shallow to trap the molecules of the liquid. But when the liquid is cooled, the molecules will eventually be trapped in those wells and form a regular hexagonal 'crystal,' thus simulating freezing of the liquid (see Figure 2.5(c)).

Both the kinetic and the potential energy of the seven molecules drop in the process and so does the entropy, because the volume becomes smaller and the 'order' increases. Indeed, the perfect disorder of the gas[6] is converted into the perfect order of the crystal as the temperature decreases and the entropy with it.

We have seen that the entropy decreases with temperature as the transition from vapor to liquid to crystalline solid occurs. Effectively, there are only seven positions in the solid phase, each occupied by one molecule so that $W = 1$ holds and $S = 0$. Observations like this were extrapolated by Hermann Walther Nernst (1864–1941) (see Nernst 1924) and Planck, and led them to proclaim the *third law of thermodynamics*:

$$\boxed{\text{3rd law}} \quad S(T, p) \xrightarrow[T \to 0]{} 0 \quad \forall p. \tag{2.18}$$

Knowing this we may calculate $S(T, p)$ by integration

$$S(T, p) = \int_0^T \frac{C_p(\tau, p)}{\tau} \, d\tau, \tag{2.19}$$

where C_p is the heat capacity. Thus the arbitrary additive constant mentioned in Section 2.1 is now known. It is listed for all materials of interest in thermo-

---

[6] Recall that the word 'gas' is a garbled form of the Greek word χάος for chaos.

dynamic handbooks, see, for example, those by d'Ans and Lax or Landolt and Börnstein.

We recall that, when the free energy $F = E - TS$ tends to a minimum, small temperatures tend to render the second term, namely, TS, unimportant. This is indeed the case, but not only because of the explicit factor T. Indeed, by the third law, S itself becomes insignificant as T drops.

In passing I note that the same decade, namely, 1900–1910, that saw the identification of the additive constant of the entropy also identified the additive constant of the energy. This energy constant is related to the mass by the famous equivalence $E = mc^2$ between mass and energy discovered by Albert Einstein (1879–1955). Unfortunately, the mass of a body, and the mass defects in chemical compounds, cannot be measured with sufficient accuracy to use this knowledge for writing tables of energy constants.

2.5 Entropy and Chemical Energies

When entropy and potential energy compete, the energy may not be the potential energy of gravitation, nor the potential energy of the van der Waals forces. It may be the potential energy of the chemical binding forces.

In a chemical reaction, when *products* appear and *educts* disappear, so do their masses, energies and entropies, including the additive constants in energies and entropies. The difference of the energy constants of the products and educts is called ΔE^R; where the superscript 'R' stands for a reference state, usually 1 atm and 25 °C, and the corresponding entropic quantity is ΔS^R. The latter is known, as was explained in Section 2.4 in connection with the third law, and the former, namely, ΔE^R, can be measured by measuring the *heat of reaction*. Both are listed in the handbooks.

An instructive reaction for study is the ammonia synthesis

$$N_2 + 3H_2 \longrightarrow 2NH_3 \qquad (2.20)$$

in which case we obtain from the handbooks[7]

$$\left.\begin{array}{l} \Delta E^R = -46.2 \text{ kJ}, \\ \Delta S^R = -89.3 \text{ J K}^{-1}, \end{array}\right\}$$

such that $\Delta F^R = \begin{cases} \Delta E^R - T\Delta S^R < 0 & \text{for } T = 298 \text{ K}, \\ \Delta E^R - T\Delta S^R > 0 & \text{for } T = 773 \text{ K}. \end{cases}$ \qquad (2.21)

Thus both the energy and the entropy drop in the formation of ammonia. In a manner of speaking the energy, which wants to decrease, favours ammonia while the entropy, which wants to increase, favours N_2 and H_2. But at 25 °C

---

[7] The thermodynamically sophisticated reader will know that ΔE^R in this section is an enthalpy and ΔF^R is a free enthalpy.

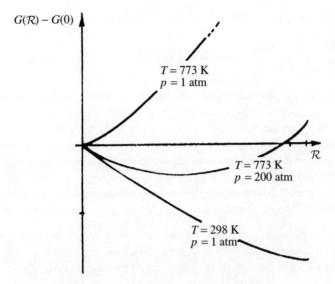

Figure 2.6. Ammonia synthesis. Free energy versus extent of reaction under different conditions.

the energy prevails, because the free energy drops. Indeed, ΔF^R is negative. Thus we might presume that a mixture of hydrogen and nitrogen is converted into ammonia, because that conversion decreases the free energy as the extent of reaction $\mathcal{R}$ proceeds from 0 to 1 (see Figure 2.6).

Unfortunately for a world that craves ammonia so as to be able to produce fertilizers and explosives, it does not happen that way. *Nothing* happens in a mixture of N_2 and H_2, because the reaction requires a catalyst to split H_2 and N_2 into their atomic components. In the Haber–Bosch synthesis the catalyst is iron and for that to be effective the temperature must be raised to 500 °C. But at that temperature the entropy prevails (see (2.21)); the minimum of the free energy lies at $\mathcal{R} = 0$ (see Figure 2.6), so that again nothing happens.

Fritz Haber (1868–1934) (see Haber 1905) was one of the first to know these relations between energy and entropy and he also knew—as we do from (2.16)—that S can be decreased by a decrease of volume, i.e. an increase of pressure. Therefore, with the help of Karl Bosch (1874–1940), an engineer, he applied a pressure of 200 bar. In that case $|\Delta S|$ is small enough for the minimum of the free energy to be shifted to $\mathcal{R} = 0.42$ (see Figure 2.6), so that ammonia could be produced in quantity.

Another nontrivial chemical reaction from the thermodynamic point of view is photosynthesis, a reaction by which plants use the CO_2 from the air and the water from the soil to produce glucose, setting oxygen free:

$$CO_2 + H_2O \longrightarrow \tfrac{1}{6}C_6H_{12}O_6 + O_2. \tag{2.22}$$

The handbooks provide the heat of reaction and the entropy of reaction as

$$\left.\begin{array}{l}\Delta E^R = +466.3 \text{ kJ}, \\ \Delta S^R = -40.1 \text{ J K}^{-1},\end{array}\right\} \text{ such that } \Delta F^R = 478.25 \text{ for } T = 298 \text{ K}. \quad (2.23)$$

Here both the growing energy and the dropping entropy work against the formation of glucose and we may ask how photosynthesis can happen at all. Part of the answer is that the necessary *increase of energy* is provided by the radiative energy of the Sun. But how do we provide for the necessary *decrease of entropy*. The answer must be that there is an accompanying process which increases the entropy so much that the system can afford to 'pay' for the negative ΔS^R of the glucose synthesis. To my knowledge there is no agreement among scientists as to the nature of the accompanying process, but there are at least two suggestions.

- Erwin Schrödinger (1887–1961) (see Schrödinger 1944) suggested that the necessary growth of entropy was to be found in the difference between the radiative entropy fluxes entering and leaving the system.

- Another suggestion is based upon the observation that plants need much more water than is required for the reaction (2.22)—100 to 1000 times as much. They evaporate that water, but evaporation is not enough, since the free energy does not change in a phase transition. However, the evaporated water mixes with the surrounding air and the positive entropy of mixing can indeed offset the negative ΔS^R of the glucose synthesis, provided the plant is watered and ventilated well. For a quantitative investigation of this 'mechanism,' we refer the interested reader to a recent paper by Klippel and Müller (1997).

2.6 Omissions

The concepts of energy and entropy, and the three laws of thermodynamics, were invented and perfected in the 19th century and they are now—and have been for 100 years—cornerstones of physics and the engineering science. They will never change! The physical, technical and philosophical implications of these discoveries were well appreciated by the founders and their contemporaries; and the 60 years of discovery in the field of thermodynamics, between 1850 and 1910, are to this day some of the most eventful years of modern science. Even now they are visited and revisited over and over again by fascinated physicists and historians of science.

I have attempted to give a flavour of these discoveries but the limited space available has forced me to omit much. Among the omissions are the entropy of an atom, degeneracy, Fermi and Bose particles, Bose–Einstein condensation, superfluidity, superconduction, electrolytes, chemical potentials, the Gibbs phase rule, phase diagrams, azeotropy, eutectics, miscibility gaps, entropic stabilization, interfacial energies and entropies, droplets and bubbles, humid air, fog and clouds, thermal convection, radiation, and more!

Entropy also plays an extremely important role in our everyday life, which is given its quality, far surpassing that of previous centuries, by technological devices for energy conversion and transport such as steam engines, power stations, car engines, refrigerators, air conditioners, heat pumps, rockets, turbines, air bags, saunas, fans, chimneys, solar panels, and more!

All this has not been mentioned above. However, all these subject may be treated by the thermodynamics of *reversible* processes—sometimes called thermo*statics*—to an excellent approximation.

Consequently, the true thermo*dynamics*—the thermodynamics of *irreversible* processes—has been neglected. That theory might be said to be still in its infancy. In Chapter 5 of this book I shall give a brief account of what has happened in *thermodynamics* in the 20th century and then I will describe the most modern branch of that theory: extended thermodynamics.

References

Alt, H. W. 1998 An evolution principle and the existence of entropy. First order systems. *Continuum Mech. Thermodyn.* **10**, 61–79.

Boltzmann, L. 1910 *Vorlesungen über die Gastheorie*, 2nd edn. J. A. Barth, Leipzig.

Carathéodory, C. 1909 Untersuchungen über die Grundlagen der Thermodynamik. *Math. Annalen* **67**.

Carnot, S. 1824 *Réflexions sur la puissance motrice du feu et sur les machines propres a développer cette puissance*. Chez Bachelier, Libraire, Paris.

Clausius, R. 1887 *Die Mechanische Wärmetheorie*, Vol. 1, 3rd edn. Vieweg & Sohn, Braunschweig.

Gibbs, J. W. 1960 *The Scientific Papers*, Vol. I. Dover, New York.

Haber, F. 1905 *Thermodynamik technischer Gasreaktionen*. R. Oldenbourg, München.

Klippel, A. and Müller, I. 1997 Plant growth—a thermodynamicist's view. *Continuum Mech. Thermodyn.* **9**, 127–142.

Maxwell, J. C. 1866 On the dynamical theory of gases. *Phil. Trans. R. Soc. Lond.* **157**, 49–88.

Müller, I. 1985 *Thermodynamics*. Pitman.

Müller, I. 2001 *Grundzüge der Thermodynamik—mit historischen Anmerkungen*, 3rd edn. Springer.

Nernst, W. 1924 *Die theoretischen und experimentellen Grundlagen des neuen Wärmesatzes*, 2nd edn. W. Knapp, Halle.

Nietzsche, F. 1872 *Kritische Gesamtausgabe*, Vol. 3, A. Abt. Nachgelassene Fragmente. Walter de Gruyter, Berlin and New York.

Planck, M. 1912 *Über neuere thermodynamische Theorien* (Nernst'sches Wärmetheorem und Quantenhypothese, Read 16 December 1911 at the German Chemical Society, Berlin). Akad. Verlagsges., Leipzig.

Schrödinger, E. 1944 *What Is Life?* Cambridge University Press.
Spengler, O. 1919 Die Entropie und der Mythos der Götterdämmerung. In *Der Untergang des Abendlandes. Umrisse einer Morphologie der Weltgeschichte*, Kapitel VI, Vol. 1, 3rd edn. C. H. Beck'sche Verlagsbuchhandlung, München.

Chapter Three

Probabilistic Aspects of Entropy

Hans-Otto Georgii
Mathematisches Institut der Universität München

> When Shannon had invented his quantity and consulted von Neumann on what to call it, von Neumann replied: 'Call it entropy. It is already in use under that name and besides, it will give you a great edge in debates because nobody knows what entropy is anyway.'
>
> (Denbigh 1990)

We give an overview of some probabilistic facets of entropy, recalling how entropy shows up naturally in various different situations ranging from information theory and hypothesis testing over large deviations and the central limit theorem to interacting random fields and the equivalence of ensembles.

3.1 Entropy as a Measure of Uncertainty

As is well known, it was Ludwig Boltzmann who first gave a probabilistic interpretation of thermodynamic entropy. He coined the famous formula

$$S = k \log W, \qquad (3.1)$$

which is engraved on his tombstone in Vienna: the entropy S of an observed macroscopic state is nothing other than the logarithmic probability of its occurrence, up to some scalar factor k (the Boltzmann constant), which is physically significant but can be ignored from a mathematical point of view. The history and physical significance of this formula are discussed in Chapter 2 by Müller. Here I will simply recall its most elementary probabilistic interpretation.

Let E be a finite set and μ a probability measure on E.[1] In the Maxwell–Boltzmann picture, E is the set of all possible energy levels for a system of particles, and μ corresponds to a specific histogram of energies describing some macrostate of the system. Assume for a moment that each $\mu(x)$, $x \in E$, is a multiple of $1/n$, i.e. μ is a histogram for n trials or, equivalently, a macrostate for a system of n particles. On the microscopic level, the system is then described

---

Opening lecture given at the 'International Symposium on Entropy' hosted by the Max Planck Institute for Physics of Complex Systems, Dresden, Germany, 26–28 June 2000.

[1] Here and throughout we assume for simplicity that $\mu(x) > 0$ for all x.

by a sequence $\omega \in E^n$, the microstate, associating with each particle its energy level. Boltzmann's idea is now the following.

> *The entropy of a macrostate μ corresponds to the degree of uncertainty about the actual microstate ω when only μ is known, and can thus be measured by $\log N_n(\mu)$, the logarithmic number of microstates leading to μ.*

Explicitly, for a given microstate $\omega \in E^n$ let[2]

$$L_n^\omega = \frac{1}{n} \sum_{i=1}^n \delta_{\omega_i} \qquad (3.2)$$

be the associated macrostate describing how the particles are distributed over the energy levels. L_n^ω is called the *empirical distribution* (or histogram) of $\omega \in E^n$. Then

$$N_n(\mu) \equiv |\{\omega \in E^n : L_n^\omega = \mu\}| = \frac{n!}{\prod_{x \in E}(n\,\mu(x))!},$$

the multinomial coefficient. In view of the n-dependence of this quantity, one should approximate a given μ by a sequence μ_n of n-particle macrostates and define the uncertainty $H(\mu)$ of μ as the $n \to \infty$ limit of the 'mean uncertainty of μ_n per particle.' Using Stirling's formula, we arrive in this way at the well-known expression for the entropy.

(3.3) Entropy as degree of ignorance. *Let μ and μ_n be probability measures on E such that $\mu_n \to \mu$ and $n\,\mu_n(x) \in \mathbb{Z}$ for all $x \in E$. Then the limit*

$$\lim_{n \to \infty} \frac{1}{n} \log N_n(\mu_n)$$

exists and is equal to

$$H(\mu) = -\sum_{x \in E} \mu(x) \log \mu(x).$$

A proof including error bounds is given in Lemma 2.3 of Csiszár and Körner (1981).

Though we have used the terms uncertainty and ignorance, the entropy $H(\mu)$ should not be considered as a subjective quantity. It simply counts the number of possibilities to obtain the histogram μ, and thus describes the *hidden multiplicity* of 'true' microstates consistent with the observed μ. It is therefore a measure of the *complexity* inherent in μ.

To summarize, in Boltzmann's picture, μ is a histogram resulting from a random phenomenon on the microscopic level, and $H(\mu)$ corresponds to the observer's uncertainty of what is really going on there.

---

[2] We write δ_x for the Dirac measure at $x \in E$.

3.2 Entropy as a Measure of Information

We will now approach the problem of measuring the 'uncertainty content' of a probability measure μ from a different side suggested by Shannon (1948). Whereas Boltzmann's view is backwards to the microscopic origins of μ, Shannon's view is ahead, taking μ as given and 'randomizing' it by generating a random signal with alphabet E and law μ. His question is, How large is the receiver's effort to recover μ from the signal? This effort can be measured by the number of yes-or-no questions to be answered on average in order to identify the signal (and thereby μ, after many independent repetitions). So it corresponds to the receiver's *a priori* uncertainty about μ. But, as observed by Shannon, this effort measures also the degree of information the receiver gets *a posteriori* when all necessary yes-or-no questions are answered. This leads to the following concept of information.

> *The information contained in a random signal with prescribed distribution is equal to the expected number of bits necessary to encode the signal.*

Specifically, a binary prefix code for E is a mapping $f : E \to \bigcup_{\ell \geq 1} \{0, 1\}^\ell$ from E into the set of all finite zero–one sequences which is decipherable, in that no codeword $f(x)$ is a prefix of another codeword $f(y)$. (Such an f can be described by a binary decision tree, the leaves of which correspond to the codewords.) Let $\#f(x)$ denote the length of the codeword $f(x)$, and $\mu(\#f)$ the expectation of the random variable $\#f$ under μ. A natural candidate for the information contained in the signal is then the minimal expected length

$$I_{\mathrm{p}}(\mu) = \inf\{\mu(\#f) : f \text{ binary prefix code for } E\}$$

of a binary prefix code for E. This quantity is already closely related to $H(\mu)$, but the relationship becomes nicer if one assumes that the random signal forms a memoryless source, in that the random letters from E are repeated independently, and one encodes signal words of length n (which are distributed according to the product measure μ^n). In this setting, $I_{\mathrm{p}}(\mu^n)/n$ is the information per signal letter, and in the limit $n \to \infty$ one obtains the following theorem.

(3.4) Source-coding theorem for prefix codes. *The information contained in a memoryless source with distribution μ is*

$$\lim_{n \to \infty} \frac{I_{\mathrm{p}}(\mu^n)}{n} = -\sum_{x \in E} \mu(x) \log_2 \mu(x) \equiv H_2(\mu) = \frac{1}{\log 2} H(\mu).$$

For a proof of a refined version, see, for example, Theorem 4.1 of Csiszár and Körner (1981); cf. Theorem 15.1 in Chapter 15.

An alternative coding scheme leading to a similar result is block coding with small error probability. A binary n-block code of length ℓ with error level $\alpha > 0$ is a mapping $f : E^n \to \{0, 1\}^\ell$ together with a decoder $\varphi : \{0, 1\}^\ell \to E^n$ such that $\mu^n(\varphi \circ f \neq \mathrm{id}) \leqslant \alpha$. Let

$$I_b(\mu^n, \alpha) = \inf\{\ell : \exists\, n\text{-block code of length } \ell \text{ at level } \alpha\}$$

be the minimal length of a binary n-block code with error level α. The following result then gives another justification of entropy.

(3.5) Source-coding theorem for block codes. *The information contained in a memoryless source with distribution μ is*

$$\lim_{n \to \infty} \frac{I_b(\mu^n, \alpha)}{n} = H_2(\mu),$$

independently of the error level $\alpha > 0$.

The proof of this result (see, for example, Theorem 1.1 of Csiszár and Körner (1981)) relies on an intermediate result which follows immediately from the weak law of large numbers. It reveals yet another role of entropy and is therefore interesting in its own right.

(3.6) Asymptotic equipartition property. *For all $\delta > 0$,*

$$\mu^n\left(\omega \in E^n : \left|\frac{1}{n}\log \mu^n(\omega) + H(\mu)\right| \leqslant \delta\right) \xrightarrow[n \to \infty]{} 1.$$

In other words, most ω have probability $\mu^n(\omega) \approx e^{-nH(\mu)}$. This may be viewed as a random version of Boltzmann's formula (3.1).

To conclude this section, let us mention that the entropy $H(\mu)$ admits several axiomatic characterizations which underline its significance as a measure of uncertainty and information; cf., for example, the discussion on pp. 25–27 of Csiszár and Körner (1981). However, compared with the previous genuine results these characterizations should rather be considered as *a posteriori* justifications.

3.3 Relative Entropy as a Measure of Discrimination

Let E still be a finite set, and consider two distinct probability measures μ_0 and μ_1 on E. Suppose we do not know which of these probability measures properly describes the random phenomenon we have in mind (which might again be a random signal with alphabet E). We then ask the following question.

> *How easy is it to distinguish the two candidates μ_0 and μ_1 on the basis of independent observations?*

This is a standard problem of statistics, and the standard procedure is to perform a test of the hypothesis μ_0 against the alternative μ_1 with error level α. In fact, if we want to use n independent observations, then we have to test the product measure μ_0^n against the product measure μ_1^n. Such a test is defined by a 'rejection region' $R \subset E^n$; if the observed outcome belongs to R, one decides in favor of the alternative μ_1, otherwise one accepts the hypothesis μ_0. There are two possible errors: rejecting the hypothesis μ_0 although it is true (first kind), and accepting μ_0 though it is false (second kind). The common practice is to keep the error probability of first kind under a prescribed level α and to choose R such that the error probability of the second kind becomes minimal. The minimum value is

$$\rho_n(\alpha; \mu_0, \mu_1) = \inf\{\mu_1^n(R^c) : R \subset E^n, \ \mu_0^n(R) \leqslant \alpha\}.$$

Consequently, it is natural to say that μ_0 and μ_1 are the easier to distinguish the smaller $\rho_n(\alpha; \mu_0, \mu_1)$ turns out to be. More precisely, we have the following.

The degree to which μ_1 can be distinguished from μ_0 on the basis of independent observations can be measured by the rate of decay of $\rho_n(\alpha; \mu_0, \mu_1)$ as $n \to \infty$.

An application of the weak law of large numbers completely similar to that in the source-coding theorem (3.5) gives the following.

(3.7) Lemma of Stein. *The measure for discriminating μ_1 from μ_0 is*

$$-\lim_{n \to \infty} \frac{1}{n} \log \rho_n(\alpha; \mu_0, \mu_1) = \sum_{x \in E} \mu_0(x) \log \frac{\mu_0(x)}{\mu_1(x)} \equiv D(\mu_0 \mid \mu_1),$$

independently of the choice of $\alpha \in \ ]0, 1[$.

This result was first published in Chernoff (1956); see also Corollary 1.2 of Csiszár and Körner (1981) or Lemma 3.4.7 of Dembo and Zeitouni (1993). $D(\mu_0 \mid \mu_1)$ is known as the *relative entropy*, *Kullback–Leibler information*, *I-divergence* or *information gain*. If μ_1 is the equidistribution on E, then

$$D(\mu_0 \mid \mu_1) = \log |E| - H(\mu_0).$$

Hence relative entropy is a generalization of entropy to the case of a nonuniform reference measure, at least up to the sign. (In view of the difference in sign, one might prefer calling $D(\mu_0 \mid \mu_1)$ the *negative* relative entropy. Nevertheless, we stick to the terminology above which has become standard in probability theory.)

Stein's lemma asserts that the relative entropy $D(\cdot \mid \cdot)$ measures the extent to which two probability measures differ. Although $D(\cdot \mid \cdot)$ is not a metric (neither being symmetric nor satisfying the triangle inequality), it can be used to introduce some kind of geometry for probability measures, and in particular some kind of projection of a probability measure on a convex set of such measures (Csiszár 1975). As we will see in a moment, these so-called I-projections play a central role in the asymptotic analysis of the empirical distributions (3.2). But first, as some motivation, let us mention a refinement of Stein's lemma for which the error probability of the first kind is not held fixed but decays exponentially at a given rate. The answer is in terms of L_n^ω and reads as follows.

(3.8) The Hoeffding Theorem. *Let $0 < a < D(\mu_1 \mid \mu_0)$, and consider the test of μ_0 against μ_1 on n observations with the rejection region $R_n = \{\omega \in E^n : D(L_n^\omega \mid \mu_0) > a\}$. Then the error probability of the first kind decays exponentially with rate a, i.e.*

$$\lim_{n \to \infty} \frac{1}{n} \log \mu_0^n(R_n) = -a,$$

and the error probability of the second kind satisfies the exponential bound

$$\mu_1^n(R_n^c) \leq \exp\left[-n \min_{\nu : D(\nu \mid \mu_0) \leq a} D(\nu \mid \mu_1)\right]$$

with optimal exponent.

Hoeffding's original paper is Hoeffding (1965); see also p. 44 of Csiszár and Körner (1981) or Theorem 3.5.4 of Dembo and Zeitouni (1993). It is remarkable that the asymptotically optimal tests R_n do not depend on the alternative μ_1. One should note that Pearson's well-known χ^2-test for the parameter of a multinomial distribution (see, for example, Rao 1973) uses a rejection region similar to R_n, the relative entropy $D(L_n^\omega \mid \mu_0)$ being replaced by a quadratic approximation.

Hoeffding's Theorem is in fact an immediate consequence of a much more fundamental result, the theorem of Sanov. This elucidates the role of relative entropy for the asymptotic behavior of the empirical distributions L_n^ω. The basic observation is the identity

$$\mu^n(\omega) = \exp[-n(D(L_n^\omega \mid \mu) + H(L_n^\omega))], \tag{3.9}$$

which holds for any probability measure μ on E and any $\omega \in E^n$. In view of our first assertion (3.3), it follows that

$$\frac{1}{n} \log \mu^n(\omega \in E^n : L_n^\omega = \nu_n) \to -D(\nu \mid \mu)$$

whenever $\nu_n \to \nu$ such that $n\nu_n(x) \in \mathbb{Z}$ for all x and n. This can be viewed as a version of Boltzmann's formula (3.1) and leads directly to the following theorem due to Sanov (1957), cf. p. 43 of Csiszár and Körner (1981) or Theorem 2.1.10 of Dembo and Zeitouni (1993).

(3.10) Sanov's large-deviation theorem. *Let μ be any probability measure on E and $\mathcal{C}$ a class of probability measures on E with dense (relative) interior, i.e. $\mathcal{C} \subset \text{cl int } \mathcal{C}$. Then*

$$\lim_{n\to\infty} \frac{1}{n} \log \mu^n(\omega \in E^n : L_n^\omega \in \mathcal{C}) = -\inf_{\nu \in \mathcal{C}} D(\nu \mid \mu).$$

Sanov's Theorem provides just a glimpse into large-deviation theory in which (relative) entropies of various kinds play a central role. (More on this can be found in Dembo and Zeitouni (1993) and Chapters 10 and 9 by den Hollander and Varadhan in this book.) Its meaning can be summarized as follows.

Among all realizations with histogram in $\mathcal{C}$, the most probable are those having a histogram closest to μ in the sense of relative entropy.

We will return to this point later in (3.20). Needless to say, Sanov's Theorem can be extended to quite general state spaces E; see Csiszár (1984) or Theorem 6.2.10 of Dembo and Zeitouni (1993).

3.4 Entropy Maximization under Constraints

The second law of thermodynamics asserts that a physical system in equilibrium has maximal entropy among all states with the same energy. Translating this into a probabilistic language and replacing entropy by the more general relative entropy, we are led to the following question.

Let $\mathcal{C}$ be a class of probability measures on some measurable space $(E, \mathcal{E})$ and μ a fixed reference measure on $(E, \mathcal{E})$. What then are the probability measures in $\mathcal{C}$ minimizing $D(\,\cdot\mid\mu)$?

The universal significance of such minimizers has been put forward by Jaynes (1957, 1982). As noticed above, they also arise in the context of Sanov's Theorem (3.10). In the present more general setting, the relative entropy can be defined by $D(\nu \mid \mu) = \sup_{\mathcal{P}} D(\nu_\mathcal{P} \mid \mu_\mathcal{P})$, where the supremum extends over all finite $\mathcal{E}$-measurable partitions $\mathcal{P}$ and $\nu_\mathcal{P}$ stands for the restriction of ν to $\mathcal{P}$. Equivalently, $D(\nu \mid \mu) = \nu(\log f)$ if ν is absolutely continuous with respect to μ with density f, and $D(\nu \mid \mu) = \infty$ otherwise; see Corollary (15.7) of Georgii (1988), for example. (For a third expression see (3.11) below.) The first definition shows in particular that $D(\,\cdot \mid \mu)$ is lower semicontinuous in the so-called τ-topology generated by the mappings $\nu \to \nu(A)$ with $A \in \mathcal{E}$. Consequently, a minimizer does exist whenever $\mathcal{C}$ is closed in this topology. If $\mathcal{C}$ is also convex, the minimizer is uniquely determined due to the strict convexity of $D(\,\cdot \mid \mu)$, and is then called the I-projection of μ on $\mathcal{C}$. We consider here only the most classical case when $\mathcal{C}$ is defined by an integral constraint. That is, writing $\nu(g)$ for the integral of some bounded measurable function $g : E \to \mathbb{R}^d$ with respect to ν, we assume that $\mathcal{C} = \{\nu : \nu(g) = a\}$ for suitable $a \in \mathbb{R}^d$. In other words, we consider the constrained variational problem

$$D(\nu \mid \mu) \stackrel{!}{=} \min, \quad \nu(g) = a.$$

In this case one can use a convex Lagrange multiplier calculus as follows.

For any bounded measurable function $f : E \to \mathbb{R}$ let $P(f) = \log \mu(e^f)$ be the log-Laplace functional of μ. One then has the variational formula

$$D(\nu \mid \mu) = \sup_f [\nu(f) - P(f)], \quad (3.11)$$

meaning that $D(\cdot \mid \mu)$ and P are convex conjugates (i.e. Legendre–Fenchel transforms) of each other; cf. Lemma 6.2.13 of Dembo and Zeitouni (1993). Let

$$J_g(a) = \inf_{\nu : \nu(g) = a} D(\nu \mid \mu)$$

be the 'entropy distance' of $\{\nu : \nu(g) = a\}$ from μ. A little convex analysis then shows that

$$J_g(a) = \sup_{t \in \mathbb{R}^d} [t \cdot a - P(t \cdot g)], \quad (3.12)$$

i.e. J_g is a partial convex conjugate of P (or, in other terms, the Cramér transform of the distribution $\mu \circ g^{-1}$ of g under μ; cf. Chapter 9 by Varadhan). Moreover, if g is nondegenerate (in the sense that $\mu \circ g^{-1}$ is not supported on a hyperplane), then J_g is differentiable on the interior $I_g = \text{int} \{J_g < \infty\}$ of its essential domain, and one arrives at the following.

(3.13) The Gibbs–Jaynes Principle. *For any nondegenerate $g : E \to \mathbb{R}^d$, $a \in I_g$ and $t = \nabla J_g(a)$, the probability measure*

$$\mu_t(dx) = Z_t^{-1} e^{t \cdot g(x)} \mu(dx)$$

on $(E, \mathcal{E})$ is the unique minimizer of $D(\cdot \mid \mu)$ on $\{\nu : \nu(g) = a\}$. Here $Z_t = e^{P(t \cdot g)}$ is the normalizing constant.

Generalized versions of this result can be found in Csiszár (1975, 1984) or Example (9.42) of Vajda (1989).

In statistical mechanics, the measures μ_t of the form above are called *Gibbs distributions*, and the preceding result (or a suitable extension) justifies that these are indeed the equilibrium distributions of physical systems satisfying a finite number of conservation laws. In mathematical statistics, such classes of probability measures are called *exponential families*. Here are some familiar examples from probability theory.

(3.14) Example. Let $E = \mathbb{R}^d$, μ the standard normal distribution on E, and for any positive definite symmetric matrix C let μ_C be the centered normal distribution with covariance matrix $\text{Cov}(\mu_C) = C$. Then μ_C minimizes the relative entropy $D(\cdot \mid \mu)$ among all centered distributions ν with covariance matrix C. Equivalently, μ_C maximizes the differential entropy

$$H(\nu) = \begin{cases} -\int dx\, f(x) \log f(x) & \text{if } \nu \text{ has Lebesgue-density } f, \\ -\infty & \text{otherwise,} \end{cases}$$

in the same class of distributions, and $H(\mu_C) = \frac{1}{2} d \log[2\pi e(\det C)^{1/d}]$.

(3.15) Example. Let $E = [0, \infty[$, $a > 0$ and μ_a be the exponential distribution with parameter a. Then $\nu = \mu_a$ minimizes $D(\nu \mid \mu_1)$, respectively maximizes $H(\nu)$, under the condition $\nu(\mathrm{id}) = 1/a$.

(3.16) Example. Let $E = \mathbb{N}$, $a > 0$ and μ_a be the Poisson distribution with parameter a. Then $\nu = \mu_a$ minimizes $D(\nu \mid \mu_1)$ under the condition $\nu(\mathrm{id}) = a$, and $D(\mu_a \mid \mu_1) = 1 - a + a \log a$.

(3.17) Example. Let $E = C[0,1]$, $a \in E$ and μ_a the image of the Wiener measure $\mu_0 = \mu$ under the shift $x \to x + a$ of E. Then $\nu = \mu_a$ minimizes $D(\nu \mid \mu)$ under the condition $\nu(\mathrm{id}) = a$, and

$$D(\mu_a \mid \mu) = \frac{1}{2} \int_0^1 \dot{a}(t)^2\, dt$$

if a is absolutely continuous with derivative $\dot{a}$, and $D(\mu_a \mid \mu) = \infty$ otherwise (see, for example, Föllmer 1988, Section II.1.1).

3.5 Asymptotics Governed by Entropy

We will now turn to the dynamical aspects of the second law of thermodynamics. As before, we will not enter into a physical discussion of this fundamental law. Rather we will show by examples that the principle of increasing entropy (or decreasing relative entropy) stands also behind a number of well-known facts of probability theory.

Our first example is the so-called ergodic theorem for Markov chains. Let E be a finite set and $P_t = e^{tG}$, $t \geq 0$, the transition semigroup for a continuous-time Markov chain on E. The generator G is assumed to be irreducible. It is well known that there is then a unique invariant distribution μ (satisfying $\mu P_t = \mu$ for all $t \geq 0$ and, equivalently, $\mu G = 0$). Let ν be any initial distribution, and $\nu_t = \nu P_t$ be the distribution at time t. Consider the relative entropy $D(\nu_t \mid \mu)$ as a function of time $t \geq 0$. A short computation (using the identities $\mu G = 0$ and $G\mathbf{1} = 0$) then gives the following result.

(3.18) Entropy production of Markov chains. *For any $t \geqslant 0$ we have*

$$\frac{\mathrm{d}}{\mathrm{d}t} D(\nu_t \mid \mu) = - \sum_{x,y \in E : x \neq y} \nu_t(y)\, \bar{G}(y,x)\, \varphi\!\left(\frac{\nu_t(x)\, \mu(y)}{\mu(x)\, \nu_t(y)} \right)$$

$$= -a(\nu_t)\, D(\tilde{\nu}_t \mid \bar{\nu}_t) \leqslant 0,$$

and in particular

$$\frac{\mathrm{d}}{\mathrm{d}t} D(\nu_t \mid \mu) < 0 \quad \text{when } \nu_t \neq \mu.$$

In the above, $\bar{G}(y,x) = \mu(x)\, G(x,y)/\mu(y)$ is the generator for the time-reversed chain, $\varphi(s) = 1 - s + s \log s \geqslant 0$ for $s \geqslant 0$,

$$a(\nu) = - \sum_{x \in E} \nu(x)\, G(x,x) > 0,$$

and the probability measures $\tilde{\nu}$ and $\bar{\nu}$ on $E \times E$ are defined by

$$\tilde{\nu}(x,y) = \nu(x)\, G(x,y)\, \frac{1 - \delta_{x,y}}{a(\nu)},$$

$$\bar{\nu}(x,y) = \nu(y)\, \bar{G}(y,x)\, \frac{1 - \delta_{x,y}}{a(\nu)},$$

$x, y \in E$. The second statement follows from the fact that 1 is the unique zero of φ, and G is irreducible. A detailed proof can be found in Chapter I of Spitzer (1971). The discrete time analogue was apparently discovered repeatedly by various authors; it appears, for example, in Renyi (1970) and on p. 98 of Kac (1959).

The entropy production formula above states that the relative entropy $D(\cdot \mid \mu)$ is a strict Lyapunov function for the fixed-time distributions ν_t of the Markov chain. Hence $\nu_t \to \mu$ as $t \to \infty$. This is the well-known ergodic theorem for Markov chains, and the preceding argument shows that this convergence result fits precisely into the physical picture of convergence to equilibrium.

Although the central limit theorem is a cornerstone of probability theory, it is often not realized that this theorem is also an instance of the principle of increasing entropy. (This is certainly due to the fact that the standard proofs do not use this observation.) To see this, let (X_i) be a sequence of i.i.d. centered random vectors in $\mathbb{R}^d$ with existing covariance matrix C, and consider the normalized sums $S_n^* = \sum_{i=1}^n X_i / \sqrt{n}$. By the very definition, S_n^* is again centered with covariance matrix C. But, as we have seen in Example (3.14), under these conditions the centered normal distribution μ_C with covariance matrix C has maximal differential entropy. This observation suggests that the relative entropy may again serve as a Lyapunov function. Unfortunately, a time-monotonicity of relative entropies seems to be so far unknown (though monotonicity along the powers of 2 follows from a subadditivity property). But the following statement is true.

(3.19) Entropic central limit theorem. *Let ν_n be the distribution of S_n^*. If ν_1 is such that $D(\nu_n \mid \mu_C) < \infty$ for some n, then $D(\nu_n \mid \mu_C) \to 0$ as $n \to \infty$.*

This theorem can be traced back to Linnik (1959), whose result was put on firm grounds by Barron (1986). The multivariate version above is due to Johnson (2000). By an inequality of Pinsker, Csiszár, Kullback and Kemperman (cf. p. 133 of Föllmer (1988) or p. 58 of Csiszár and Körner (1981)), it follows that $\nu_n \to \mu_C$ in total variation norm (which is equal to the L^1-distance of their densities).

A similar result holds for products of i.i.d. random elements X_i of a compact group G. Let μ_G denote the normalized Haar measure on G, and let ν_n be the distribution of $\prod_{i=1}^n X_i$, i.e. the n-fold convolution of the common distribution of the X_i. A recent result of Johnson and Suhov (2000) then implies that $D(\nu_n \mid \mu_G) \downarrow 0$ as $n \uparrow \infty$, provided $D(\nu_n \mid \mu_G)$ is ever finite. Note that μ_G is the measure of maximal entropy (certainly if G is finite or a torus), and that the convergence here is again monotone in time.

Our third example is intimately connected to Sanov's Theorem (3.10). Suppose again (for simplicity) that E is finite, and let μ be a probability measure on E. Let $\mathcal{C}$ be a closed convex class of probability measures on E such that int $\mathcal{C} \neq \emptyset$. We consider the conditional probability

$$\mu_{\mathcal{C}}^n = \mu^n(\,\cdot\, \mid \{\omega \in E^n : L_n^\omega \in \mathcal{C}\})$$

under the product measure μ^n given that the empirical distribution belongs to the class $\mathcal{C}$. (By Sanov's Theorem, this condition has positive probability when n is large enough.) Do these conditional probabilities converge to a limit? According to the interpretation of Sanov's Theorem, the most probable realizations ω are those for which $D(L_n^\omega \mid \mu)$ is as small as possible under the constraint $L_n^\omega \in \mathcal{C}$. But we have seen above that there exists a unique probability measure $\mu_* \in \mathcal{C}$ minimizing $D(\,\cdot\, \mid \mu)$, namely, the I-projection of μ on $\mathcal{C}$. This suggests that, for large n, $\mu_{\mathcal{C}}^n$ concentrates on configurations ω for which L_n^ω is close to μ_*. This and even more is true, as was shown by Csiszár (1984).

(3.20) Csiszár's conditional limit theorem. *For closed convex $\mathcal{C}$ with non-empty interior,*

$$\mu^n(\,\cdot\, \mid \{\omega \in E^n : L_n^\omega \in \mathcal{C}\}) \to \mu_*^{\mathbb{N}} \quad \text{as } n \to \infty,$$

where μ_* is the I-projection from μ on $\mathcal{C}$.

Note that the limit is again determined by the maximum entropy principle. It is remarkable that this result follows from purely entropic considerations. Writing $\nu_{\mathcal{C},n} = \mu_{\mathcal{C}}^n(L_n^{\cdot})$ for the mean conditional empirical distribution (which by symmetry coincides with the one-dimensional marginal of $\mu_{\mathcal{C}}^n$), Csiszár

(1984) observes that

$$-\frac{1}{n}\log\mu^n(\omega \in E^n : L_n^\omega \in \mathcal{C}) = \frac{1}{n}D(\mu_\mathcal{C}^n \mid \mu^n)$$
$$= \frac{1}{n}D(\mu_\mathcal{C}^n \mid (\nu_{\mathcal{C},n})^n) + D(\nu_{\mathcal{C},n} \mid \mu)$$
$$\geq \frac{1}{n}D(\mu_\mathcal{C}^n \mid \mu_*^n) + D(\mu_* \mid \mu).$$

The inequality can be derived from the facts that $\nu_{\mathcal{C},n} \in \mathcal{C}$ by convexity and μ_* is the I-projection of μ on $\mathcal{C}$. Now, by Sanov's Theorem, the left-hand side tends to $D(\mu_* \mid \mu)$, whence $(1/n)D(\mu_\mathcal{C}^n \mid \mu_*^n) \to 0$. In view of the superadditivity properties of relative entropy, it follows that for each $k \geq 1$ the projection of $\mu_\mathcal{C}^n$ onto E^k converges to μ_*^k, and one arrives at (3.20).

The preceding argument is completely general: Csiszár's original paper (Csiszár 1984) deals with the case when E is an arbitrary measurable space. In fact, some modifications of the argument even allow the replacement of the empirical distribution L_n^ω with the so-called empirical process; this will be discussed below in (3.27).

3.6 Entropy Density of Stationary Processes and Fields

Although we have already occasionally considered sequences of i.i.d. random variables, our main concern so far has been the entropy and relative entropy of (the distribution of) a single random variable with values in E. In this final section we will recall how the ideas described so far extend to the set-up of stationary stochastic processes, or stationary random fields, and our emphasis here is on the nonindependent case.

Let E be a fixed state space. For simplicity we assume again that E is finite. We consider the product space $\Omega = E^{\mathbb{Z}^d}$ for any dimension $d \geq 1$. For $d = 1$, Ω is the path space of an E-valued process, while for larger dimensions Ω is the configuration space of an E-valued random field on the integer lattice. In each case, the process or field is determined by a probability measure μ on Ω. We will assume throughout that all processes or fields are *stationary*, respectively *translation invariant*, in the sense that μ is invariant under the shift-group $(\vartheta_x)_{x \in \mathbb{Z}^d}$ acting on Ω in the obvious way.

In this setting it is natural to consider the entropy or relative entropy *per time*, respectively *per lattice site*, rather than the (total) entropy or relative entropy. (In fact, $D(\nu \mid \mu)$ is infinite in all interesting cases.) The basic result on the existence of the entropy density is the following. In its statement, we write $\Lambda \uparrow \mathbb{Z}^d$ for the limit along an arbitrary increasing sequence of cubes exhausting $\mathbb{Z}^d$, μ_Λ for the projection of μ onto E^Λ, and ω_Λ for the restriction of $\omega \in \Omega$ to Λ.

(3.21) The Shannon–McMillan–Breiman Theorem. *For any stationary μ on Ω, there exists the entropy density*

$$h(\mu) = \lim_{\Lambda \uparrow \mathbb{Z}^d} |\Lambda|^{-1} H(\mu_\Lambda),$$

and for the integrands we have

$$-\lim_{\Lambda \uparrow \mathbb{Z}^d} |\Lambda|^{-1} \log \mu_\Lambda(\omega_\Lambda) = h(\mu(\,\cdot\, | \, \mathcal{I})(\omega))$$

for μ-almost ω and in $L^1(\mu)$. Here $\mu(\,\cdot\, | \, \mathcal{I})(\omega)$ is a regular version of the conditional probability with respect to the σ-algebra $\mathcal{I}$ of shift-invariant events in Ω.

For a proof we refer to Section 15.2 of Georgii (1988) (and the references therein), and Section I.3.1 of Föllmer (1988). In the case of a homogeneous product measure $\mu = \alpha^{\mathbb{Z}^d}$ we have $h(\mu) = H(\alpha)$.

In view of Boltzmann's interpretation (3.3) of entropy, $h(\mu)$ is a measure of the lack of knowledge about the process or field per time, respectively per site. Also, the L^1-convergence result of McMillan immediately implies an asymptotic equipartition property analogous to (3.6), whence $h(\mu)$ is also the optimal rate of a block code, and thus the *information per signal* of the stationary source described by μ (provided we take the logarithm to base 2).

What about the existence of a *relative* entropy per time or per site? Here we need to assume that the reference process has a nice dependence structure, which is also important in the context of the maximum entropy problem.

Let $f : \Omega \to \mathbb{R}$ be any function depending only on the coordinates in a finite subset Δ of $\mathbb{Z}^d$. Such a function will be called local. A probability measure μ on Ω is called a *Gibbs measure* for f if its conditional probabilities for observing a configuration ω_Λ in a finite region $\Lambda \subset \mathbb{Z}^d$, given a configuration ω_{Λ^c} outside of Λ, are almost surely given by the formula

$$\mu(\omega_\Lambda \mid \omega_{\Lambda^c}) = Z_\Lambda(\omega_{\Lambda^c})^{-1} \exp\left[\sum_{x:(\Delta+x)\cap\Lambda\neq\emptyset} f(\vartheta_x \omega)\right],$$

where $Z_\Lambda(\omega_{\Lambda^c})$ is the normalization constant. Since f is local, each Gibbs measure μ is Markovian in the sense that the conditional probabilities above only depend on the restriction of ω_{Λ^c} to a bounded region around Λ. (This assumption of finite range could be weakened, but here is no place for this.) The main interest in Gibbs measures comes from their use for describing systems of interacting spins in equilibrium, and the analysis of phase transitions; a general account can, for example, be found in Georgii (1988). (To make the connection with the definition given there let the potential Φ be defined as in Lemma (16.10) of this reference.) In the present context, Gibbs measures simply show up because of their particular dependence properties. We now can state the following counterpart to (3.21).

(3.22) Ruelle–Föllmer Theorem. *Suppose μ is a Gibbs measure for some local function f, and ν is translation invariant. Then the relative entropy density*

$$d(\nu \mid \mu) = \lim_{\Lambda \uparrow \mathbb{Z}^d} |\Lambda|^{-1} D(\nu_\Lambda \mid \mu_\Lambda) \tag{3.23}$$

exists and is equal to $p(f) - h(\nu) - \nu(f)$, where

$$p(f) = \max_\nu [h(\nu) + \nu(f)] = \lim_{\Lambda \uparrow \mathbb{Z}^d} |\Lambda|^{-1} \log Z_\Lambda(\omega_{\Lambda^c}), \tag{3.24}$$

the so-called pressure of f, is the counterpart of the log-Laplace functional appearing in (3.13).

The second identity in (3.24) is often called the *variational formula*; it dates back to Ruelle (1967). Föllmer (1973) made the connection with relative entropy; for a detailed account see also Theorem (15.30) of Georgii (1988) or Section I.3.3 of Föllmer (1988). An example of a nonGibbsian μ for which $d(\cdot \mid \mu)$ fails to exist was constructed by Kieffer and Sokal (see van Enter et al. 1993, pp. 1092–1095). As in (3.21), there is again an $L^1(\nu)$ and ν-almost sure convergence behind (3.23) (Föllmer 1973). In the case $f = 0$ when the unique Gibbs measure μ is equal to $\alpha^{\mathbb{Z}^d}$ for the equidistribution α on E, the Ruelle–Föllmer Theorem (3.22) reduces to (3.21).

Since $D(\nu \mid \mu) = 0$ if and only if $\nu = \mu$, the preceding result leads us to ask what one can conclude from the identity $d(\nu \mid \mu) = 0$. The answer is the following celebrated variational characterization of Gibbs measures first derived by Lanford and Ruelle (1969). Simpler proofs were given later by Föllmer (1973) and Preston (1976, Theorem 7.1); cf. Section 15.4 of Georgii (1988) or Theorem (I.3.39) of Föllmer (1988).

(3.25) The variational principle. *Suppose ν is stationary. Then ν is a Gibbs measure for f if and only if $h(\nu) + \nu(f)$ is equal to its maximum value $p(f)$.*

Physically speaking, this result means that the stationary Gibbs measures are the minimizers of the free energy density $\nu(-f) - h(\nu)$, and therefore describe a physical system with interaction f in thermodynamic equilibrium.

It is now easy to obtain an analogue of the Gibbs–Jaynes Principle (3.13). Let $g : \Omega \to \mathbb{R}^d$ be any vector-valued local function whose range $g(\Omega)$ is not contained in a hyperplane. Then for all $a \in \mathbb{R}^d$ we have in analogy to (3.12)

$$j_g(a) \equiv - \sup_{\nu : \nu(g) = a} h(\nu) = \sup_{t \in \mathbb{R}^d} [t \cdot a - p(t \cdot g)],$$

which together with (3.25) gives us the following result (cf. Georgii 1993, Section 4.3).

(3.26) The Gibbs–Jaynes Principle for the entropy density. *Suppose $a \in \mathbb{R}^d$ is such that j_g is finite on a neighborhood of a, and let ν be translation invariant. Then $h(\nu)$ is maximal under the constraint $\nu(g) = a$ if and only if ν is a Gibbs measure for $t_a \cdot g$, where $t_a = \nabla j_g(a)$.*

The next topic to be discussed is the convergence to stationary measures of maximal entropy density. The preceding Gibbs–Jaynes Principle suggests that an analogue of Csiszár's conditional limit theorem (3.20) might hold in the present setting. This is indeed the case, as was proved by Deuschel et al. (1991), Georgii (1993) and Lewis et al. (1995) using a suitable large-deviation theorem for Gibbs measures (cf. Chapter 9 by Varadhan). We state the result only in the most interesting particular case.

(3.27) The equivalence of ensembles. *Let $\mathcal{C} \subset \mathbb{R}^d$ be closed and such that*

$$\inf j_g(\mathcal{C}) = \inf j_g(\operatorname{int} \mathcal{C}) = j_g(a)$$

for a unique $a \in \mathcal{C}$ having the same property as in (3.26). For any cube Λ in $\mathbb{Z}^d$ let $\nu_{\Lambda,\mathcal{C}}$ be the uniform distribution on the set

$$\left\{ \omega \in E^\Lambda : |\Lambda|^{-1} \sum_{x \in \Lambda} g(\vartheta_x^{\mathrm{per}} \omega_\Lambda) \in \mathcal{C} \right\},$$

where $\vartheta_x^{\mathrm{per}}$ is the periodic shift of E^Λ defined by viewing Λ as a torus. (The assumptions imply that this set is nonempty when Λ is large enough.) Then, as $\Lambda \uparrow \mathbb{Z}^d$, each (weak) limit point of the measures $\nu_{\Lambda,\mathcal{C}}$ is a Gibbs measure for $t_a \cdot g$.

In statistical mechanics, the equidistributions of the type $\nu_{\Lambda,\mathcal{C}}$ are called *microcanonical Gibbs distributions*, and 'equivalence of ensembles' is the classical term for their asymptotic equivalence with the (grand canonical) Gibbs distributions considered before. A similar result holds also in the context of point processes, and thus applies to the classical physical models of interacting molecules (Georgii 1995).

Finally, we want to mention that the entropy approach (3.18) to the convergence of finite-state Markov chains can also be used for the time-evolution of translation-invariant random fields. For simplicity let $E = \{0, 1\}$ and thus $\Omega = \{0, 1\}^{\mathbb{Z}^d}$. We define two types of continuous-time Markov processes on Ω which admit the Gibbs measures for a given f as reversible measures. These are defined by their pregenerator G acting on local functions g as

$$Gg(\omega) = \sum_{x \in \mathbb{Z}^d} c(x, \omega)[g(\omega^x) - g(\omega)]$$

or

$$Gg(\omega) = \sum_{x, y \in \mathbb{Z}^d} c(xy, \omega)[g(\omega^{xy}) - g(\omega)],$$

respectively. Here $\omega^x \in \Omega$ is defined by $\omega_x^x = 1 - \omega_x$, $\omega_y^x = \omega_y$ for $y \neq x$, and ω^{xy} is the configuration in which the values at x and y are interchanged. Under mild locality conditions on the rate function c, the corresponding Markov processes are uniquely defined. They are called *spin-flip* or *Glauber processes* in the first case, and *exclusion* or *Kawasaki processes* in the second case. The Gibbs measures for f are reversible stationary measures for these processes as soon as the rate function satisfies the detailed balance condition that $c(x, \omega) \exp[\sum_{z:x\in\Delta+z} f(\vartheta_z\omega)]$ does not depend on ω_x, respectively an analogous condition in the second case. The following theorem is due to Holley (1971a,b); for streamlined proofs and extensions see Moulin-Ollagnier and Pinchon (1977), Georgii (1979) and Wick (1982).

(3.28) The Holley Theorem. *For any translation-invariant initial distribution ν on Ω, the negative free energy $h(\nu_t) + \nu_t(f)$ is strictly increasing in t as long as the time-t distribution ν_t is no Gibbs measure for f. In particular, ν_t converges to the set of Gibbs measures for f.*

This result is just another instance of the principle of increasing entropy. For similar results in the nonreversible case see Künsch (1984) and Maes et al. (2000) and Chapter 12 by Maes in this book. Further aspects of Glauber and exclusion processes are discussed in Chapter 11 by Olivieri and in Chapter 9 by Varadhan.

Let me conclude by noting that the results and concepts of this section serve also as a *paradigm for ergodic theory*. The set Ω is then replaced by an arbitrary compact metric space with a μ-preserving continuous $\mathbb{Z}^d$-action $(\vartheta_x)_{x\in\mathbb{Z}^d}$. The events in a set $\Lambda \subset \mathbb{Z}^d$ are those generated by the transformations $(\vartheta_x)_{x\in\Lambda}$ from a generating partition of Ω. The entropy density $h(\mu)$ then becomes the well-known Kolmogorov–Sinai entropy of the dynamical system $(\mu, (\vartheta_x)_{x\in\mathbb{Z}^d})$. Again, $h(\mu)$ can be viewed as a measure of the inherent randomness of the dynamical system, and its significance comes from the fact that it is invariant under isomorphisms of dynamical systems. Measures of maximal Kolmogorov–Sinai entropy again play a key role. It is quite remarkable that the variational formula (3.24) also holds in this general setting, provided the partition functions $Z_\Lambda(\omega_{\Lambda^c})$ are properly defined in terms of f and the topology of Ω. $p(f)$ is then called the topological pressure, and $p(0)$ is the so-called topological entropy describing the randomness of the action $(\vartheta_x)_{x\in\mathbb{Z}^d}$ in purely topological terms. All this is discussed in detail in Chapter 16 by Young in this book.

Acknowledgement

I am grateful to A. van Enter, O. Johnson, R. Lang and H. Spohn for a number of comments on a preliminary draft.

References

Barron, A. R. 1986 Entropy and the central limit theorem. *Ann. Prob.* **14**, 336–342.

Chernoff, H. 1956 Large sample theory: parametric case. *Ann. Math. Statist.* **23**, 493–507.

Csiszár, I. 1975 *I*-divergence geometry of probability distributions and minimization problems. *Ann. Prob.* **3**, 146–158.

Csiszár, I. 1984 Sanov property, generalized *I*-projection and a conditional limit theorem. *Ann. Prob.* **12**, 768–793.

Csiszár, I. and Körner, J. 1981 *Information Theory, Coding Theorems for Discrete Memoryless Systems*. Akdaémiai Kiadó, Budapest.

Dembo, A. and Zeitouni, O. 1993 *Large Deviation Techniques and Applications*. Jones and Bartlett, Boston, MA, and London.

Denbigh, K. 1990 How subjective is entropy. In *Maxwell's Demon, Entropy, Information, Computing* (ed. H. S. Leff and A. F. Rex), pp. 109–115. Princeton University Press.

Deuschel, J.-D., Stroock, D. W. and Zessin, H. 1991 Microcanonical distributions for lattice gases. *Commun. Math. Phys.* **139**, 83–101.

van Enter, A. C. D., Fernandez, R. and Sokal, A. D. 1993 Regularity properties and pathologies of position-space renormalization-group transformations: scope and limitations of Gibbsian theory. *J. Statist. Phys.* **72**, 879–1167.

Föllmer, H. 1973 On entropy and information gain in random fields. *Z. Wahrscheinlichkeitsth.* **26**, 207–217.

Föllmer, H. 1988 Random fields and diffusion processes. In *École d'Été de Probabilités de Saint Flour XV–XVII, 1985–87* (ed. P. L. Hennequin), Lecture Notes in Mathematics, Vol. 1362, pp. 101–203. Springer.

Georgii, H.-O. 1979 *Canonical Gibbs Measures*, Lecture Notes in Mathematics, Vol. 760. Springer.

Georgii, H.-O. 1988 *Gibbs Measures and Phase Transitions*. Walter de Gruyter, Berlin and New York.

Georgii, H.-O. 1993 Large deviations and maximum entropy principle for interacting random fields on $\mathbb{Z}^d$. *Ann. Prob.* **21**, 1845–1875.

Georgii, H.-O. 1995 The equivalence of ensembles for classical systems of particles. *J. Statist. Phys.* **80**, 1341–1378.

Hoeffding, W. 1965 Asymptotically optimal tests for multinomial distributions. *Ann. Math. Statist.* **36**, 369–400.

Holley, R. 1971a Pressure and Helmholtz free energy in a dynamic model of a lattice gas. *Proc. Sixth Berkeley Symp. Prob. Math. Statist.* **3**, 565–578.

Holley, R. 1971b Free energy in a Markovian model of a lattice spin system. *Commun. Math. Phys.* **23**, 87–99.

Jaynes, E. T. 1957 Information theory and statistical mechanics. *Phys. Rev.* **106**, 620–630, **108**, 171 190.

Jaynes, E. T. 1982 On the rationale of maximum entropy methods. *Proc. IEEE* **70**, 939–952.

Johnson, O. 2000 Entropy inequalities and the central limit theorem. *Stoch. Proc. Appl.* **88**, 291–304.

Johnson, O. and Suhov, Y. 2000 Entropy and convergence on compact groups. *J. Theor. Prob.* **13**, 843–857.

Kac, M. 1959 *Probability and Related Topics in Physical Sciences*, Lectures in Applied Mathematics, Proceedings of the Summer Seminar, Boulder, CO, 1957, Vol. I. Interscience, London and New York.

Künsch, H. 1984 Nonreversible stationary measures for infinite interacting particle systems. *Z. Wahrscheinlichkeitsth.* **66**, 407–424.

Lanford III, O. E. and Ruelle, D. 1969 Observables at infinity and states with short range correlations in statistical mechanics. *Commun. Math. Phys.* **13**, 194–215.

Lewis, J. T., Pfister, C.-E. and Sullivan, W. G. 1995 Entropy, concentration of probability and conditional limit theorems. *Markov Process. Rel. Fields* **1**, 319–386.

Linnik, Yu. V. 1959 An information-theoretic proof of the central limit theorem with the Lindeberg condition. *Theor. Prob. Appl.* **4**, 288–299.

Maes, C., Redig, F. and van Moffaert, A. 2000 On the definition of entropy production via examples. *J. Math. Phys.* **41**, 1528–1554.

Moulin-Ollagnier, J. and Pinchon, D. 1977 Free energy in spin-flip processes is nonincreasing. *Commun. Math. Phys.* **55**, 29–35.

Preston, C. J. 1976 *Random Fields*. Lecture Notes in Mathematics, Vol. 534. Springer.

Rao, C. R. 1973 *Linear Statistical Inference and its Applications*. Wiley.

Renyi, A. 1970 *Probability Theory*. North-Holland, Amsterdam.

Ruelle, D. 1967 A variational formulation of equilibrium statistical mechanics and the Gibbs phase rule. *Commun. Math. Phys.* **5**, 324–329.

Sanov, I. N. 1957 On the probability of large deviations of random variables (in Russian). *Mat. Sbornik* **42**, 11–44. (English translation *Selected Translations in Mathematical Statistics and Probability I*, 1961, pp. 213–244.)

Shannon, C. E. 1948 A mathematical theory of communication. *Bell Syst. Technol. J.* **27**, 379–423, 623–657. (Reprinted *Key Papers in the Development of Information Theory* (ed. D. Slepian), IEEE Press, New York (1974).)

Spitzer, F. 1971 Random fields and interacting particle systems. Notes on Lectures Given at the 1971 MAA Summer Seminar Williamstown, MA, Mathematical Association of America.

Vajda, I. 1989 *Theory of Statistical Inference and Information*. Kluwer, Dordrecht.

Wick, W. D. 1982 Monotonicity of the free energy in the stochastic Heisenberg model. *Commun. Math. Phys.* **83**, 107–122.

PART 2
Entropy in Thermodynamics

Chapter Four

Phenomenological Thermodynamics and Entropy Principles

Kolumban Hutter and Yongqi Wang

Darmstadt University of Technology

There is no unified approach to irreversible thermodynamics in the phenomenological theories of continuum thermodynamics. The different approaches are based on different forms of the second law. Depending upon which basic underlying principles are postulated, the entropy principle yields different implications on the nonequilibrium quantities for these to fulfil the irreversibility requirements.

We review the most popular approaches and demonstrate that one may prefer or reject a certain entropy principle simply by the results it generates. We draw heavily on the work of Hutter (1977b) and Wang and Hutter (1999c).

4.1 Introduction

This chapter concentrates on a certain aspect of thermodynamics that was fashioned by applied mathematicians and engineers during the last 30–40 years of the 20th century and was coined 'rational thermodynamics' by Truesdell. It commenced in the 1940s with the 'thermodynamics of irreversible processes' of Eckart and Meixner, merged into 'rational thermodynamics' in 1964 when Coleman and Noll introduced an entirely new axiomatic approach, and then diverted into various almost separate versions of rational thermodynamics. The two most important versions are the second law in the form of the Clausius–Duhem inequality and its exploitation according to Coleman and Noll and the entropy principle of Müller with the method of exploitation introduced by Liu. These two versions of the entropy principle applied in rational thermodynamics are the topic of this chapter. There are natural intellectual links to the chapters written in this book by Professor Ingo Müller. In particular, Chapter 2 outlines the basics that lie behind these versions of the second law. The other chapters in this book dealing with methods are obviously also connected, albeit less directly.

To some extent, modern continuum thermodynamics amounts to a collection of 'thermodynamic theories' sharing common premises and a common methodology. There are theories of elastic materials, of viscous materials, of materials with memory, of mixtures, and so on. It is generally the case that, in the context of each theory, one considers all processes (compatible with classical conservation laws) that bodies composed of the prescribed material might admit. Moreover, there exist for the theory some universal physical principles that have been abstracted from experience. One can therefore reduce the generality of the constitutive relations of dependent material variables by relying upon these principles. The most important of these principles is the second law of thermodynamics.

It is known that the second law of thermodynamics is not a unique statement defined by precise rules. On the contrary, there are various versions of the second law and, likewise, also various degrees of generality at which these are exploited. All these laws express some notion of irreversibility and the implications drawn from them necessarily differ from each other.

Mathematicians interested in continuum thermodynamics are generally not aware of the differences in the various postulations of the second law of thermodynamics. Virtually the same is true for many continuum mechanicians; in particular, it is surprising how uncritical and undiscriminating many continuum mechanicians handle the second law. It appears that they have superficially learned how Coleman and Noll apply the Clausius–Duhem inequality and use it as a machine to generate inferences with little contemplation on whether the deduced results, let alone the form of the postulates used to reach them, make physical sense. In this chapter we will summarize various thermodynamic theories, such as ordinary and extended irreversible thermodynamics as well as rational thermodynamics, and explain how their basic postulates of the entropy principle differ from one another and then demonstrate that they yield different results. Specifically, we compare two different formulations of rational thermodynamics. It is these results which allow us to favour one set of basic postulates over another. The two entropy principles are the

(i) generalized Clausius–Duhem–Coleman–Noll approach (CDCN), and

(ii) Müller's entropy principle with Liu's suggestion of its exploitation (ML).

Our demonstration of the essential steps in these two principles will include only the most important mathematical steps and omit significant details that would detract from the main ideas. The reader can fill in these details by reading the pertinent literature. It is shown that wherever it has been applied, the implications of the Müller–Liu version of the second law, which appears to be the most general form of it, are farther reaching than those from other postulations.

4.2 A Simple Classification of Theories of Continuum Thermodynamics

Thermodynamics is a field theory with the ultimate objective to determine the independent field variables of this theory. In a thermodynamic theory of fluids these fields are usually the density and velocity fields as well as the field of the empirical temperature, and it is well known that the necessary field equations are obtained from the balance laws of mass, momentum, moment of momentum and energy, if these balance laws are complemented by *known* constitutive equations. The latter relate stress, energy flux and internal energy to the above-mentioned variables. We have assumed here that these constitutive laws are known, but this is not so; the self-comprehension of thermodynamics is more restrictive. Indeed, it is the purpose of thermodynamics to *reduce* postulated constitutive equations by means of a well-defined law and using well-defined rules. This law is what we usually call the second law of thermodynamics. We would, however, like to caution the reader to use this expression with care, because there are many second laws of thermodynamics. In other words, there are various postulates to quantify the irreversibility of physical processes. These different postulates, axioms, or second laws of thermodynamics are not necessarily equivalent. Nonetheless, the suitability of a set of thermodynamic axioms can be based on objective reasoning. First, a set of thermodynamic axioms should be as general as possible, and, second, it should in the course of its exploitation lead to as many *reasonable* implications as possible. Furthermore, and in this regard, we are motivated physically rather than mathematically, a set of thermodynamic postulates or axioms should produce results which can be counterchecked by the methods of statistical physics. The reason for this latter requirement is our belief that statistical mechanics is based upon an axiomatic structure which is not only general, but also simultaneously different enough from phenomenological theories that a phenomenological theory must necessarily be attributed a higher content of physical truth when it leads to results which coincide with those of statistical mechanics. This principle, of course, must be understood in the sense of natural philosophy and not of the theory of cognition. In particular, its goodness depends on its being demonstrated correct through experiments.

Thermodynamics is an extension and outgrowth of *thermostatics* of the 19th century. The latter is a theory for physical processes that are reversible and for which the motion, that is the balance law of momentum, plays no role. The concept of entropy is part of thermostatics, ranging from Clausius' ideas of entropy as a measure of conversion of heat to Carathéodory's version of entropy as a thermodynamic potential. Boltzmann's *kinetic theory of gases* offers a confirmation of the existence of entropy in thermodynamics where entropy is related to the probability of a distribution of particles. The works of Clausius (1854, 1887), Duhem (1891) and Carathéodory (1925, 1955) have motivated entropy inequalities in thermodynamics and, in particular, the Clausius–Duhem inequality, while the kinetic theory provides an estimate of the range of validity of the Clausius–Duhem inequality. A detailed description of thermostatics and kinetic theory can be found in Müller (1967b, 1971, 1985). In thermostatics and kinetic theory, entropy and absolute temperature are derived quantities.

While in some thermodynamic theories the existence of quantities, which in the thermostatic limit can be identified with entropy or absolute temperature, is *a priori* assumed, we attempt in other theories not to introduce one or the other of these quantities, but tries to prove their existence in due course with the exploitation of the theory.

Thermostatics is concerned with *reversible* processes, i.e. processes whose time evolution can be reversed, thereby leading to another possible thermostatic process. The first theories dealing with irreversible processes go back to Eckart (1940a,b,c) and Meixner (1943). Their theory was applied by many workers in the 1940s and 1950s (see, for example, de Groot and Mazur 1963; Meixner 1969a,b), and it is now known as *irreversible thermodynamics*, although all other thermodynamic theories equally aim at describing irreversible processes. In irreversible thermodynamics the absolute temperature is assumed to be a primitive quantity whose existence has been proved in thermostatics and is therefore taken over into thermodynamics. Regarding entropy, the irreversible thermodynamicist appears to assume its *a priori* existence with some hesitation. The unbiased assumption of its existence apparently is only sufficient to consider neighboring states of thermostatic equilibria for which the existence of equilibrium entropy is assured. This assumption forces the irreversible thermodynamicist to deal with thermodynamic processes which deviate only slightly from a thermostatic equilibrium. Such a restriction leads to the well-known *linearity* of the constitutive equations of this theory.

Irreversible thermodynamics can be subdivided into two groups. One of these is *ordinary irreversible thermodynamics*, whose chief exponents include, among others, Eckart (1940a,b,c), Meixner (1943) and de Groot and Mazur (1963). This theory as extended by Müller (1967a) is known as the *extended irreversible thermodynamic* theory. As with most thermodynamic theories, irreversible thermodynamics derives all its consequences from an inequality. It is this inequality that expresses the irreversibility content of the second law. Most thermodynamic postulates can be motivated by using the entropy inequality:

$$\rho\dot{\eta} + \text{div}\,\boldsymbol{\phi} - \rho s^\eta = \rho\pi^\eta \geqslant 0. \tag{4.1}$$

Here ρ denotes mass density, η specific entropy, $\boldsymbol{\phi}$ entropy flux, s^η the entropy supply, and π^η the entropy production. As most continuum mechanicians do, irreversible thermodynamics uses (4.1) under the simultaneous assumption that

$$\boldsymbol{\phi} = \frac{\boldsymbol{q}}{\theta} \quad \text{and} \quad s^\eta = \frac{s^\varepsilon}{\theta}, \tag{4.2}$$

where $\boldsymbol{q}$ denotes the heat flux vector, s^ε the energy supply density, and θ the absolute temperature. The first goes back to Duhem (1891), the second is due to Truesdell (1957). Substituting (4.2) into (4.1), the well-known Clausius–Duhem inequality is formed, namely,

$$\rho\dot{\eta} + \text{div}\left(\frac{\boldsymbol{q}}{\theta}\right) - \rho\frac{s^\varepsilon}{\theta} = \rho\pi^\eta \geqslant 0. \tag{4.3}$$

Starting from (4.3), the ordinary irreversible thermodynamicist uses a further basic assumption: the Gibbs equation of thermostatics that is derived for slow thermostatic processes to describe the complete differential of the entropy with time is postulated to hold also for all those thermodynamic processes which take place in the neighborhood of a thermostatic equilibrium.[1] For instance, for a simple heat-conducting fluid, the Gibbs equation may be written as[2]

$$d\eta = \frac{1}{\theta}\left(d\varepsilon - \frac{p}{\rho^2}d\rho\right) \quad \text{with } \eta = \eta(\rho, \theta), \tag{4.4}$$

where ε denotes the internal energy and p the hydrostatic pressure, which are considered to be functions of ρ and θ as prescribed by the caloric and thermal equations of state, respectively. Exploiting (4.3) and (4.4), we can obtain some restrictions for constitutive relations. Had we assumed a viscous heat-conducting fluid, phenomenological assumptions would have had to be made for stress and heat flux as *linear* functions of strain rate and temperature gradient. Once these laws are applied to the equations of balance of mechanics and thermodynamics, a system of parabolic equations emerges that predicts infinite speeds for the propagation of heat and shear stress. This phenomenon has been called a paradox.[3] Obviously, the reason for this linearity lies in the restriction that ordinary thermodynamics of irreversible processes borrows the Gibbs equation from thermostatics whose functional form could in a general thermodynamic process be more general than proved correct in thermostatics, as, for example, suggested in (4.5) below.

Discounting Maxwell (1866), who implicitly corrected the Fourier law in the context of the kinetic theory of gases, Cattaneo (1948) was the first to propose a remedy. He was concerned with heat conduction and based his arguments on molecular kinetics. He formulated a modification of Fourier's law, which is now called the Cattaneo equation, in which the heat flux at a certain time depends on the temperature not only at the present time but also at an earlier time. Thus, Cattaneo arrived at a hyperbolic differential equation for the propagation of temperature. Müller (1967a) made an attempt to derive modified versions of the laws of Fourier and Navier–Stokes from the principles of thermodynamics of

---

[1] Rather obviously, this may be a good assumption for thermodynamic processes that do not vary too rapidly and in which all fields differ but little across a particle of the body. This assumption is called the *principle of local equilibrium*.

[2] Clausius was the first to introduce the concept of entropy and establish the Gibbs equation with regard to a Carnot cycle from the viewpoint of thermostatics (see Clausius 1854, 1887; Mendoza 1960).

[3] It is fair to say that few scientists take note of this paradox, because the laws of Fourier and Navier–Stokes are perfectly well suited for most of practical problems of engineers and physicists. Only a few mathematicians and theoretical physicists are concerned. However, the desire to have hyperbolic equations in thermodynamics and, hence, finite speeds of propagation has been a driving motivation for the development of extended thermodynamics. For details see Müller and Ruggeri (1993).

irreversible processes. This theory has been called *extended* irreversible thermodynamics. It is structurally different from ordinary irreversible thermodynamics in two ways. First, the Gibbs equation of thermostatics can also be extended to dynamical processes in a slightly different way. It is indeed possible to assume that the nonequilibrium entropy here can also depend upon quantities which vanish in thermodynamic equilibrium. Such quantities are the heat flux q and the momentum flux t. Hence a generalized Gibbs equation is attained

$$d\eta = \frac{1}{\theta}\left(d\varepsilon - \frac{p}{\rho^2}d\rho - \boldsymbol{A} \cdot d\boldsymbol{q} - \boldsymbol{B} \cdot d\boldsymbol{t} - Cd(\operatorname{tr} \boldsymbol{t})\right), \tag{4.5}$$

where A, B and C are vector-, tensor- and scalar-valued functions of the variables ρ, ε, q, t and $\operatorname{tr} t$ still to be determined. In a second step of extended irreversible thermodynamics Müller assumes a modified entropy flux

$$\boldsymbol{\phi} = \frac{\boldsymbol{q}}{\theta} + \boldsymbol{k}, \tag{4.6}$$

where k is a constitutive quantity.[4] With these statements irreversible thermodynamics is exploited. An important constitutive result of heat flux for a heat-conducting fluid allows us to remove the peculiar deficiency of classical irreversible thermodynamics, namely, the fact that thermal pulses propagate at infinite speed. A large number of papers have appeared using Müller's methodology and are, in particular, devoted to gases. For a review article of this theory, see Jou et al. (1988).

Three symptoms appear to be typical not only for this extended irreversible thermodynamic theory, but partly also for ordinary irreversible thermodynamics.

(i) The entropy inequality is exploited without explicitly accounting for the field equations.

(ii) Thermodynamic processes are restricted to all those which lie in the neighborhood of a thermostatic equilibrium. The theory therefore cannot be extended to fully nonlinear processes.

(iii) For the determination of k, global arguments are left aside.

Irreversible thermodynamics is therefore defective at least in these special points. It was not until Coleman and Noll (1963) and, later, Müller (1967a) first in a restricted sense and then in full generality (Müller 1971) that a rational method for the exploitation of the entropy inequality free from this defect was found. In contrast to irreversible thermodynamics these theories exhibiting

---

[4] Those assumptions can be shown, however, to have specific counterparts in the kinetic theory of gases at the level of Grad's 13-moment approximation (Grad 1949). While the extended theory is not properly invariant under a change of frame, the kinetic theory of gases provides the arguments by which that flaw may be remedied (Müller and Ruggeri 1993).

this generality are summarized as *rational thermodynamics*. Both, irreversible thermodynamics as well as rational thermodynamics can be motivated from thermostatics. Thereby, however, some basic differences must be observed. The main difference must be sought in the fact that in rational thermodynamics the second law is not interpreted as a restriction to the processes a body can possibly experience, but rather as a restriction to its constitutive equation. This was the fundamental new interpretation introduced in 1963 by Coleman and Noll. In other words, a body must be so conditioned that there is no possibility of violating the balance laws of mass, momentum, moment of momentum, energy and *the second law of thermodynamics*, by whatever processes that might occur in this body. In irreversible thermodynamics these conditions are only satisfied for linear processes.

Rational thermodynamics can also be subdivided into two types: *entropy theories* and *entropy free thermodynamics*. We shall, however, not discuss the latter, and refer the interested reader to Day (1972) and Meixner (1969b). The starting point for all entropy theories is the entropy inequality (4.1), whose form, however, varies from author to author. In the following we need not discuss the version of Green and Laws (1972) as well as Green and Naghdi (1972) for the deficiencies, in which one assumes that the entropy inequality in an integral form only holds for the *entire* body and not for its parts. Such an assumption is equivalent to the statement that entropy is not an *additive* quantity, an assumption that is contradictory to results obtained in statistical mechanics.[5] On the other hand the fundamental assumptions behind the two entropy theories—the theory using the Coleman–Noll approach and Müller's theory—are so different that it is worthwhile comparing the two formulations in detail.

4.3 Comparison of Two Entropy Principles

In this section we will explain how the two exploitations of the entropy principle in rational thermodynamics are made and what postulates are underlying them. We then demonstrate how they differ from one another.

4.3.1 Basic Equations

Consider a field theory for a number of field variables $u = (u_i, u_d)$ defined over the body. Let u_i be the independent fields, i.e. those field variables for which the theory provides field equations. Moreover, let u_d be the dependent

---

[5] The additivity property of the entropy in all those theories in which entropy is used as a primitive quantity is tacitly assumed, where the same additivity assumption is also made for all other variables. It is conceptually displeasing to require the additivity property for all the fields of a physical theory except the entropy. The concept of localization should apply to all variables or to none.

field variables which are functionally expressed in terms of the independent fields. Let s^F, s^ε be source terms, arbitrary known functions defined over the body and over time.[6]

Any continuum mechanical field theory consists of the following statements.

Balance laws

$$\mathcal{F}(u_\mathrm{i}, u_\mathrm{d}) - s^F = \mathbf{0}, \qquad f^\varepsilon(u_\mathrm{i}, u_\mathrm{d}) - s^\varepsilon = 0. \tag{4.7}$$

These are, for instance, the balance laws of momentum, angular momentum and energy, but can in electromagneto-mechanical applications also include the Maxwell equations. In (4.7) we have singled out one scalar-type equation, the one with the superscript ε, from the others; this is the energy equation. $\mathcal{F}$ and f^ε denote functional differential operators involving differentiations of space and time.

Constraint relations and source-free balance laws It is often the case that the field variables are subjected to constraint conditions which are either kinematic or thermomechanical in nature. These constraint conditions are also expressible as functional relations between the field variables $(u_\mathrm{i}, u_\mathrm{d})$,

$$\mathcal{C}(u_\mathrm{i}, u_\mathrm{d}) = \mathbf{0}. \tag{4.8}$$

For example, an incompressible material is kinematically constrained by the equation $\det F = 1$, where F is the deformation gradient, or by $\det C = 1$, where $C = F^\mathrm{T} F$. The corresponding equation (4.8) is

$$\mathrm{tr}(\dot F F^{-1}) = 0 \quad \text{or} \quad \dot\gamma = 0, \tag{4.9}$$

where γ is the mass density. Another example of a constraint condition is the *saturation condition* in a porous mixture of a solid and a fluid, if volume fractions are field variables of the theory. It states that the water fills the entire pore space. If ν_f and ν_s are the fluid and solid volume fractions of a binary mixture continuum, then the constraint condition requires $\nu_\mathrm{f} + \nu_\mathrm{s} = 1$, or

$$\frac{\mathrm{d}}{\mathrm{d}t}(\nu_\mathrm{f} + \nu_\mathrm{s}) = 0 \quad \Longrightarrow \quad \sum_\alpha (\acute\nu_\alpha - u_\alpha \cdot \mathrm{grad}\, \nu_\alpha) = 0, \tag{4.10}$$

where $\acute\nu_\alpha$ is the material time derivative with respect to the constituent velocity v_α, and $u_\alpha = v_\alpha - v$ the constituent diffusion velocity in the mixture with

---

[6] The notation we are using here is mathematically fairly sloppy, but we prefer this to a more rigorous presentation, which would require the introduction of a large number of function spaces with particular properties, which vary from case to case anyhow. Bold variables have vector character, and may for convenience be thought of as being n-tuples (for some n) and are defined as elements of a vector space of some dimension. The variables u_i and u_d are such continuously differentiable vector-valued functions defined over the body $\mathcal{B}$ under consideration and as functions of time. $\mathcal{F}$, $\mathcal{C}$, $\mathcal{M}$ and f^ε are three vector-valued and a scalar-valued functional, and s^F, s^ε are functions defined over $\mathcal{B}$ and of time having the same dimension as $\mathcal{F}$ and f^ε.

the mixture velocity v. α is an identifier of the constituents, and $(4.10)_2$ is also valid for a mixture with an arbitrary number of constituents.

Source-free balance relations are also of the form expressed by (4.8). The conservation equation of mass is of such a form, as are the constituent balance laws of mass in a mixture of a finite number of constituents. In these latter laws production terms can enter due either to phase changes or to chemical reactions. These production terms are no source terms as their origin is within the body and not external.

It is often the case that authors introduce external source terms in evolution equations of the type (4.8) to make them of type (4.7). In most situations the reason is mathematical and we will see the implications in a moment, but there is often no justification on physical grounds to introduce such source terms. For instance, to add an arbitrary source term to a balance law of mass is physically not justifiable, because there are physically no situations where it would arise. Neither can balance relations for hidden variables have such external source terms simply because they express something about the microstructure of the body which is entirely internal to the body. Equations like this are the 'equilibrated force balances' in granular theories, the 'balance laws for configurational forces' used in connection with phase changes, the 'spin balances' in polar theories such as micropolar, micromorphic and liquid crystal theories, etc.

Constitutive relations The constitutive relations are functional relations between the dependent fields u_d and the independent fields u_i. When the fields u_d are expressed as functional relations of u_i they read[7]

$$u_d = \mathcal{M}(u_i). \tag{4.11}$$

It is for these relations that we have divided the field variables into u_i and u_d. Examples are an equation of the stress tensor in terms of the strain tensor in an elastic constitutive relation, or the heat flux expressed as being affine to the temperature gradient.

In most continuum thermodynamic theories it is stipulated that the balance laws and constitutive relations together define a well-posed problem; in other words, with appropriate initial and boundary conditions these equations are supposed to yield unique functions of space and time for the field variables, at least for some finite nonzero interval of time. When constraint conditions are added, additional variables enter the theory, representing the constraint stresses or constraint fields that must be applied to guarantee the maintenance of the

---

[7] The constitutive postulates (4.11) should be widely understood as functional relations of the independent variables u_i themselves and their temporal and spatial derivatives to certain orders, which constitute different classes of constitutive postulates.

constraint conditions. These additional fields are not contained in u_i and u_d of (4.11).

Combining (4.7), (4.8) and (4.11) yields

$$\left.\begin{aligned} \mathcal{F}(u_i, \mathcal{M}(u_i)) - s^F &= \mathbf{0}, \\ f^\varepsilon(u_i, \mathcal{M}(u_i)) - s^\varepsilon &= 0, \\ \mathcal{C}(u_i, \mathcal{M}(u_i)) &= \mathbf{0}. \end{aligned}\right\} \quad \left.\begin{aligned} \mathbb{F}(u_i) - s &= \mathbf{0}, \end{aligned}\right\} \qquad (4.12)$$

These equations are called *field equations*. Any set of u_i that satisfies equations (4.12) is called a *thermodynamic process*.[8] In reality the constitutive functions (4.11) are not arbitrary, they should obey universal physical principles, i.e. one can reduce the generality of these functions by relying upon these physical principles. The most important of these principles is the second law of thermodynamics, which we now introduce in the form of the *entropy principle*.

There exists a specific entropy η, entropy flux $\boldsymbol{\phi}$, entropy production density π^η and entropy supply density s^η, which obey a balance law. The quantities η, π^η and $\boldsymbol{\phi}$ belong to u_d and s^η is a source, but this is not emphasized at this stage. The second law of thermodynamics requires that the following general inequality is satisfied,

$$\pi^\eta := \mathcal{H}(\eta, \boldsymbol{\phi}) - s^\eta \geqslant 0, \qquad (4.13)$$

which can explicitly be expressed as in (4.1). Now, any process satisfying (4.13) represents a so-called physically admissible process. The entropy inequality, however, need not hold for arbitrary fields u_i, but only for thermodynamic processes, i.e. solutions of the field equations. The *working principle* is therefore either that all thermodynamic processes must satisfy (4.13), or that all fields which satisfy (4.12) must in addition satisfy (4.13). We must point out that as long as η, $\boldsymbol{\phi}$, s^η are not related to any of the quantities in (4.12) or in previously stated equations the second law is an empty statement. *Various second laws differ by the method of how this link is made.*

4.3.2 Generalized Coleman–Noll Evaluation of the Clausius–Duhem Inequality

Most modern thermodynamicists set at the beginning of their investigations the Clausius–Duhem inequality in the form (4.13) (or the explicit form (4.3)). Explicitly, this is tantamount to an *a priori assignment* of the entropy flux and entropy supply according to (4.2) and establishes the above-mentioned link. Coleman and Noll (1963) have formalized this as follows.

---

[8] There are situations in which the constitutive relations (4.11) are of the form $\mathcal{M}(u_i, u_d) = \mathbf{0}$ and not explicitly solvable for u_d. In those circumstances the union of (4.7), (4.8) and (4.11) form the field equations and their solution for u_i is called a thermodynamic process. For simplicity of the arguments we shall restrict considerations to the case (4.12).

(i) The measure of the coldness of a body particle is unquestionably the absolute temperature θ as proved suitable in thermostatics.

(ii) There exist *a priori postulates* by which the entropy supply rate density s^η and the entropy flux $\boldsymbol{\phi}$ are connected to some field variables of equations (4.7) and (4.8). For instance, in the classical Clausius–Duhem inequality one postulates the relations

$$s^\eta = \frac{s^\varepsilon}{\theta}, \qquad \boldsymbol{\phi} = \boldsymbol{\phi}^{\mathcal{M}}(\boldsymbol{u}_i) = \frac{\boldsymbol{q}}{\theta}(\boldsymbol{u}_i), \qquad (4.14)$$

where s^ε and $\boldsymbol{q}$ represent the energy supply density and energy flux density vector, respectively. Most authors use (4.14) if they apply the CDCN procedure for the exploitation of the entropy inequality. In mixture theories, however, supporters of the CD approach generally recognize that entropy flux and heat flux need not necessarily be collinear. In those instances entropy and energy balance statements are formally written down for each constituent and (4.14) is replaced by

$$s_\mathfrak{a}^\eta = \frac{s_\mathfrak{a}^\varepsilon}{\theta_\mathfrak{a}}, \qquad \boldsymbol{\phi}_\mathfrak{a} = \boldsymbol{\phi}_\mathfrak{a}^{\mathcal{M}}(\boldsymbol{u}_i) = \frac{\boldsymbol{q}_\mathfrak{a}(\boldsymbol{u}_i)}{\theta_\mathfrak{a}}, \qquad \mathfrak{a} \in (1,\ldots,N), \qquad (4.15)$$

where $\mathfrak{a}$ is a counting index for the number of constituents. Again, these are *a priori* assignments.

(iii) η is a constitutive quantity with constitutive relation

$$\eta = \eta^{\mathcal{M}}(\boldsymbol{u}_i). \qquad (4.16)$$

The second law of thermodynamics is here expressed as a statement concerning the entropy balance as a whole; it requires its entropy production to be nonnegative. Finally, notice that in (4.15) we have assumed each constituent to possess its own temperature. Of course, we may also specialize these relations for constituents having the same temperatures. In the above, the *a priori* assignments are given as (4.14) and (4.15) and the link is established by (4.14)–(4.16).

It is basic in the Coleman–Noll approach that the entropy inequality be *identically satisfied for all thermodynamic processes*. Hereby, a thermodynamic process is understood to be any time-dependent solution of field variables satisfying the balance laws (4.7), possible constraint relations (4.8) as well as the constitutive relations (4.11). Furthermore, a constitutive equation for the entropy must be given as stated in (iii).

Combination of the energy equation (4.7)$_2$ and the entropy inequality (4.13) by use of the postulate (4.14) yields

$$\underbrace{\mathcal{H}(\eta^{\mathcal{M}}(\boldsymbol{u}_i), \boldsymbol{\phi}^{\mathcal{M}}(\boldsymbol{u}_i)) - \frac{1}{\theta} f_\varepsilon(\boldsymbol{u}_i, \mathcal{M}(\boldsymbol{u}_i))}_{\mathbb{H}(\boldsymbol{u}_i)} \geqslant 0 \quad \text{or} \quad \mathbb{H}(\boldsymbol{u}_i) \geqslant 0, \qquad (4.17)$$

which should be satisfied for all thermodynamic processes. This form of the entropy inequality no longer contains any source terms.

A clear formulation of the fundamental approach to the exploitation of the second law is due to Coleman and Noll (1963). In their point of view it is important to realize that they regard the source terms s^F, s^ε in the balance laws (4.7) (or (4.12)$_1$) to be external fields *that can be assigned arbitrarily*. It follows that for whatever dynamical process that might occur there are always externally applied body force and energy supply distributions which guarantee that balance laws of momentum and energy be identically satisfied. Day (1972) and Truesdell (1969) pointed out the following metaphysical 'principle' in this regard.

> [T]he fact that practical difficulties prevent us from varying body force and energy supply arbitrarily in the laboratory and thereby prevent us from varying the motion and the temperature arbitrarily, does not affect our argument any more than our inability to produce arbitrary forces acting on mass points prevents us from calculating, on the basis of Newton's law of motion, the force required to produce a given, but arbitrary motion of a mass point.

As a consequence, in exploiting the entropy inequality as an identity only source-free balance laws, e.g. the balance of mass as well as possible constraint relations must be taken care of. It is vital that the reader be fully aware of this understanding, because it will be essentially (and physically) different in Müller–Liu's approach later on. In the CDCN approach one assumes all balance equations (4.7) contain free-source terms, therefore only relations (4.8) constrain the independent fields u_i in the exploitation of the inequality (4.17).

As has been pointed out by Truesdell and Noll (1965), constraints are maintained by reaction forces and there are infinitely many different systems of reaction forces which suffice to maintain a given constraint. However, they also assert that the simplest reaction forces imaginable are those which do not perform any work. This requirement can be implemented by formally introducing arbitrary reactive stresses and reducing their form such that the above workless requirement is identically satisfied. When exploiting the entropy inequality, the work done by the presumed reactive stresses corresponding to the given constraints is artificially added to the balance equation of energy and then the entropy inequality is analyzed by the Coleman–Noll approach. The emerging result is the same, and obviously the reactive stresses do not produce entropy. This process can be described as follows, for example, for the saturation condition in a mixture (see Passman et al. 1984).

Let π be the reaction force for the saturation constraint (4.10) and let the power of working w_π done by the constraint be defined as the product of the reaction force π times the time rate of change of ν_a for all the constituents at any spatial position, namely,

$$w_\pi = \pi \sum_a (\dot{v}_a - u_a \cdot \operatorname{grad} v_a). \tag{4.18}$$

Other internal constraints, such as density preservation of any constituent, may be handled in much the same way as described here. For example, the reaction to the balance law of mass of a single continuum

$$\dot{\rho} + \rho \operatorname{div} v = 0 \tag{4.19}$$

is a thermodynamic pressure p, and the work associated with this constraint is

$$w_p = p(\dot{\rho} + \rho \operatorname{div} v). \tag{4.20}$$

We will see that in the resulting constitutive relations obtained later such reaction forces appear automatically. In reality, with the concept of Lagrange multipliers, such a process is equivalent to considering the balance law of mass and the saturation conditions as constraints for the entropy inequality in the following step.

It has been shown by Liu (1972) that satisfaction of (4.17) for all fields constrained by (4.8) is equivalent to satisfying the inequality

$$\mathbb{H}(u_i) - \lambda_\mathcal{C} \cdot \mathcal{C}(u_i, \mathcal{M}(u_i)) \geqslant 0, \quad \forall u_i, \tag{4.21}$$

for unconstrained fields u_i, where $\lambda_\mathcal{C}$ represent Lagrange multipliers, corresponding to the reaction forces associated with the different constraint conditions. Inequality (4.21) is the resulting generalized entropy inequality, written in the Clausius–Duhem form and exploited by the Coleman–Noll approach.

If all differentiations contained in $\mathbb{H}$ and $\mathcal{C}$ are performed, this inequality can be written in the form

$$a(u_i) \cdot (Du_i) + b(u_i) \geqslant 0, \tag{4.22}$$

where Du_i represent new emerging temporal and spatial derivatives of the independent variables u_i,[9] which are not included in the constitutive relations (4.11), hence inequality (4.22) is linear in Du_i, and Du_i may assume any value we please. This recognition is probably the major step in Coleman–Noll's achievement of the Clausius–Duhem inequality. Since the inequality must hold for all fields u_i and the variables Du_i can hence take any values, it could be violated unless

$$a(u_i) = \mathbf{0}, \qquad b(u_i) \geqslant 0. \tag{4.23}$$

We recall that the main purpose of the entropy principle is to derive restrictions upon the constitutive relation (4.11). With relations (4.23)$_1$ the following results can be obtained:

---

[9] Some of the variables Du_i may and generally do arise in the balance laws (4.7), but since source terms are present, these equations can always be fulfilled by selecting the external sources accordingly. Thus these equations do not influence inequality (4.22).

- reduced dependences of the constitutive relations (e.g. for internal energy, entropy, pressure and other thermodynamic potentials),
- thermodynamic canonical relations,
- the Gibbs relation.

Defining thermodynamic *equilibrium* to be a time-independent process with vanishing thermodynamic quantities such as temperature gradient, velocity gradient, etc., the entropy production π^η assumes its minimum in equilibrium. Hence, a set of constitutive relations in equilibrium can also be obtained from the residual inequality $(4.23)_2$.

The Gibbs equation obtained here is more general than that of irreversible thermodynamics because the entropy and internal energy may depend on variables which vanish in thermostatic equilibrium; see, for example, Hutter (1977b) for a heat-conducting simple fluid, in which the Gibbs relation becomes in equilibrium the classical Gibbs equation of thermostatics (4.4). Thus, in this respect the Coleman–Noll approach is at least as general as the extended irreversible thermodynamic theory with the additional advantage that *processes may be as nonlinear as we please* and, furthermore, that *the Gibbs equation is a proven and not an assumed statement*. This should be emphasized, because earlier it was said that the Gibbs relation holds true for slow and linear processes in the vicinity of thermostatic equilibrium. The Gibbs relation is a proven statement which here holds true for nonlinear processes subject to fast changes if needed. Such 'extensions' have already been tacitly used by, for example, gas dynamicists before this proof was given. Nevertheless, the theory is still less general than one would desire, and in particular it contains some drawbacks that should be discussed. First, it is in general desirable that $\dot{\theta}$ be an independent variable in the constitutive relations. Otherwise, the Gibbs equation has the same form as that of thermostatics (4.4), irrespective of whether the thermodynamic process under consideration is slow or fast, linear or nonlinear. Hence under such conditions this theory is in a way more restrictive than the extended theory of irreversible thermodynamics. Its advantage is again that it proves the Gibbs relation for any thermodynamic process no matter what its complexity. If processes are linear in the sense that constitutive relations are linear equations in the variables that vanish in equilibrium, this theory does not produce results beyond those generated by irreversible thermodynamics.[10] Secondly, it is also very desirable that $\dot{\theta}$ be not an independent constitutive variable. If $\dot{\theta}$ is an independent constitutive variable, the evolution equation for temperature (the energy equation) is elliptic, implying that thermal disturbances propagate with infinite speed. If $\dot{\theta}$ is missing as a variable in the constitutive relations,

---

[10] However, for more complex bodies, rational thermodynamics gives results that are different from those of extended linear irreversible thermodynamics. Materials with fading memory are of this class.

the energy equation becomes parabolic, and then the same paradox obtains. Of course, this does not mean that the Coleman–Noll theory rules out the possibility of a finite speed of propagation for thermal disturbances. For that purpose the heat flux must be represented as a functional of the entire history of the temperature gradient, as illustrated by Gurtin and Pipkin (1969). A review of a large number of papers dealing with this is given by Joseph and Preziosi (1989).

We close by mentioning some important points pertinent to the CDCN approach.

- When there are no constraints, only the energy equation has an influence on the results implied by (4.23).

- To preserve the property that all balance equations contain free-source terms, authors often invent source terms without physical motivation, e.g. for mass balances, structure balance laws, etc. In such cases the results obtained from the Coleman–Noll exploitation of the entropy inequality are doubtful.

- When $\dot{\theta}$ as well as θ is an independent constitutive variable in the constitutive relations (4.11), this approach is *a priori* in doubt because the existence of absolute temperature is questionable under those circumstances except in equilibrium.

- When mixtures with distinct constituent temperatures are considered, the method is equally in doubt for the same reason.

4.3.3 Müller–Liu's Entropy Principle[11]

In the CDCN approach, the flux and the supply of entropy are related *a priori* to the flux and supply of heat. Some examples have suggested that, in general, the entropy flux should not be assumed to be collinear with heat flux using the inverse absolute temperature as the constant of proportionality. This has been demonstrated for mixtures, electro-mechanical interactions, kinetic theory of gases, etc.[12] (see, for example, Hutter 1975, 1977a; Leslie 1968; Liu and Müller 1972; Müller 1985). Moreover, the supposition of the existence of arbitrary-source terms in most balance laws cannot really physically be justified. In order

---

[11] The entropy principle of Müller and its exploitation according to Liu is restricted here to systems not encompassing extended thermodynamics. This is so because we are interested in a comparison with the Clausius–Duhem inequality and the Coleman–Noll approach. For that the reader is directed to Chapter 5 by Müller. This is also the reason why no comparison with higher-order statistical models is made.

[12] Generally, noncollinearity can be observed in two different situations: for mixtures, and if the balance laws of mass, momentum and energy are complemented by additional balance laws (Maxwell equations, spin balance, balance law of an internal variable, materials with microstructure).

to relax these assumptions, Müller (1967b, 1971) proposed an entropy principle in which the entropy and its flux are both *a priori* unrestricted constitutive quantities. Liu (1972) introduced Lagrange multipliers to consider the influences of all balance laws (4.7) and constraint conditions (4.8) in the entropy inequality (4.13), by which the exploitation of the general entropy inequality is much facilitated. It is at this point that Müller and Liu's approach differs substantially from that of Coleman and Noll.

The basic axioms of this approach are as follows.

- θ is an empirical temperature.[13] The absolute temperature, if it exists, is a derived concept, as it is in classical thermostatics.

- The specific entropy η and the entropy flux ϕ are general constitutive relations and, except for this, no *a priori* postulates are introduced,

$$\eta = \eta^M(u_i), \qquad \phi = \phi^M(u_i). \qquad (4.24)$$

- Source terms do not affect the material behaviour.

This shows that the link between the field equations and the second law is weaker than in the Coleman–Noll approach.

To satisfy the entropy inequality (4.13) for all thermodynamic processes, all field equations (4.12) serve as constraints for the inequality (4.13). It follows that

$$\mathcal{H}(\eta^M(u_i), \phi^M(u_i)) - \Lambda \cdot \mathbb{F}(u_i) - \lambda_{\mathcal{C}} \cdot \mathcal{C}(u_i, \mathcal{M}(u_i)) + (\Lambda \cdot s - s^\eta) \geqslant 0, \qquad (4.25)$$

where $\Lambda = (\lambda, \lambda^\varepsilon)$, $\lambda_{\mathcal{C}}$ represent Lagrange multipliers. The third assumption above requires

$$s^\eta = \Lambda \cdot s, \qquad (4.26)$$

so that the entropy supply s^η is known as soon as Λ is determined. By evaluation of the entropy inequality (4.25) for a given constitutive class, the following variables or relations can be obtained:

- Lagrange multipliers Λ as functions of u_i;

- $\lambda_{\mathcal{C}}$ may be new emerging independent variables, i.e. reaction forces subject to constraints;

- reduced dependences of constitutive relations;

- thermostatic equilibrium relations for constitutive quantities;

- the Gibbs relation;

---

[13] It can be shown that in some simple fluids the absolute temperature is a derived quantity, which is a function of the empirical temperature (see, for example, Hutter 1977b; Müller 1985).

- thermodynamic potential relations.

It is important to emphasize that these results differ from those of the classical evaluation of the entropy inequality in the following respects.

- This second law holds for open as well as for closed systems.

- Reaction forces belonging to specific constraints appear automatically in the resulting constitutive relations. For example, if the material is compressible, the Lagrange multiplier with regard to the mass balance is a dependent quantity determined by the free energy. On the other hand, if the material is density preserving, this multiplier is an independent field variable, which corresponds to the reaction of the incompressible constraint and must be specified by the field equations.

- Results are in many cases the same as for the CDCN approach, but not when theories are complex. As a rule, differences are likely to occur when structural variables enter the formulation, such as Cosserat continua, liquid crystals, gradient theories, porous media.

- Experience shows that when results between the two entropy principles differ those obtained with this principle are generally physically better founded.

In particular, we note that entropy, internal and free energies may depend on nonequilibrium variables yielding a different Gibbs relation than obtained with the CDCN approach. As a rule, the differences occur primarily in thermodynamic nonequilibrium, but not exclusively so. For instance, in the theory of liquid crystals the orientation field of the rodlike molecules in equilibrium is determined by the entropy flux contribution that is not collinear to the heat flux vector. If the Clausius–Duhem inequality were true, the orientation field in thermodynamic equilibrium would be arbitrary, whence chaotic. We would never be able to read on our laptop screen what we write if the screen is a liquid crystal display. In granular media of elongated particles (rice), the situation must be very much the same.

The disadvantages of Müller's theory must all be sought in the complexity of the algebraic manipulations one gets into, once the constitutive theory becomes too complicated. The number of equations that must be fulfilled in the course of the exploitation of the entropy principle grows strongly with the number of independent vector- and tensor-valued constitutive variables. Instead of a full exploitation with all the steps of necessity and sufficiency that should be carried through, one is forced to introduce ad hoc assumptions which still allow conclusions of sufficiency. Such is actually always the case when constitutive equations are restricted to special forms (say linearity or the rule of equipresence is violated).

4.4 Concluding Remarks

In this chapter we have briefly summarized various theories of thermodynamics, from thermostatics, thermodynamics of irreversible processes to modern rational thermodynamics.[14] Thermostatics is identified as a special case of thermodynamics and is concerned with reversible processes. It can only deal with slow thermodynamic processes. There is an obvious weak point in the thermodynamics of irreversible processes as it proceeds from the Gibbs equation of thermostatics, even though in extended irreversible thermodynamics this relation has been relaxed to allow the entropy to depend on nonequilibrium variables. Within the class of rational thermodynamics, an attempt was made to characterize the various different postulates of two modern forms of the entropy principle—the Coleman–Noll approach of the Clausius–Duhem inequality (CDCN) and Müller's entropy principle with Liu's suggestion of its exploitation (ML)—to compare their differences. CDCN makes *a priori* postulates about the entropy flux and entropy supply and assumes external source terms in (most) balance laws, so that they do not affect the exploitation of the entropy principle. Müller and Liu postulate the entropy flux to be a general constitutive variable and treat all field equations as constraints for the exploitation of the entropy principle. With this reasoning the ML version of the entropy principle is clearly generally preferable. Explicit support for this is given by the following facts.

- Hutter et al. (1994) demonstrated for a saturated mixture of viscous fluids that the emerging constitutive statements, if obtained with the CDCN exploitation of the entropy principle (Ehlers 1989), are extremely restricted. There exists no solution for a simple gravity-driven shearing flow of viscous constituents. However, in this case, if the mixture theory is derived using the ML approach of the entropy principle (Svendsen and Hutter 1995), this nonexistence of the solution can be avoided. It is then tempting to favour the ML approach of the entropy principle over the CDCN on account of the generated results, not simply by the basic postulates.

- As has been shown in Wang and Hutter (1999c), those two entropy principles yield different constitutive relations for granular materials with/without fluid.

- CDCN should definitely be abandoned in the following classes of models: *polar continua* (solids, anisotropic fluids, liquid crystals) because the spin balance has no free-source terms; *structured continua*; *mixtures* (as has long been known); *coupled field theories*; etc.

---

[14] Extended thermodynamics and rational extended thermodynamics were not discussed at all (Müller and Ruggeri 1993).

Acknowledgement

This work was supported by the German Research Foundation (DFG) through the special collaborative research project SFB 298 'Deformation and failure of metallic and granular media.'

References

Carathéodory, C. 1925 Über die Bestimmung der Energie und der absoluten Temperatur mit Hilfe von reversiblen Prozessen. *Sitzungsberichte der preußischen Akademie der Wissenschaften, Math. Phys. Klasse*, pp. 39–47.

Carathéodory, C. 1955 *Gesammelte Mathematische Schriften*, Band II. Beck'sche Verlagsbuchhandlung, München.

Cattaneo, C. 1948 Sulla conduzione del calore. *Atti Sem. Mat. Fis. Univ. Modena* **3**, 83–101.

Clausius, R. 1854 Über eine veränderte Form des zweiten Hauptsatzes der mechanischen Wärmetheorie. *Poggendroff's Annln Phys.* **93**, 481–506.

Clausius, R. 1887 *Die Mechanische Wärmetheorie*, Vol. 1, 3rd edn. Vieweg & Sohn, Braunschweig.

Coleman, B. D. and Noll, W. 1963 The thermodynamics of elastic materials with heat conduction and viscosity. *Arch. Ration. Mech. Analysis* **13**, 167–178.

Day, W. A. 1972 *The Thermodynamics of Simple Materials with Fading Memory*. Springer Tracts in Natural Philosophy, Vol. 22. Springer.

de Groot, S. R. and Mazur, P. 1963 *Non-Equilibrium Thermodynamics*. North-Holland, Amsterdam.

Duhem, P. 1891 *Hydrodynamique, Elasticité, Acoustique* (2 vols). Paris, Hermann.

Eckart, C. 1940a The thermodynamics of irreversible processes. I. The simple fluid. *Phys. Rev.* **58**, 267–269.

Eckart, C. 1940b The thermodynamics of irreversible processes. II. Fluid mixtures. *Phys. Rev.* **58**, 269–275.

Eckart, C. 1940c The thermodynamics of irreversible processes. III. Relativistic theory of the simple fluid. *Phys. Rev.* **58**, 919–924.

Ehlers, W. 1989 Poröse Medien, ein kontinuumsmechanisches Modell auf der Basis der Mischungstheorie. Habilitation, Universität-Gesamthochschule Essen.

Goodman, M. A. and Cowin, S. C. 1972 A continuum theory for granular materials. *Arch. Ration. Mech. Analysis* **44**, 249–266.

Grad, H. 1949 *On the Kinetic Theory of Rarefied Gases*. Communications in Pure and Applied Mathematics, vol. 2, Wiley.

Green, A. E. and Laws, N. 1972 On a global entropy production inequality. *Q. J. Mech. Appl. Math.* **25**, 1–11.

Green, A. E. and Naghdi, P. M. 1972 On continuum thermodynamics. *Arch. Ration. Mech. Analysis* **48**, 352–378.

Gurtin, M. E. and Pipkin, A. C. 1969 A general theory of heat conduction with finite wave speed. *Arch. Ration. Mech. Analysis* **31**, 113–126.

Hutter, K. 1975 On thermodynamics and thermostatics of viscous thermoelastic solids in the electromagnetic fields. A Lagrangian formulation. *Arch. Ration. Mech. Analysis* **54**, 339–366.

Hutter, K. 1977a A thermodynamic theory of fluids and solids in the electromagnetic fields. *Arch. Ration. Mech. Analysis* **64**, 269–289.

Hutter, K. 1977b The foundation of thermodynamics, its basic postulates and implications. A review of modern thermodynamics. *Acta Mech.* **27**, 1–54.

Hutter, K., Jöhnk, K. and Svendsen, B. 1994 On interfacial transition conditions in two phase gravity flow. *Z. Angew. Math. Phys.* **45**, 746–762.

Joseph, D. D. and Preziosi, L. 1989 Heat waves. *Rev. Mod. Phys.* **61**, 41–73.

Jou, D., Casas-Vázquez, J. and Lebon, G. 1988 Extended irreversible thermodynamics. *Rep. Prog. Phys.* **51**, 1105–1179.

Leslie, F. M. 1968 Some constitutive equations for liquid crystals. *Arch. Ration. Mech. Analysis* **28**, 265–283.

Liu, I.-S. 1972 Method of Lagrange multipliers for exploitation of the entropy principle. *Arch. Ration. Mech. Analysis* **46**, 131–148.

Liu, I.-S. and Müller, I. 1972 On the thermodynamics and thermostatics of fluids in electromagnetic fields. *Arch. Ration. Mech. Analysis* **46**, 149–176.

Liu, I.-S. and Müller, I. 1984 Thermodynamics of mixtures of fluids. In *Rational Thermodynamics* (ed. C. Truesdell), pp. 264–285. Springer.

Maxwell, J. C. 1866 On the dynamical theory of gases. *Phil. Trans. R. Soc. Lond.* **157**, 49–88.

Meixner, J. 1943 Zur Thermodynamik der irreversiblen Prozesse. *Z. Phys. Chem.* **538**, 235–263.

Meixner, J. 1969a Thermodynamik der Vorgänge in einfachen fluiden Medien und die Charakterisierung der Thermodynamik irreversibler Prozesse. *Z. Phys.* **219**, 79–104.

Meixner, J. 1969b Processes in simple thermodynamic materials. *Arch. Ration. Mech. Analysis* **33**, 33–53.

Mendoza, E. (ed.) 1960 *Reflections on the Motive Power of Fire by Sadi Carnot and Other Papers on the Second Law of Thermodynamics by E. Clapeyron and R. Clausius*. Dover, New York.

Müller, I. 1967a Zum Paradox der Wärmeleitungstheorie. *Z. Phys.* **198**.

Müller, I. 1967b On the entropy inequality. *Arch. Ration. Mech. Analysis* **26**, 118–141.

Müller, I. 1971 Die Kältefunktion, eine universelle Funktion in der Thermodynamik viskoser wärmeleitender Flüssigkeiten. *Arch. Ration. Mech. Analysis* **40**, 1–36.

Müller, I. 1985 *Thermodynamics*. Pitman.

Müller, I. and Ruggeri, T. 1993 *Extended Thermodynamics*. Springer Tracts in Natural Philosophy, Vol. 37. Springer.

Nunziato, J. W. and Passman, S. L. 1981 A multiphase mixture theory for fluid-saturated granular materials. In *Mechanics of Structured Media A* (ed. A. P. S. Selvadurai), pp. 243–254. Elsevier, Amsterdam.

Passman, S. L., Nunziato, J. W. and Walsh, E. K. 1984 A theory of multiphase mixtures. In *Rational Thermodynamics* (ed. C. Truesdell), pp. 286–325. Springer.

Svendsen, B. and Hutter, K. 1995 On the thermodymics of a mixture of isotropic materials with constraints. *Int. J. Engng Sci.* **33**, 2021–2054.

Truesdell, C. 1957 Sulle basi della termomeccanica. *Rend. Accad. Lincei* **22**(8), 33–88, 158–166. (English translation in *Rational Mechanics of Materials*, Gordon & Breach, New York, 1965.)

Truesdell, C. 1969 *Rational Thermodynamics*. McGraw-Hill, New York.

Truesdell, C. and Noll, W. 1965 *The Non-Linear Field Theories of Mechanics*. Flügge's Handbuch der Physik, Vol. III/3. Springer.

Wang, Y. and Hutter, K. 1999a Shearing flows in a Goodman–Cowin type granular material—theory and numerical results. *Particulate Sci. Technol.* **17**, 97–124.

Wang, Y. and Hutter, K. 1999b A constitutive model for multi-phase mixtures and its application in shearing flows of saturated soil–fluid mixtures. *Granular Matter* **1**(4), 163–181.

Wang, Y. and Hutter, K. 1999c Comparison of two entropy principles and their applications in granular flows with/without fluid. *Arch. Mech.* **51**, 605–632.

Chapter Five

Entropy in Nonequilibrium

Ingo Müller

Technical University of Berlin

While thermo*statics*—the thermodynamics of equilibrium—is cut and dried, and fully accepted, this cannot be said for thermo*dynamics* proper, i.e. the theory of nonequilibrium processes. It is true that there is the reliable theory of the 'thermodynamics of irreversible processes,' TIP to most people, but that theory is only valid in the immediate neighborhood of equilibrium, nor does it claim to be more general.

When gradients are steep and rates of change are large we need a theory that can treat processes far from equilibrium, like shock waves or light scattering in rarefied gases, and that is what extended thermodynamics does. Extended thermodynamics owes a lot to the kinetic theory of gases, but basically it is a macroscopic theory, in which the problems of closure of the equations of state and of boundary values for high moments figure prominently. For the resolution of both problems entropic arguments have been pressed into service: the maximum entropy principle, or the principle of minimum entropy production, or the minmax principle of entropy production. None of these is universally accepted and perhaps they are not here to stay. But they seem to belong to an account of the uses of entropy, if only to exhibit the aspects of entropy that are currently being discussed.

This contribution forms the second part of what was originally a single paper on entropy. The first part is now Chapter 2 of this book, and references to the equations in that chapter will have the prefix 2.

5.1 Thermodynamics of Irreversible Processes and Rational Thermodynamics for Viscous, Heat-Conducting Fluids

The objective of irreversible thermodynamics is the determination of the five fields of

$$\left.\begin{array}{rl} \text{mass density} & \rho(x,t), \\ \text{velocity} & v_i(x,t), \\ \text{temperature} & T(x,t), \end{array}\right\} \quad (5.1)$$

in all points x of a fluid and at all times t.

For this purpose we need field equations and these are based upon the equations of balance of mechanics and thermodynamics, namely, the conservation laws of mass and momentum and the equation of balance of internal energy[1]

$$\left.\begin{array}{l} \dfrac{d\rho}{dt} + \rho\dfrac{\partial v_j}{\partial x_j} = 0, \\[4pt] \rho\dfrac{dv_i}{dt} - \dfrac{\partial t_{ij}}{\partial x_j} = \rho f_i, \\[4pt] \rho\dfrac{du}{dt} + \dfrac{\partial q_j}{\partial x_j} = t_{ij}\dfrac{\partial v_i}{\partial x_j} + r. \end{array}\right\} \quad (5.2)$$

While these are five equations, the correct number for five fields, they are not field equations for the fields (5.1). Indeed, T does not even appear in equations (5.2), and instead they contain new quantities, namely, the

$$\left.\begin{array}{r} \text{(symmetric) stress tensor} \quad t_{ij}, \\ \text{heat flux} \quad q_j, \\ \text{specific internal energy} \quad u. \end{array}\right\} \quad (5.3)$$

The specific body force f_i and the density of energy absorption r are assumed to be given functions of x and t. In order to close the system (5.2) we must therefore find relations between t_{ij}, q_j, u and the fields $\rho(x, t)$, $v_i(x, t)$, $T(x, t)$—the so-called constitutive relations, or state functions, or material equations.

In the thermodynamics of irreversible processes,[2] a theory universally known as TIP, such relations are derived in a heuristic manner from an entropy inequality that is based upon the Gibbs equation of equilibrium thermodynamics, namely, (2.6), or, for a unit of mass,

$$\frac{ds}{dt} = \frac{1}{T}\left(\frac{du}{dt} + p\frac{d(1/\rho)}{dt}\right). \quad (5.4)$$

p and u are supposed to be dependent on ρ and T in the manner dictated by the thermal and caloric equations of state of equilibrium thermodynamics. This reference to equilibrium thermodynamics is often called the *principle of local equilibrium*.

Elimination of du/dt and $d\rho/dt$ between the Gibbs equation (5.4) and the equations of balance of mass and (internal) energy (5.2) and a little rearrangement of terms leads to the equation[3]

$$\rho\frac{ds}{dt} + \frac{\partial}{\partial x_i}\left(\frac{q_i}{T}\right) = q_i\frac{\partial(1/T)}{\partial x_i} + \frac{1}{T}t_{\langle ij\rangle}\frac{\partial v_{\langle i}}{\partial x_{j\rangle}} + \frac{1}{T}(\tfrac{1}{3}t_{ii} + p)\frac{\partial v_j}{\partial x_j}, \quad (5.5)$$

---

[1] We follow the convention of thermodynamics by which small letters characterize specific, i.e. mass-related quantities. Thus u and s are the specific internal energy and the specific entropy. We also use the summation convention, which requires summation over repeated indices.

[2] TIP was first properly formulated by Carl Henry Eckart (1902–1973) (see Eckart 1940a,b,c).

[3] t_{ii} is the trace of the stress, while $t_{\langle ij\rangle}$ denotes its deviatoric part,

ENTROPY IN NONEQUILIBRIUM

which may be interpreted as an equation of balance of entropy. That interpretation implies that

$$\left.\begin{array}{l}\hat{h}_i =: \dfrac{q_i}{T} \quad \text{is the nonconvective entropy flux (\emph{assumption of Duhem})} \\ \qquad\qquad \text{Pierre Maurice Marie Duhem (1861--1916),} \\[4pt] \Sigma =: -\dfrac{q_i}{T^2}\dfrac{\partial T}{\partial x_i} + \dfrac{1}{T}t_{\langle ij\rangle}\dfrac{\partial v_{\langle i}}{\partial x_{j\rangle}} + \dfrac{1}{T}(\tfrac{1}{3}t_{ii}+p)\dfrac{\partial v_n}{\partial x_n} \\ \qquad\qquad \text{is the density of entropy production.} \end{array}\right\} \quad (5.6)$$

Inspection shows that the entropy production density is a sum of products of *thermodynamic fluxes* (on the left) and *thermodynamic forces* (on the right):

$$\left.\begin{array}{ll} \text{heat flux } q_i, & \text{temperature gradient } \dfrac{\partial T}{\partial x_i}, \\[6pt] \text{deviatoric stress } t_{\langle ij\rangle}, & \text{deviatoric velocity gradient } \dfrac{\partial v_{\langle i}}{\partial x_{j\rangle}}, \\[6pt] \text{dynamic pressure } \pi \equiv -\tfrac{1}{3}t_{ii}-p, & \text{divergence of velocity } \dfrac{\partial v_n}{\partial x_n}. \end{array}\right\} \quad (5.7)$$

The entropy production density is postulated to be nonnegative. Assuming only linear relations between forces and fluxes, TIP theory ensures the nonnegative entropy production by constitutive equations—*phenomenological equations* in the jargon of TIP—of the type

$$\left.\begin{array}{lll} q_i = -\kappa \dfrac{\partial T}{\partial x_i} & \kappa \geqslant 0 & \text{thermal conductivity,} \\[6pt] t_{\langle ij\rangle} = 2\mu \dfrac{\partial v_{\langle i}}{\partial x_{j\rangle}} & \mu \geqslant 0 & \text{shear viscosity,} \\[6pt] \pi = -\lambda \dfrac{\partial v_i}{\partial x_i} & \lambda \geqslant 0 & \text{bulk viscosity.} \end{array}\right\} \quad (5.8)$$

Along with the thermal and caloric equations of state $p = p(\rho, T)$ and $u = u(\rho, T)$, equations (5.8) represent the constitutive equations of TIP. They are known as the laws of Fourier and Navier–Stokes, because these scientists read the relations off from the phenomena of heat conduction and fluid friction—hence phenomenological equations—long before there was any thermodynamic theory to base them on.

The 1960s saw the emergence of a variant of TIP, so-called rational thermodynamics, first formulated by Coleman and Noll (1963) (see also Coleman and Mizel 1964). That theory is slightly more systematic than TIP in that it *postulates* an entropy inequality, the Clausius–Duhem inequality,

$$\rho \frac{ds}{dt} + \frac{\partial}{\partial x_i}\left(\frac{q_i}{T}\right) - \frac{r}{T} \geqslant 0 \qquad (5.9)$$

and *derives* the Gibbs equation for the equilibrium part of the specific entropy. Thus in rational thermodynamics there is a nonequilibrium part of the specific entropy, at least *a priori*, so that rational thermodynamics does not need to rely on the principle of local equilibrium, which is to its credit. However, rational thermodynamics needs a specific form, namely, r/T, for the supply of entropy, and it makes use of the specific form, namely, q_i/T, of the nonconvective entropy flux as if that were a fundamental law of nature. There are other awkward points about rational thermodynamics, the most obvious one being that, for the exploitation of the Clausius–Duhem inequality (5.9), it needs arbitrarily adjustable supplies $f_i(x, t)$ and $r(x, t)$ in the field equations (5.2).

Rational thermodynamics was promoted by its proponents with considerable fanfare as a breakthrough in nonequilibrium thermodynamics making TIP hopelessly obsolete. Naturally, there was a reaction, actually more like an overreaction, by the adherents of TIP, in particular Woods (1973). The truth is that both theories are more similar than either group would like to admit. Thus for a linearly viscous, heat-conducting fluid both arrive at the same conclusion, namely, the constitutive relations (5.7) of Navier–Stokes and Fourier. And neither theory is much good beyond that! Thus, for instance, both theories fail for nonNewtonian fluids.

5.2 Kinetic Theory of Gases, the Motivation for Extended Thermodynamics

5.2.1 A Remark on Temperature

Extrapolation of thermodynamics of fluids away from a Navier–Stokes–Fourier theory is precarious and fraught with uncertainty. We need some guidelines and fortunately there is the kinetic theory of gases to guide us.

But even so, we need to be subtle where, unintentionally, we were lax before; and one subtle point concerns the temperature. Indeed, if we think about it, we realize that temperature is defined twice, in different ways:

- as the quantity that is continuous at the boundary between two bodies,[4] and
- as a measure for the mean kinetic energy of the particles.

The former is called the thermodynamic temperature and we shall henceforth reserve the letter T for it. The latter might be called the kinetic temperature and we shall denote it by θ. The kinetic theory knows only θ.

We shall prove that both temperatures differ, except in equilibrium. And for that argument we need to refer to entropy again, or rather to entropy and entropy flux.

---

[4] That definition is sometimes called the zeroth law of thermodynamics.

5.2.2 Entropy Density and Entropy Flux

The kinetic theory of gases and the Boltzmann equation have provided the 'mechanical' interpretation of entropy as described in Section 2.2 (see (2.10)). But this does not exhaust the heuristic value of the kinetic theory—and of the kinetic expressions for the entropy and its flux—for the formulation of a proper theory of nonequilibrium thermodynamics.

First of all, according to (2.10), the local form of the entropy density reads

$$\rho s = \int (-k \ln f) f \, d\mathbf{p} \qquad (5.10\,a)$$

and the corresponding entropy flux

$$h_i = \int \left(-\frac{k}{m} p_i \ln f\right) f \, d\mathbf{p}. \qquad (5.10\,b)$$

Thus, if the distribution function f were known, we should be able to calculate the entropy and its flux.

5.2.3 13-Moment Distribution. Maximization of Nonequilibrium Entropy

Nonequilibrium in a viscous, heat-conducting gas means that the state of a volume element is characterized by 13 quantities, namely, mass density ρ, momentum density ρv_i, specific internal energy $u = \frac{3}{2}k\theta$, deviatoric stress $t_{\langle ij \rangle}$, and heat flux q_i. All of these quantities are moments of the distribution function. We have[5]

$$\left.\begin{aligned}
\rho &= \int mf \, dp, \\
\rho v_i &= \int mc_i f \, dp, \\
\rho u &= \int \tfrac{1}{2} mC^2 f \, dp, \\
t_{\langle ij \rangle} &= \int mC_i C_j f \, dp, \\
q_i &= \int \tfrac{1}{2} mC^2 C_i f \, dp.
\end{aligned}\right\} \qquad (5.11)$$

These 13 fields are those that occur in the conservation laws of mass, momentum and energy (see (5.2)), and that is why they represent a natural choice as nonequilibrium variables.

---

[5] c_i is the velocity of an atom, while $C_i = c_i - v_i$ is its relative velocity, relative to the velocity of the volume element.

We approximate the nonequilibrium distribution function by expanding f in terms of Hermite polynomials. This was first done by Harold Grad (1923–1983) (see Grad 1958), and we obtain Grad's 13-moment distribution function

$$f_G = f_E \left(1 - \frac{1}{2p} t_{\langle ij \rangle} \left(\frac{m}{k\theta} C_i C_j - \delta_{ij} \right) - \frac{1}{p} \frac{m}{k\theta} q_i C_i \left(1 - \frac{1}{5} \frac{m}{k\theta} C^2 \right) \right), \quad (5.12)$$

where f_E is the equilibrium distribution function, the Maxwell distribution. With the Grad distribution it is easy to calculate the entropy density ρs and the nonconvective entropy flux $\hat{h}_i$ from (5.10). We obtain

$$\rho s = \rho s_E - \frac{t_{\langle ij \rangle} t_{\langle ij \rangle}}{4 p \rho \theta} - \frac{q_i q_i}{5 p^2 \theta}, \quad \hat{h}_i = \frac{q_i}{\theta} + \frac{2}{5 p \theta} t_{\langle ij \rangle} q_j. \quad (5.13)$$

Thus, neither is the nonequilibrium entropy s equal to its equilibrium value s_E, nor is the nonconvective entropy flux $\hat{h}_i$ parallel to the heat flux q_i. Therefore, neither the principle of local equilibrium nor the Duhem assumption for the entropy flux is true. To be sure, both are 'approximately true' in the sense that the departures of (5.13) from these assumptions are nonlinear in the heat flux q_i and the deviatoric stress $t_{\langle ij \rangle}$.

There is a possible connection between Grad's 13-moment method and the entropy in the form $(5.10)_1$. Müller (1985) noticed[6] that the distribution function (5.11) may be derived from the maximization of entropy in the form $\int (-k \ln f) f \, d\boldsymbol{p}$ under the constraints of prescribed values of ρ, ρv_i, ρu, $t_{\langle ij \rangle}$ and q_i (cf. (5.11)). It is true that this procedure, which came to be called the 'maximum entropy principle,' leads to an exponential distribution, but linearization in the nonequilibrium values $t_{\langle ij \rangle}$ and q_i provides the Grad distribution. The maximum entropy principle implies that equilibrium—the state of maximum entropy—is not just approached along any path; rather there is an optimal path in some sense, i.e. a path along 'relatively maximal entropies,' i.e. maximal for given nonequilibrium values of $t_{\langle ij \rangle}$, q_i.

(In recent years mathematicians have cast doubt upon the maximum entropy principle, because there are circumstances, and the 13-moment theory is one of them, in which it leads to a nonintegrable distribution function (see Junk 2000; Levermore 1996). This matter is currently under investigation (see, for example, Junk 2002; Müller et al. 2002).)

5.2.4 Balance Equations for Moments

Multiplication of the Boltzmann equation (5.10) by $c_{i_1} c_{i_2} \cdots c_{i_n}$ for $n = 0, 1, \ldots, N$ and integration over all momenta $\boldsymbol{p}$ provide a hierarchy of equations of balance for moments of increasing tensorial order of the distribution function. The generic moment has the form

$$F_{i_1 \cdots i_n} = \int m c_{i_1} \cdots c_{i_n} f \, d\boldsymbol{p}$$

---

[6] This had already been observed by Kogan (1967) in another context.

ENTROPY IN NONEQUILIBRIUM 85

and the equations of balance are[7]

$$\begin{aligned}
\frac{\partial F}{\partial t} + \frac{\partial F_n}{\partial x_n} &= 0, \\
\frac{\partial F_{i_1}}{\partial t} + \frac{\partial F_{i_1 n}}{\partial x_n} &= 0, \\
\frac{\partial F_{i_1 i_2}}{\partial t} + \frac{\partial F_{i_1 i_2 n}}{\partial x_p} &= -\frac{1}{\tau} F_{\langle i_1 i_2 \rangle}, \\
\frac{\partial F_{i_1 i_2 i_3}}{\partial t} + \frac{\partial F_{i_1 i_2 i_3 n}}{\partial x_n} &= -\frac{1}{\tau} F_{i_1 i_2 i_3}, \\
&\vdots \\
\frac{\partial F_{i_1 i_2 \cdots i_N}}{\partial t} + \frac{\partial F_{i_1 i_2 \cdots i_N n}}{\partial x^n} &= -\frac{1}{\tau}(F^E_{i_1 i_2 \cdots i_N} - F_{i_1 i_2 \cdots i_N}).
\end{aligned} \quad (5.14)$$

The system is never closed; closure can, however, be achieved by calculating the last flux, namely, $F_{i_1 i_2 \cdots i_N n}$, in terms of $F, F_{i_1}, \ldots, F_{i_1 i_2 \cdots i_N}$ by use of the distribution function that maximizes the nonequilibrium entropy under the constraint of fixed values for $F, F_{i_1}, \ldots, F_{i_1 i_2 \cdots i_N}$.

5.2.5 Moment Equations for 13 Moments. Stationary Heat Conduction

For 13 moments, the system (5.14) is, in closed form,

$$\begin{aligned}
\frac{\partial \rho}{\partial t} + \frac{\partial \rho v_k}{\partial x_k} &= 0, \\
\frac{\partial \rho v_i}{\partial t} + \frac{\partial \rho v_i v_k + p\delta_{ik} - t_{\langle ik \rangle}}{\partial x_k} &= 0, \\
\frac{\partial \frac{3}{2} p}{\partial t} + \frac{\partial \frac{3}{2} p v_k + q_k}{\partial x_k} + (p\delta_{ik} - t_{\langle ik \rangle})\frac{\partial v_i}{\partial x_k} &= 0, \\
\frac{\partial t_{\langle ij \rangle}}{\partial t} + \frac{t_{\langle ij \rangle} v_k - \frac{4}{5} q_{\langle i} \delta_{j \rangle k}}{\partial x_k} + 2t_{\langle k\langle i} \frac{v_{j\rangle}}{\partial x_k} \boxed{-2p\frac{\partial v_{\langle i}}{\partial x_{j\rangle}}} &= -\frac{1}{\tau} t_{\langle ij \rangle}, \\
\frac{\partial q_i}{\partial t} + \frac{\partial q_i v_k - (k/m)\theta t_{\langle ik \rangle}}{\partial x_k} - \frac{t_{\langle ij \rangle}}{\rho}\frac{\partial t_{\langle jk \rangle} - 3p\delta_{jk}}{\partial x_k} + \frac{2}{5} q_i \frac{\partial v_k}{\partial x_k} + \frac{7}{5} q_k \frac{\partial v_i}{\partial x_k} & \\
+ \frac{2}{5} q_k \frac{\partial v_k}{\partial x_i} + (-\frac{5}{6} t_{\langle ik \rangle} + \boxed{\frac{5}{6} p \delta_{ik}})\frac{\partial (k/m)\theta}{\partial x_k} &= -\frac{1}{\tau} q_i .
\end{aligned} \quad (5.15)$$

The Navier–Stokes–Fourier equations are shown in frames. Here we have denoted all moments by their canonical letters, like ρ, v_i, θ and $t_{\langle ij \rangle}$, q_i for

---

[7] For simplicity we use the so-called BGK equation (see Bhatnagar et al. 1954), in which the right-hand side of the Boltzmann equation is replaced by the simple collision term $(1/\tau)(f_E - f)$. τ is a relaxation time, it has the order of magnitude of the mean time of free flight of the atoms.

deviatoric stress and heat flux. We recognize the five conservation laws of mass, momentum and energy, common to all thermodynamic theories, and note that all 13 equations are completely explicit. The last two equations contain, within the frames, the phenomenological equations of Navier–Stokes and Fourier by which the deviatoric stress is proportional to the deviatoric part of the velocity gradient and the heat flux is proportional to the temperature gradient.

The other terms in the last two equations of (5.15) include rates and gradients of stress and heat flux and we expect that those terms are negligible, if rates are slow and if gradients are small; in such a case the Navier–Stokes–Fourier theory may suffice. But for rapid rates and steep gradients the full equations (5.15) with all terms are needed.

As a case in point we investigate the simplest problem of nonequilibrium thermodynamics: stationary radial heat conduction in a gas at rest in a cylindrical tube with $r_i = 10^{-3}$ m and $r_e = 10^{-2}$ m as inner and outer radii, respectively.[8] The inner wall is heated by $q = 10^4$ W m^{-2} and the homogeneous pressure p is chosen to be 10^2 Pa, appropriate for a rarefied gas in which $\tau = 10^{-5}$ s holds. The temperature field—*kinetic* temperature—has the form

$$\frac{k}{m}\theta(r) - \frac{k}{m}\theta(r_e)$$
$$= \begin{cases} -\frac{1}{5}\frac{qr_i}{\tau p}\ln\frac{(28\tau r_i/25p) + r^2}{(28\tau r_i/25p) + r_e^2} & \text{as solution of (5.15),} \\ -\frac{2}{5}\frac{qr_i}{\tau p}\ln\frac{r}{r_e} & \text{for the Navier–Stokes–Fourier theory.} \end{cases}$$
(5.16)

Figure 5.1 shows these two solutions. As expected, the two solutions coincide for large radii, where the gradient is small and they differ for small radii, where the gradient is steep.

To complete the solution of the field equations (5.15), we note that the heat flux only has a radial component and that the deviatoric stress comes out as[9]

$$t^{\langle ij \rangle} = \begin{bmatrix} -\frac{4}{5}\frac{\tau q r_i}{r^2} & 0 & 0 \\ 0 & \frac{4}{5}\frac{\tau q r_i}{r^4} & 0 \\ 0 & 0 & 0 \end{bmatrix},$$
(5.17)

while in the Navier–Stokes–Fourier case the deviatoric stress vanishes, since the gas is at rest.

---

[8] This case was recently studied by Müller and Ruggeri (2003).

[9] We write contravariant components appropriate to the cylindrical coordinates (r, ϑ, z).

ENTROPY IN NONEQUILIBRIUM

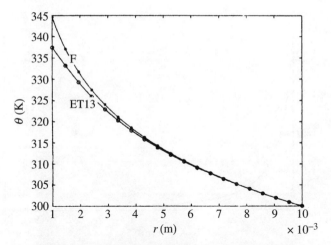

Figure 5.1. Kinetic temperature in a cylindrical tube. Fourier theory (F) and 13-moment theory (ET13).

5.2.6 Kinetic and Thermodynamic Temperatures

Therefore, the entropy flux (5.13)$_2$ only has an r-component, and that component is proportional to q^r:

$$\hat{h}^r = \underbrace{\frac{1}{\theta}\left(1 - \frac{8}{25}\frac{\tau q r_i}{p}\frac{1}{r^2}\right)}_{1/T} q^r. \tag{5.18}$$

We denote the factor of proportionality by $1/T$, as indicated in (5.18), where T is the *thermodynamic* temperature. The motivation and a discussion of this introduction of T in nonequilibrium thermodynamics follows below. First, we note that by (5.18) we have

$$T = \theta \bigg/ \left(1 - \frac{8}{25}\frac{\tau q r_i}{p}\frac{1}{r^2}\right), \tag{5.19}$$

whence we conclude that, in general, the thermodynamic temperature and the kinetic temperature differ. Both are equal, however,

- in equilibrium, where q vanishes, and
- in a dense gas, where τ is very small.

The quotient T/θ is represented in Figure 5.2 for the same data that were used to calculate the graphs of Figure 5.1. Inspection shows that the difference between the temperatures is small where the gradients are small enough for the Fourier law to hold and that the difference is big in the range of steep gradients.

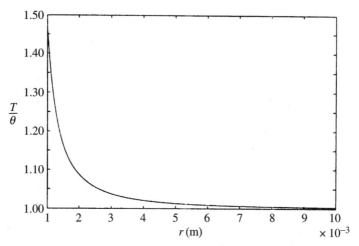

Figure 5.2. Kinetic temperature θ and thermodynamic temperature T for cylindrical heat conduction in the 13-moment theory.

It is not Clausius–Duhem nostalgia that makes us identify the thermodynamic temperature by

$$\hat{h}^r = \left(\frac{1}{T}\right) q^r.$$

Rather, the motivation lies in the *zeroth law of thermodynamics*. We proceed to discuss the rationale.

Temperature is the oldest of all thermodynamic concepts and it was intuitively clear to the pioneers, such as Galileo Galilei (1564–1642) 400 years ago, that the temperature was continuous at the wall of a thermometer. Much later, when thermodynamics was systematically scrutinized, this intuitive assumption became known as the *zeroth law of thermodynamics*:

> 0th law   temperature is continuous at an ideal interface between two bodies, typically the interface between a thermometer and a body whose temperature is measured.

(To be sure, this temperature was originally an empirical temperature, ϑ (say), but then, along with the work of Clausius and Carathéodory, it became known that the T in the second law (see (2.5)) is a universal, monotone function of ϑ — independent of material—and therefore T is also continuous at the interface between two bodies, along with ϑ. After a proposition of William Thomson (Lord Kelvin) (1824–1907) T itself may thus serve as a measure of temperature. T became known as the *absolute* temperature, or the *Kelvin* temperature, or the *thermodynamic* temperature.)

It seems eminently reasonable to assume that the ideal interface, at which temperature is continuous, does not contribute to the thermodynamic processes

ENTROPY IN NONEQUILIBRIUM   89

on either side. In particular, there should be no entropy production in the ideal interface and this means that the normal component of the entropy flux is continuous at the interface. As a matter of course the heat flux is also continuous at the interface, ideal or not. Therefore, if the entropy flux is proportional to the heat flux, as it is in our case (see (5.18)), the factor of proportionality must also be continuous. By the zeroth law, that factor must then be a universal function of the thermodynamic temperature and we set it equal to $1/T$ so as to make the relation $\hat{h}^i = (1/T)q^i$ compatible with the second law.

The continuity of T guarantees the measurability of temperature and its controllability. Thus, while it is impossible to prescribe the kinetic temperature on the boundary of a gas, it is possible to prescribe the thermodynamic temperature, because of its continuity.

Three cautionary remarks are in order here.

(i) While the zeroth law is certainly true at the interface between two bodies in equilibrium, it is often used with excellent results for heat conduction problems, e.g. heat conduction through a multilayered wall. However, it is not at all certain whether the zeroth law holds for strong nonequilibria, or in the case of a gas for a strong degree of rarefaction.

(ii) Equation $(5.13)_2$ shows that in the 13-moment theory $\hat{h}_i$ and q_i are connected by a tensor, which might be called the coldness tensor. In a general case, a case without planar or cylindrical symmetry, this means that it is only the normal component of this tensor that is continuous at the ideal wall. (A similar circumstance can be found in connection with the chemical potential, which in general must be replaced by a tensor, the chemical potential tensor, or the Eshelby tensor.)

(iii) If more and more moments are taken into account, the entropy flux will no longer be essentially determined by the heat flux; there will be terms with higher moments, but no heat flux. Such cases have not yet been investigated systematically and it is entirely uncertain whether and how thermodynamics can be extrapolated to treat them.

5.2.7 Moment Equations for 14 Moments. Minimum Entropy Production

It is generally assumed that more moments are better than fewer moments, and we shall see that rule confirmed in Section 5.3 in connection with light scattering. Therefore, the next natural step beyond 13 moments is 14 moments, namely, $F, F_{i_1}, \ldots, F_{i_1 i_2}, F_{i_1 ee} F_{iiee}$. The distribution function is determined by maximization of the entropy $(5.10)_1$ under constraints, and the closed system of equations of balance is shown as follows,

$$\begin{aligned}
&\frac{\partial \rho}{\partial t} + \frac{\partial \rho v_k}{\partial x_k} = 0, \\
&\frac{\partial \rho v_i}{\partial t} + \frac{\partial \rho v_i v_k + p\delta_{ik} - t_{\langle ik \rangle}}{\partial x_k} = 0, \\
&\frac{\partial \frac{3}{2}p}{\partial t} + \frac{\partial \frac{3}{2}pv_k + q_k}{\partial x_k} + (p\delta_{ik} - t_{\langle ik \rangle})\frac{\partial v_i}{\partial x_k} = 0, \\
&\frac{\partial t_{\langle ij \rangle}}{\partial t} + \frac{t_{\langle ij \rangle}v_k - \frac{4}{5}q_{\langle i}\delta_{j\rangle k}}{\partial x_k} + 2t_{\langle k\langle i}\frac{v_{j\rangle}}{\partial x_k} \boxed{-2p\frac{\partial v_{\langle i}}{\partial x_{j\rangle}}} = -\frac{1}{\tau}t_{\langle ij \rangle}, \\
&\frac{\partial q_i}{\partial t} + \frac{\partial q_i v_k - (k/m)\theta t_{\langle ik \rangle}}{\partial x_k} + \frac{1}{6}\frac{\partial \Delta}{\partial x_i} - \frac{t_{\langle ij \rangle}}{\rho}\frac{\partial t_{\langle jk \rangle} - 3p\delta_{jk}}{\partial x_k} + \frac{2}{5}q_i\frac{\partial v_k}{\partial x_k} \\
&\quad + \frac{7}{5}q_k\frac{\partial v_i}{\partial x_k} + \frac{2}{5}q_k\frac{\partial v_k}{\partial x_i} + (-\frac{5}{6}t_{\langle ik \rangle} + \boxed{\frac{5}{6}p\delta_{ik}})\frac{\partial (k/m)\theta}{\partial x_k} = -\frac{1}{\tau}q_i, \\
&\frac{\partial \Delta}{\partial t} + \frac{\partial \frac{28}{3}(k/m)\theta q_k + \Delta v_k}{\partial x_k} + 8\frac{q_i}{\rho}\frac{\partial t_{\langle ik \rangle} - p\delta_{ik}}{\partial x_k} - 20\frac{k}{m}\theta\frac{\partial q_k}{\partial x_k} \\
&\quad + \frac{4}{3}\left(-6t_{\langle ik \rangle}\frac{k}{m}\theta + \Delta\delta_{ik}\right)\frac{\partial v_i}{\partial x_k} = -\frac{1}{\tau}\Delta,
\end{aligned}$$

(5.20)

where, once again, the canonical notation for the variables has been employed. Of course, the last moment F_{iiee} has no name, apart from being the double trace of the fourth moment, nor a canonical notation; in equation (5.20) its nonequilibrium part is denoted by Δ and we note that Δ must satisfy a specific differential equation. Once again, as in equation (5.15), the classical constitutive relations of Navier–Stokes and Fourier are indicated by the frames.

The problem of stationary, now planar, heat conduction in a gas at rest has been considered by Barbera et al. (1999) and by Au et al. (2000). All fields depend only on $x_1 = x$, and $x > 0$ is the domain of interest. The solution based on the BGK equation has an explicit form, namely,

$$v = 0, \quad p = \text{const.}, \quad t_{\langle ij \rangle} = 0, \quad q_i = \begin{pmatrix} q = \text{const.} \\ 0 \\ 0 \end{pmatrix}, \quad (5.21)$$

and

$$\begin{aligned}
\frac{k}{m}\theta(x) &= \frac{k}{m}\theta(0) - \frac{1}{15p}\left[\left(\Delta(0) - \frac{56}{5}\frac{q^2}{p}\right)(e^{x/\lambda} - 1) + \frac{56}{5}\frac{q^2}{p}\frac{x}{\lambda}\right], \\
\Delta(x) &= \left(\Delta(0) - \frac{56}{5}\frac{q^2}{p}\right)e^{x/\lambda} + \frac{56}{5}\frac{q^2}{p}.
\end{aligned}$$

(5.21cont.)

ENTROPY IN NONEQUILIBRIUM

λ stands for $\frac{28}{5}\tau q/p$, and is of the order of magnitude of the mean free path of the atoms in the gas. We conclude from (5.21) that $\theta(x)$ is essentially a linear function of x, as expected, while $\Delta(x)$ is essentially constant. Both, however, have a boundary layer of the approximate thickness of a mean free path. The boundary layer will occur at $x \gtrsim 0$, if the heat flux is negative, which will be the case here.

The entropic quantities, specific entropy s, nonconvective entropy flux $\hat{h}^i$, and entropy production density Σ in this case read

$$\left. \begin{array}{l} s = -\dfrac{5}{2}\dfrac{k}{m}\ln\dfrac{\rho}{p^{3/5}} + c - \dfrac{1}{4}\dfrac{k}{m}\dfrac{t_{\langle ij\rangle}t_{\langle ij\rangle}}{p^2} - \dfrac{1}{5}\dfrac{k}{m}\dfrac{\rho}{p^3}q_i^2 - \dfrac{1}{240}\dfrac{k}{m}\dfrac{\rho^2}{p^4}\Delta^2, \\[2mm] h^i = \underbrace{\dfrac{1}{\theta}\left(1 - \dfrac{1}{15p}\dfrac{\Delta}{k/m\theta}\right)}_{1/T} q^i, \\[2mm] \Sigma = \dfrac{2}{5}\dfrac{k/m}{\tau p}\left\{\dfrac{q^2}{(k/m\theta)^2} + \dfrac{1}{48}\dfrac{\Delta^2}{(k/m\theta)^3}\right\}. \end{array} \right\} \quad (5.22)$$

Once again, as explained in the previous section, we identify the factor of proportionality between the heat flux and the entropy flux with the thermodynamic temperature T (see (5.22)$_2$).

The equations (5.21) need boundary conditions for p, and q and for $\theta(0)$, and $\Delta(0)$ before the solution is complete. We assume that p is given and proceed to show how q, $\theta(0)$, and $\Delta(0)$ may be determined from the entropic quantities (5.22).

We consider a layer of gas with boundaries at $x = 0$ and $x = L$ and we recall that the thermodynamic temperatures may be prescribed on the boundaries, namely, $T(0)$ and $T(L)$. Thus we have

$$\left. \begin{array}{l} T(0) = \theta(0)\left(1 - \dfrac{1}{15p}\dfrac{\Delta(0)}{k/m\theta(0)}\right)^{-1}, \\[2mm] T(L) = \theta(L)\left(1 - \dfrac{1}{15p}\dfrac{\Delta(L)}{k/m\theta(L)}\right)^{-1}, \end{array} \right\} \quad (5.23)$$

which represent two equations for the two unknowns $\theta(0)$ and $\Delta(0)$ in (5.21). (Note that $\theta(L)$ and $\Delta(L)$ are given by (5.21) in terms of $\theta(0)$ and $\Delta(0)$.)

There remains the heat flux q to be determined, actually the boundary value of q, since q is constant. For practical reasons there is no way to control, or adjust, q on the boundary, when the temperatures are already prescribed there. We must conclude that the layer of gas will adjust q so as to 'suit itself' and here is where the principle of minimal entropy production comes into play. Let us consider the following.

It is an everyday experience that a body, if left to itself, will tend to approach equilibrium, i.e. the state of minimal entropy production, in which in fact the

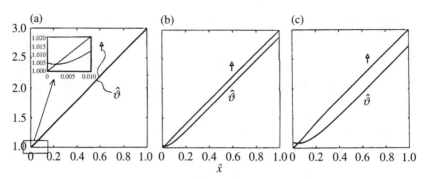

Figure 5.3. Temperature fields according to minmax principle:
(a) dense gas, (b) intermediate density, (c) rarefied gas.

entropy production is zero. But, if constant boundary conditions—here the boundary conditions $T(0)$ and $T(L)$—prevent equilibrium from being established, it seems eminently plausible to assume that the body will approach a stationary state which is as close to equilibrium as possible, meaning perhaps that it will tend to minimize the entropy production. This is an idea that goes back to Ilya Prigogine (1917–) (Glansdorff and Prigogine 1971) and, for all its plausibility, *it does not work*. Indeed, Barbera (1999) has recently shown that the 'principle of minimal entropy production' contradicts the equations of balance of mass, momentum and energy for stationary heat conduction in a Navier–Stokes–Fourier fluid at rest and for a Couette flow of such a fluid.

So great is the intrinsic suggestiveness of the principle of minimal entropy production, however, that it remains memorable. Thus Struchtrup and Weiss (1998) have recently employed it to determine the value of q chosen by the body in our present 14-moment heat conduction problem. And again, the principle provides a nonsensical result for the case of a dense gas, where we are certain that the solution is a linear increase of temperature. Instead of that linear graph, minimal entropy production requires a nearly constant temperature field throughout most of the layer and a very steep drop on the side with the lower temperature. Therefore, Struchtrup and Weiss have modified the principle of minimal entropy production and they replace it by the *minmax principle*. Their idea is that the entropy production *density* must be as small as possible everywhere in the layer, and they satisfy that condition, in the present case, by requiring that the gas adjusts q to the value that makes the maximum of the entropy production density minimal.

With the entropy production density given explicitly by (5.22)$_3$ and (5.21) as a function of x and q, the minmax principle is easily exploited and the result is shown in Figure 5.3. That result makes good sense. For a dense gas, where everything is crystal clear to begin with, both the thermodynamic and the kinetic temperatures are linear functions to the naked eye. And for a rarefied gas the two temperatures differ. The thermodynamic function is still linear while the kinetic function shows a marked boundary layer as predicted by (5.21)$_2$.

5.3 Extended Thermodynamics

5.3.1 Paradoxes

The original motive for the development of extended thermodynamics was the 'paradox of heat conduction,' so-called by Carlo Cattaneo (1911–1979) (see Cattaneo 1948) in 1948. By this paradox Cattaneo meant the fact that the heat equation is parabolic and thus implies an infinite speed of propagation of disturbances in temperature.

Upon reflection it was clear to Cattaneo that the Fourier law was to blame for the paradox. The Fourier law sets the heat flux proportional to the temperature gradient and opposite in direction. Thus it reads

$$q_i = -\kappa \frac{\partial T}{\partial x_i} \quad (\kappa \geqslant 0).$$

Indeed, if Fourier's law is used to eliminate the heat flux from the balance of energy

$$\rho c_v \frac{dT}{dt} + \frac{\partial q_i}{\partial x_i} = 0,$$

we obtain the heat equation in the form

$$\frac{dT}{dt} = \frac{\kappa}{\rho c_v} \Delta T,$$

which is the prototype of all parabolic equations.

Therefore, the Fourier law must be modified. Before explaining how Cattaneo found this modification, I should like to describe how we understand Fourier's law in terms of molecular arguments. For that purpose, we focus our attention on a small volume element of a gas, a volume element of the linear dimensions of a mean free path. We establish a temperature gradient downwards so that the bottom is hotter than the top (see Figure 5.4). The temperature determines the mean kinetic energy of the atoms so that the atoms at the bottom are faster than those at the top. Therefore, if, in the course of the thermal motion, two atoms are exchanged between top and bottom, this exchange is tantamount to an upwards energy flux through the middle surface. Therefore, this energy flux is opposite to the temperature gradient just as postulated by Fourier's law and, obviously, proportional to the temperature difference between bottom and top. This is how the kinetic theory of gases interprets the heat flux and Fourier's law.

In order to modify this argument, Cattaneo considered a situation in which the temperature gradient changes rapidly. In that case the energy transport carried by the two atoms across the middle surface depends on the temperature gradient

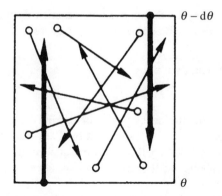

Figure 5.4. On the molecular interpretation of Fourier's law.

half a flight time *prior* to the transition, i.e. the heat flux depends on the recent past of the temperature gradient. Thus, making a Taylor expansion in time of the temperature gradient, Cattaneo came up with a modified Fourier law. It reads

$$q_i = -\kappa \left(\frac{\partial T}{\partial x_i} - \tau \frac{d}{dt}\left(\frac{\partial T}{\partial x_i}\right) \right).$$

If this modified Fourier law was used to eliminate the heat flux from the energy balance, Cattaneo obtained a new heat equation which, ironically, is still parabolic.

Cattaneo does not say much about this, but in his paper there follows a bit of creative mathematics involving the approximation of an operator, namely,

$$\left(1 - \tau \frac{d}{dt}\right)^{-1} \approx 1 + \tau \frac{d}{dt}$$

at the end of which argument he comes up with a rate-type equation for the heat flux, which we now call the *Cattaneo equation*. It reads

$$\tau \frac{dq_i}{dt} + q_i = -\kappa \frac{\partial T}{\partial x_i}. \tag{5.24}$$

Combining this new equation with the balance of energy we now obtain a hyperbolic equation for the temperature; it is, in fact, the telegraph equation in the form

$$\tau \frac{d^2 T}{dt^2} + \frac{dT}{dt} = \frac{\kappa}{\rho c_v} \Delta T. \tag{5.25}$$

And, if we calculate the speed of propagation dictated by this equation for a gas under normal conditions, we obtain a speed of the order of magnitude of the speed of sound

$$V = \pm \sqrt{\frac{\kappa}{\rho c_v \tau}} \approx 10^2\text{--}10^3 \text{ m s}^{-1}.$$

ENTROPY IN NONEQUILIBRIUM

Thus the paradox was resolved and that is as far as Cattaneo went. However, there is more than heat conduction to be considered. Indeed, the paradox of heat conduction is not alone. There are similar paradoxes of diffusion and of the propagation of shear waves. These are due to deficiencies of Fick's law and the Navier–Stokes law. The Navier–Stokes law between the stress and velocity gradients is quite analogous to Fourier's law, if stress is substituted for heat flux and the velocity gradient is substituted for the temperature gradient. It reads

$$t_{\langle ij \rangle} = 2\mu \frac{\partial v_{\langle i}}{\partial x_{j \rangle}}.$$

And the paradox of infinitely fast shear waves may be resolved by replacing the Navier–Stokes expression by a rate law for stress which is in turn analogous to the Cattaneo equation, namely,

$$\tau \frac{d t_{\langle ij \rangle}}{dt} + t_{\langle ij \rangle} = 2\mu \frac{\partial v_{\langle i}}{\partial x_{j \rangle}}.$$

5.3.2 Formal Structure

The paradoxes of heat conduction and shear waves are only the tip of the iceberg; their resolution, however, clearly indicates that the Navier–Stokes–Fourier theory is inadequate for processes with rapid rates and steep gradients and that it must be replaced by a theory with hyperbolic field equations.

That new theory was developed in the last decades of the 20th century and is called *extended thermodynamics*. The theory is strongly motivated by the kinetic theory of gases and combines all the best features of thermostatics, the thermodynamics of irreversible processes and rational thermodynamics. Its formal structure is beautifully simple and I shall now describe it. See also Müller and Ruggeri (1993, 1998) for an account of extended thermodynamics up to 1998.

The object of thermodynamics, extended or ordinary, is the determination of certain fields, say n of them, which we combine in a field vector u whose components depend on the event x_D. Capital indices run from 0 to 3, and x_0 is the time, while x_d ($d = 1, 2, 3$) denote the spatial coordinates of the event. For the determination of the fields u we need field equations and these are based upon the equations of balance of mechanics and thermodynamics, which read, in compact form,

$$F_{A,A} = \Pi.$$

F_A are 4-fluxes and the comma denotes differentiation. The n components of F_0 are called densities, while F_a denotes the spatial components of the fluxes. Π is the n-vector of productions.

In order to arrive at field equations for the fields u, we need constitutive equations for F_A and Π which relate the 4-fluxes F_A and the productions Π

to the fields u in a materially dependent manner by what we call constitutive relations. In extended thermodynamics these *constitutive relations* are local in space-time, such that F_A and Π at one event depend only on the values of the fields u at that event. We have

$$F_A = \hat{F}_A(u) \quad \text{and} \quad \Pi = \hat{\Pi}(u).$$

(The local character of the constitutive relations distinguishes extended thermodynamics from ordinary thermodynamics, where all kinds of gradients and time derivatives may occur among the variables.)

If the constitutive functions $\hat{F}_A$ and $\hat{\Pi}$ are explicitly known, we may eliminate the fluxes F_A and the productions Π between the constitutive relations and the equations of balance. In this way we come up with an explicit set of field equations. Every solution of those we call a *thermodynamic process*.

Thus, in the last few paragraphs I have described thermodynamics and, in particular, extended thermodynamics in a nutshell. However, there is a catch: there is not a single material for which we know the constitutive functions $\hat{F}_A$ and $\hat{\Pi}$ explicitly, and therefore the thermodynamicist spends most of the time trying to determine these functions, or at least trying to reduce their generality and, if possible, come up with only a few coefficients for the experimenter to measure. This is the subject of constitutive theory.

There are three main tools of constitutive theory: the entropy inequality, the requirement of convexity, and the principle of relativity.

The entropy inequality requires that the equation of balance of entropy

$$h_{A,A} = \Sigma \geqslant 0$$

holds for all thermodynamic processes. h_A is the entropy 4-flux and Σ is the entropy production, both are constitutive quantities so that we have

$$h_A = \tilde{h}_A(u) \quad \text{and} \quad \Sigma = \tilde{\Sigma}(u).$$

The requirement of convexity, ironically enough, states that the entropy density h_0 is a concave function of u, such that

$$\frac{\partial^2 h_0}{\partial u \partial u}\text{-negative definite.}$$

The principle of relativity requires that the field equations have the same form in all Galilei frames. This is a powerful restrictive principle on the constitutive functions, but we shall not exploit it in this chapter.

The entropy principle, comprising the entropy inequality and the convexity requirement, provides the possibility of casting the system of field equations into a symmetric hyperbolic form. Indeed, in the process of exploiting the entropy inequality we come up with a privileged set of fields which may be chosen instead of the generic field u, so that the field equations are symmetric hyperbolic for that privileged field. I proceed to explain how the privileged field is identified and how its properties are recognized. For that purpose we have to exploit the entropy inequality.

The key to the exploitation of the entropy inequality is the observation that this inequality need not hold for all fields, but only for those that are thermodynamic processes, i.e. solutions of the field equations. In a manner of speaking, therefore, the field equations provide constraints for the fields that must satisfy the entropy inequality. According to a lemma proved by Liu (1972), we may get rid of this constraint by using Lagrange multipliers. Indeed Liu proved that the more complex inequality,

$$h_{A,A} - \Lambda(F_{A,A} - \Pi) \geq 0$$

with Lagrange multipliers Λ, must hold for all fields rather than only for thermodynamic processes. The Lagrange multipliers Λ, according to Liu's proof, are constitutive quantities, depending on u locally in space-time. In fact, the components of Λ represent the privileged fields as we shall presently show.

From the new inequality we have that

$$\left(\frac{\partial h_A}{\partial u} - \Lambda \frac{\partial F_A}{\partial u}\right) u_{,A} - \Lambda \cdot \Pi \geq 0$$

must hold for all $u_{,A}$, and hence

$$dh_A = \Lambda \, dF_A \quad \text{and} \quad \Lambda \Pi \geq 0. \tag{I}$$

If, without essential loss of generality, we set $u = F_0$, we conclude

$$\frac{\partial h_0}{\partial u} = \Lambda \quad \text{and hence} \quad \frac{\partial^2 h_0}{\partial u \partial u} = \frac{\partial \Lambda}{\partial u} \text{-negative definite,}$$

since h_0 is concave. Therefore, the relation $\Lambda = \Lambda(u)$ is globally invertible and we may use the Lagrange multipliers as the fields, instead of the generic fields u. If this is done, we may write (I)$_1$ in the form

$$dh'_A = F_A \, d\Lambda, \quad \text{where } h'_A = \Lambda F_A - h_A. \tag{II}$$

Thus

$$F_A = \frac{\partial h'_A}{\partial \Lambda} \tag{III}$$

holds and $F_{A,A} - \Pi$ may be written in the form

$$\frac{\partial F_A}{\partial \Lambda} \Lambda_{,A} = \Pi \quad \text{or, by (III),} \quad \frac{\partial^2 h'_A}{\partial \Lambda \partial \Lambda} \Lambda_{,A} = \Pi. \tag{IV}$$

Equation (IV)$_2$ represents a symmetric hyperbolic system, because all four matrices are symmetric and since h'_0 is concave in the fields Λ. The latter results from the concavity of h_0 in the fields u, since, by (II)$_2$ and still with $u = F_0$, we have $h'_0 = \Lambda u - h_0$, so that h'_0 is the Legendre transform of h_0; and Legendre transformations preserve convexity or, in our case, concavity.

Symmetric hyperbolicity is a desirable feature for the field equations to have, because it implies well-posedness of initial-value problems and, above all, finite characteristic speeds.

The characteristic speeds are the speeds of acceleration waves, i.e. propagating singular surfaces on which the fields, now the privileged fields Λ, are continuous but their derivatives are not. Indeed, we have

$$[\Lambda_{,a}] = n_a \delta \Lambda \quad \text{and} \quad [\Lambda_{,0}] = -V \delta \Lambda,$$

so that the jump of the gradient is in the normal direction and the jump of the time derivative determines the speed of the wave. $\delta \Lambda$ is called the amplitude of the wave; it represents the jump of the normal component of the gradient. Subtracting the field equations (IV)$_2$ before and behind the wave we thus obtain a homogeneous linear algebraic system for the amplitudes, namely,

$$\left(-\frac{\partial^2 h'_0}{\partial \Lambda \partial \Lambda} V + \frac{\partial^2 h'_a}{\partial \Lambda \partial \Lambda} n_a \right) \delta \Lambda = 0.$$

This system has nontrivial solutions only if the determinant vanishes, and that condition determines the characteristic speeds, n of them, not necessarily all different, but all finite, because of the symmetric hyperbolic character of the equations.

5.3.3 Pulse Speeds

All of the above formal structure is quite synthetic. We have chosen neither specific fields nor a specific material. This will now change, because now we focus attention on a monatomic ideal gas and as fields we consider the moments of the distribution function, a set of fields of increasing tensorial rank, up to rank N, namely,

$$u = \begin{bmatrix} F \\ F_{i_1} \\ F_{i_1 i_2} \\ \vdots \\ F_{i_1 i_2 \cdots i_N} \end{bmatrix}.$$

If we take the fluxes to be moments as well, we obtain a set of balance equations of the same type as the moment equations in the kinetic theory of gases (see (5.14)). Special cases include the 13-moment system and the 14-moment system shown in equations (5.15) and (5.20). These have already been closed by constitutive relations that are subject to the entropy principle. Weiss (1990) has the linearized field equations for any N ready at the touch of a button. He has calculated the characteristic speeds and, in particular, the biggest characteristic speed, which we call the *pulse speed*. For the 20-moment case this speed is 1.8

Table 5.1. Pulse speeds V_{max} (see Weiss 1990).

Number of moments	Number of equations (5.21)	Highest order of moments	V_{max}/c_0
4	2	1	0.774 596 67
10	4	2	1.341 640 79
20	6	3	1.808 229 48
35	9	4	2.212 999 46
56	12	5	2.574 958 74
84	16	6	2.905 078 11
120	20	7	3.210 352 45
165	25	8	3.495 557 91
220	30	9	3.764 123 72
286	36	10	4.018 608 47
364	42	11	4.260 980 14
455	49	12	4.492 790 23
560	56	13	4.715 287 16
680	64	14	4.929 492 84
816	72	15	5.136 256 17
969	81	16	5.336 291 3
1 140	90	17	5.530 205 69
1 330	100	18	5.718 521 12
1 540	110	19	5.901 689 62
1 771	121	20	6.080 105 85
2 024	132	21	6.254 116 73
2 300	144	22	6.424 029 19
2 600	156	23	6.590 116 27
2 925	169	24	6.752 622 13
3 276	182	25	6.911 766 15
3 654	196	26	7.067 746 31
4 060	210	27	7.220 741 98
4 495	225	28	7.370 916 29
4 960	240	29	7.518 418 07
5 456	256	30	7.663 383 62
5 984	272	31	7.805 938 04
6 545	289	32	7.946 196 54
7 140	306	33	8.084 265 49
7 770	324	34	8.220 243 31
8 436	342	35	8.354 221 29
9 139	361	36	8.486 284 32
9 880	380	37	8.616 511 44
10 660	400	38	8.744 976 44
11 480	420	39	8.871 748 33
12 341	441	40	8.996 891 71
13 244	462	41	9.120 467 22
14 190	484	42	9.242 531 84
15 180	506	43	9.363 139 18

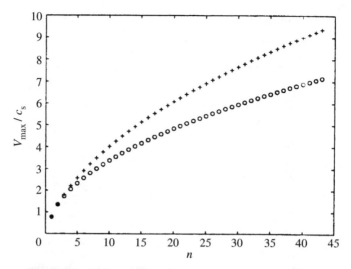

Figure 5.5. Pulse speed versus rank N. Crosses, actual values by Weiss (1990); circles, lower bound by Boillat and Ruggeri (1997).

times the regular sound speed. For $N = 20$, the 1771-moment case, that ratio is 6.1 and for $N = 40$, the 12 341-moment case, it is 9 (see Table 5.1).

The graph of Figure 5.5 shows the pulse speed versus the number of moments; it increases monotonically and there is the clear indication that it tends to infinity as N tends to infinity. This conjecture has in fact recently been proved by Boillat and Ruggeri (1997), who found a lower bound for the pulse speed which tends to infinity as $\sqrt{N - 1/2}$.

In a manner of speaking, if we think of the original motive of extended thermodynamics, which was the quest for finite speeds, this is an anticlimax. It is true that we now have finite speeds for every finite number of equations, but the speeds tend to infinity as the number of equations increases. To my thinking this is a satisfactory result. But, if you still regard this as an anticlimax, we must say that at this point extended thermodynamics has long since outgrown the original motive of finite speeds. It has instead become a predictive theory for processes with steep gradients and rapid changes, as is particularly obvious from the application of extended thermodynamics to light scattering.

Before we go into that, two remarks are appropriate. First of all, in a relativistic theory the limit of the pulse speed for infinitely many moments is the speed of light rather than infinity. This was also proved by Boillat and Ruggeri (1999). Secondly, the pulse speed is only the maximum characteristic speed; there are many more characteristic speeds.

In fact, for $N = 40$ we have 12 341 moments and 441 one-dimensional equations; therefore, more than 400 longitudinal waves, all with their speeds. Weiss has listed them all; some are smaller than the speed of sound, but most

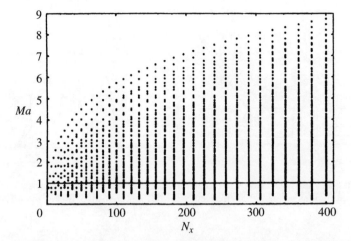

Figure 5.6. Characteristic speeds versus rank N (Weiss 1990).

are larger (see Figure 5.6) and there arise the following questions: where are all these waves, why do we not see them, or, better, why do we not hear them? The short answer is that all these modes of propagation are quickly damped out, but I do not go into this in this chapter.

5.3.4 Light Scattering

Light scattering is our paradigm for the usefulness and practicability of extended thermodynamics. Incoming light of a single frequency is scattered on the density fluctuations of a gas in thermodynamic equilibrium (see Figure 5.7). The scattered light consists mostly of light of the same frequency as the incoming light, but neighboring frequencies are also represented in the scattering spectrum. The spectral distribution of the scattered light is characterized by three distinctive peaks in a dense gas, with the central one at the frequency of the incident wave. For smaller pressures the lateral peaks become less well pronounced and in a rarefied gas there remains only a single bump in the middle.

The widths of the peaks determine viscosity and heat conductivity of the gas, the relative height of the peaks determines the specific heats and the distance of the peaks determines the compressibility. It is a fair question at this point to ask how and why the microscopic density fluctuations of the gas in equilibrium can possibly determine a macroscopic property of nonequilibrium like the viscosity.

This is where the Onsager hypothesis comes into the picture. Lars Onsager (1903–1976), in a flash of insight, postulated—and that postulate was beautifully confirmed—that the mean regression of a *microscopic* fluctuation in its temporal development is identical to the relaxation of a *macroscopic* deviation from equilibrium. Thus the microscopic fluctuation, reflected in the spectral

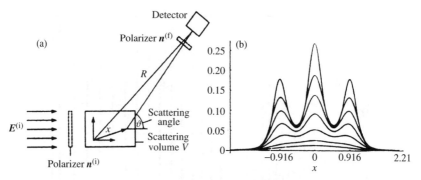

Figure 5.7. (a) Light scattering apparatus (schematic). (b) Spectral distribution of scattered light as a function of frequency for different densities (measured by Au and Wehr (1997)).

distribution, is related to the predictions of the *macroscopic* field equation. Therefore, according to Onsager's hypothesis it is possible to calculate the spectral distribution from our field equations. What we have to do is

- Fourier transform the equations spatially;
- Laplace transform them temporally;
- solve them algebraically; and then
- average the solutions over all initial conditions.

In this manner we should be able to calculate the measured spectral distribution, *provided we have the correct field equations.*

And indeed in a dense gas, where Navier–Stokes–Fourier provides the correct field equations, we obtain the three-peak distribution of Figure 5.7. But in a rarefied gas we need extended thermodynamics, because the results of the Navier–Stokes–Fourier theory are not satisfactory. For even a moderate degree of rarefaction the Navier–Stokes–Fourier theory is not good enough. The dots in Figure 5.8 show the measured values of the spectral distribution and the calculated values are represented by the curves marked ET20, 35, 56, and 84, corresponding to the respective theories of so many moments.

None of these fit well and now it might seem time to adjust parameters in order to force a fit. But extended thermodynamics is free of parameters; in extended thermodynamics we have a *theory of specific theories* with only one parameter: the number of equations. And indeed, if we raise that number to 120, we obtain a perfect fit between theory and experiment and *that fit remains unchanged for higher numbers.*

Thus we observe a convergence and it follows that at the given pressure we need not go beyond ET120. Let me emphasize that point by saying that our 'theory of theories' carries the possibility of determining its own range of

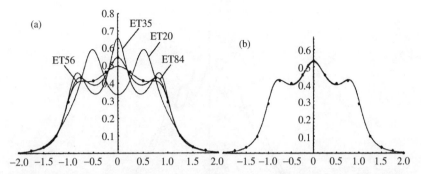

Figure 5.8. Spectral distributions in a moderately rarefied gas. Dots represent measurements by Clark (1975). (a) ET20, 35, 56, and 84. (b) ET120, 165, 220, and 286. All curves calculated by Weiss (1990); see also Weiss and Müller (1995) or Müller and Ruggeri (1993, 1998).

validity—something that is usually said a theory cannot possibly do. Here, if we have two successive theories which provide the same results, the lower one is good enough; and we may say this *without carrying out a single experiment*.

All this is most satisfactory, but there is also disappointment. Indeed, one might have hoped that 13 or 14 moments would bring about a great improvement over Navier–Stokes–Fourier and a good representation of experimental results. Instead, we need hundreds of moments for even a moderately rarefied gas. This may be disappointing, but that is what nature requires. It might have been nicer to have to use fewer moments, but nature does not permit that.

For lower pressures we need even more moments. Indeed for a very rarefied gas even ET210 through ET256 disagree, although their disagreement is kept to a narrow strip (see Figure 5.9). Inside this strip runs a Gaussian distribution which is the ultimate spectral distribution for the case when the pressure tends to zero, and it reflects the Maxwellian distribution of the gas. That can be proved analytically, albeit only for the BGK model of the Boltzmann equation (see, for example, Weiss and Müller 1995).

5.4 A Remark on Alternatives

Extended thermodynamics has been quite successful for light scattering where boundary and initial values are unimportant. The theory is also useful for the calculation of shock structures and for understanding the formation of a shock structure in a shock tube experiment (see Au et al. 2001) because in these cases the boundary and initial conditions are those of equilibrium, and they are known. Problems arise whenever initial and boundary values must be specified for unconventional quantities, like moments of rank higher than three. The principles of minimum entropy production, described in Section 5.2.7, may

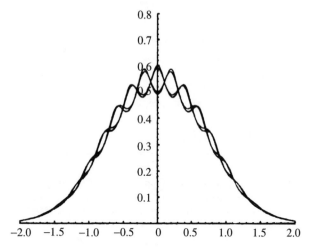

Figure 5.9. Spectral distribution for ET210, 225, 240, and 256 in a very rarefied gas.

help, but it is as yet uncertain whether they are true and whether the results which they predict can be confirmed.

In the meantime, with the increasing power of computers, numerical schemes have come forth, like the direct simulation Monte Carlo method (see, for example, Bird 1994), or kinetic schemes for the calculation of slip velocities and temperature jumps at the boundary of a gas (see Struchtrup 2003). Quantities like entropy, its flux and its production play no role in such schemes, or only a minor role.

References

Au, J. and Wehr, L. R. 1997 Light scattering experiment and extended thermodynamics. *Continuum Mech. Thermodyn.* **9**, 155–164.

Au, J., Müller, I. and Ruggeri, T. 2000 Temperature jumps at the boundary of a rarefied gas. *Continuum Mech. Thermodyn.* **12**, 19–30.

Au, J. D., Torrilhon, M. and Weiss, W. 2001 The shock tube study in extended thermodynamics. *Phys. Fluids* **13**, 2423–2432.

Barbera, E. 1999 On the principle of minimal entropy production for Navier–Stokes–Fourier fluids. *Continuum Mech. Thermodyn.* **11**, 327–330.

Barbera, E., Müller, I. and Sugiyama, M. 1999 On the temperature of a rarefied gas in non-equilibrium. *Meccanica* **34**, 103–113.

Bhatnagar, P. L., Gross, E. P. and Krook, M. 1954 A model for collision processes in gases. *Phys. Rev.* **94**, 511.

Bird, G. 1994 *Molecular Gas Dynamics and the Direct Simulation of Gas Flows*. Clarendon Press, Oxford.

Boillat, G. and Ruggeri, T. 1997 Moment equations in the kinetic theory of gases and wave velocities. *Continuum Mech. Thermodyn.* **9**, 205–212.

Boillat, G. and Ruggeri, T. 1999 Relativistic gas: moment equations and maximum wave velocity. *J. Math. Phys.* **40**.

Cattaneo, C. 1948 Sulla conduzione del calore. *Atti Sem. Mat. Fis. Univ. Modena* **3**, 83–101.

Clark, N. A. 1975 Inelastic light scattering from density fluctuations in dilute gases. The kinetic-hydrodynamic transition in a monatomic gas. *Phys. Rev.* A **12**.

Coleman, B. D. and Mizel, V. J. 1964 Existence of caloric equations of state in thermodynamics. *J. Chem. Phys.* **40**, 1116.

Coleman, B. D. and Noll, W. 1963 The thermodynamics of elastic materials with heat conduction and viscosity. *Arch. Ration. Mech. Analysis* **13**, 167–178.

Eckart, C. 1940a The thermodynamics of irreversible processes. I. The simple fluid. *Phys. Rev.* **58**, 267–269.

Eckart, C. 1940b The thermodynamics of irreversible processes. II. Fluid mixtures. *Phys. Rev.* **58**, 269–275.

Eckart, C. 1940c The thermodynamics of irreversible processes. III. Relativistic theory of the simple fluid. *Phys. Rev.* **58**, 919–924.

Glansdorff, P. and Prigogine, I. 1971 *Thermodynamic Theory of Structure, Stability and Fluctuations*. Wiley Interscience, London.

Grad, H. 1958 *Principles of the Kinetic Theory of Gases*. Handbuch der Physik XII. Springer.

Junk, M. 2000 Maximum entropy for reduced moment problems. *Math. Models Meth. Appl. Sci.* **10**, 1001–1026.

Junk, M. 2002 Maximum entropy moment systems and Galilean invariance. *Continuum Mech. Thermodyn.* **14**, 563–576.

Kogan, M. N. 1967 In *Proc. 5th Symp. on Rarefied Gas Dynamics*, 1, Suppl. 4. Academic.

Levermore, C. D. 1996 Moment closure hierarchies for kinetic theories. *J. Statist. Phys.* **83**.

Liu, I.-S. 1972 Method of Lagrange multipliers for exploitation of the entropy principle. *Arch. Ration. Mech. Analysis* **46**, 131–148.

Müller, I. 1985 *Thermodynamics*. Pitman.

Müller, I. and Ruggeri, T. 1993 *Extended Thermodynamics*. Springer Tracts in Natural Philosophy, Vol. 37.

Müller, I. and Ruggeri, T. 1998 *Rational Extended Thermodynamics*, 2nd edn. Springer.

Müller, I. and Ruggeri, T. 2003 Stationary heat conduction in radially symmetric situations—an application of extended thermodynamics. *J. Non-Newtonian Fluids*. (In the press.)

Müller, I., Reitebuch, D. and Weiss, W. 2002 Extended thermodynamic, consistent in order. *Continuum Mech. Thermodyn.* **15**, 113–146.

Struchtrup, H. 2003 Heat transfer in the transition regime: solution of boundary value problems for Grad's moment equations via kinetic schemes. *Phys. Fluids*. (In the press.)

Struchtrup, H. and Weiss, W. 1998 Maximum of the local entropy production becomes minimal in stationary processes. *Phys. Rev. Lett.* **80**, 5048–5051.

Weiss, W. 1990 Zur Hierarchie der erweiterten Thermodynamik. Dissertation, TU Berlin.

Weiss, W. and Müller, I. 1995 Light scattering and extended thermodynamics. *Continuum Mech. Thermodyn.* **7**, 123–178.

Woods, L. C. 1973 The bogus axioms of continuum mechanics. *Bull. Math. Applic.* **9**, 40.

Chapter Six

Entropy for Hyperbolic Conservation Laws

C. M. Dafermos

Brown University

This contribution describes how the notion of entropy was transplanted from its natural habitat, namely, classical thermodynamics, to the theory of hyperbolic systems of conservation laws and then discusses its multifaceted role as a stabilizing agent in the context of that theory.

6.1 Introduction

Thermodynamic entropy is the oldest among the various notions of entropy discussed in this book. It made its first appearance in the 1850s, in the work of Clausius, and there is an aura of mystery about it: a physical quantity which is not directly observable and yet pervades the whole of nature, regulating the direction of the flow of time and responsible for the ageing and eventual demise of the Universe. The aim of this chapter is to sketch how the notion of thermodynamic entropy spread from its natural habitat, which is continuum physics, to the theory of a class of partial differential equations called hyperbolic conservation laws.

In continuum physics, the state of the medium is characterized by a number of fields, such as density, velocity, temperature, etc., whose time evolution is governed by a system of field equations called conservation laws, as they result from the coupling of the conservation laws of physics with constitutive assumptions on the material properties of the medium. The system of conservation laws is supplemented by the Clausius–Duhem inequality, which expresses the second law of thermodynamics: at any point of space-time, the temporal rate of entropy change exceeds the sum of the rates of entropy flux and entropy supply at that point.

By the choice of constitutive relations, the entropy inequality must be identically satisfied for as long as the fields of the state variables remain smooth; not necessarily so, however, when discontinuities develop. This is particularly crucial for media with 'elastic' response for which the system of conservation laws is quasilinear hyperbolic. For such systems, solutions starting out from smooth initial values spontaneously develop jump discontinuities which propagate as shock waves. Thus one has to deal throughout with weak solutions.

The difficulty is compounded further by the fact that, in the framework of weak solutions, uniqueness is lost. The entropy inequality is then called to serve as an arbiter for singling out physically relevant solutions and for testing the stability of such solutions. In order to get a taste of this process, the reader will be introduced, in Section 6.2, to one-dimensional, isothermal thermoelasticity, which offers the simplest example of a hyperbolic system of conservation laws, arising in physics, which is supplemented by an entropy inequality.

The general theory of hyperbolic systems of conservation laws in one space dimension was formulated by Lax (1957), by distilling experience accumulated over the years from a thorough study of the special system of conservation laws of gas dynamics. A brief outline of the present status of this theory is presented in Section 6.3.

Section 6.4 introduces the notion of entropy in the context of general hyperbolic systems of conservation laws in one space dimension, and discusses some of its basic properties. The remainder of that section, together with Section 6.5, present a variety of results that shed light, from different angles, on the role of entropy as an instrument for stabilizing weak solutions.

Our restricting the discussion to a single space dimension is done in order to simplify the notation: notions introduced and issues raised here are relevant regardless of the dimension of the spatial variable. Naturally, only a bird's eye view of the theory may be sketched in a few pages, and few references are cited in the bibliography. Detailed treatment, together with extensive lists of references, may be found in several books (see, for example, Bressan 2000; Dafermos 2000; Holden and Risebro 2002; LeFloch 2002; Serre 1999, 2000; Smoller 1994). In particular, my book (Dafermos 2000) provides a thorough discussion of all the topics touched on in this chapter, emphasizing the intimate relation of the analytical theory with continuum physics.

6.2 Isothermal Thermoelasticity

In one-dimensional continuum physics, the reference space and the physical space are both copies of $\mathbb{R}$. The typical point of the reference space is denoted by x, while the typical point of the physical space is denoted by χ. The reference configuration of a one-dimensional body is an interval (a, b) of the reference space. A *placement* of the body is a bilipschitz homeomorphism $\chi = \chi(x)$ from its reference configuration (a, b) to the physical space.

A motion of the body over a time interval (s, τ) is a Lipschitz function $\chi = \chi(x, t)$, defined for $x \in (a, b)$, $t \in (s, \tau)$, such that for each fixed time t the map $\chi(\cdot, t)$ is a placement. A motion induces the kinematic fields of *strain* and *velocity*,

$$u(x, t) = \partial_x \chi(x, t), \quad v(x, t) = \partial_t \chi(x, t), \qquad (6.1)$$

which are bounded and measurable. Motions are assumed to be merely Lipschitz continuous, rather than smooth, because in continuous media with elastic response jump discontinuities develop spontaneously in the strain and velocity fields.

The aim of continuum physics is to monitor the evolution of certain fields with physical significance along motions of the body. Here we shall employ the *Lagrangian formulation*, in which these fields are viewed as functions of (x, t).

Continuum physics rests on a system of *conservation laws* and a set of *constitutive relations*. The former set the physical framework of the continuum theory, such as mechanics, thermomechanics, electrodynamics, etc., while the latter specify the material properties of the continuous medium, e.g. elastic, viscoelastic, magnetoelastic, and so on. Certain conservation laws may be of purely kinematic nature, for instance,

$$\partial_t u - \partial_x v = 0, \tag{6.2}$$

which expresses the integrability condition induced by (6.1). However, conservation laws typically express the balance of conserved 'extensive' physical quantities. In the Lagrangian formulation of continuum mechanics, with reference density $\rho_0 = 1$, there is only one conserved quantity, namely, *momentum*,

$$\partial_t v - \partial_x \sigma = 0, \tag{6.3}$$

where σ is the *stress*. In contrast, in continuum thermomechanics *energy* is conserved, in addition to momentum. Moreover, the list of conservation laws of momentum and energy is supplemented by the *Clausius–Duhem inequality*, which expresses the second law of thermodynamics. In *isothermal thermomechanics*, it is assumed that heat is supplied to the body at such a rate that energy is balanced while the temperature is kept constant. Therefore, in that case the conservation law of energy is discarded and only the conservation law of momentum (6.3) has to be accounted for, as in mechanics. However, in contrast to (pure) mechanics, the second law of thermodynamics is still present in isothermal thermomechanics, in the guise of the inequality

$$\partial_t(\psi + \tfrac{1}{2}v^2) - \partial_x(v\sigma) \leqslant 0, \tag{6.4}$$

which follows from the Clausius–Duhem inequality under isothermal conditions. In (6.4) ψ stands for the *Helmholtz free energy*.

In the tradition of continuum physics, the conservation laws (6.2) and (6.3) constitute a system of evolution equations from which the state variables (u, v), and thereby the motion χ, are to be determined. On the other hand, the inequality (6.4) plays a dual role. First, it is supposed to hold identically for any smooth motion that satisfies (6.2) and (6.3); this is to be accomplished by judicious choice of the constitutive relations.[1] Secondly, it provides an admissibility condition for weak solutions of (6.2), (6.3), weeding out motions with discontinuities that may conserve momentum but are thermodynamically inadmissible. To

---

[1] The reader may find details and elaborations in Chapter 5 by Müller, especially Section 5.1, as well as Chapter 4 by Hutter and Wang, especially Section 4.3.

see this scheme in a concrete situation, consider the case of *isothermal thermoelasticity*, in which stress and the Helmholtz free energy are determined solely by the strain u, that is $\sigma = \sigma(u)$, $\psi = \psi(u)$. For smooth motions, (6.4) together with (6.2) and (6.3) yield

$$[\psi'(u) - \sigma(u)]\partial_x v \leq 0, \tag{6.5}$$

which may hold identically if and only if

$$\sigma(u) = \psi'(u). \tag{6.6}$$

It should be re-emphasized that, even though (6.6) holds, (6.4) is not necessarily satisfied by weak solutions of (6.2), (6.3). In the context of weak solutions, (6.4) remains as a criterion of thermodynamic admissibility. This is the simplest manifestation of a situation that is ubiquitous in continuum physics, and has motivated the formulation of the general theory of hyperbolic systems of conservation laws that will be outlined in the following sections.

6.3 Hyperbolic Systems of Conservation Laws

A *conservation law* in one space dimension is a quasilinear, first-order system in divergence form:

$$\partial_t U(x, t) + \partial_x F(U(x, t)) = 0. \tag{6.7}$$

The (unknown) *state vector* U takes values in $\mathbb{R}^n$ and the *flux* F is a given, smooth function from $\mathbb{R}^n$ to $\mathbb{R}^n$. The system is called *strictly hyperbolic* whenever, for any fixed $U \in \mathbb{R}^n$, the Jacobian matrix $DF(U)$ has n real distinct eigenvalues (characteristic speeds)

$$\lambda_1(U) < \cdots < \lambda_n(U) \tag{6.8}$$

and thereby a set of linearly independent eigenvectors $R_1(U), \ldots, R_n(U)$.

Simple examples we shall use as models in our discussion include the scalar *Burgers equation*

$$\partial_t u + \partial_x(\tfrac{1}{2}u^2) = 0, \tag{6.9}$$

and the system of conservation laws of isothermal thermoelasticity

$$\left.\begin{array}{l}\partial_t u - \partial_x v = 0,\\ \partial_t v - \partial_x \sigma(u) = 0,\end{array}\right\} \tag{6.10}$$

introduced in the previous section. Note that (6.10) is strictly hyperbolic provided $\sigma'(u) > 0$.

The difficulty in the analytic study of solutions of (6.7) stems from the fact that when F is nonlinear the characteristic speeds $\lambda_i(U)$ are not constant. It turns out that the effect of nonlinearity is particularly pronounced in the *genuinely*

ENTROPY FOR HYPERBOLIC CONSERVATION LAWS

nonlinear case, where each characteristic speed λ_i varies in the direction of its associated eigenvector R_i:

$$D\lambda_i(U) R_i(U) \neq 0, \quad U \in \mathbb{R}^n, \quad i = 1, \ldots, n. \tag{6.11}$$

The relevance of this condition was first identified by Lax (1957). Notice that (6.9) is genuinely nonlinear, and so is (6.10) when $\sigma''(u) \neq 0$.

The Cauchy problem for (6.7), with smooth initial data

$$U(x, 0) = U_0(x), \quad -\infty < x < \infty, \tag{6.12}$$

always has a locally defined classical solution.

Theorem 6.1. *Assume (6.7) is strictly hyperbolic and $U_0(\cdot)$ is a C^1 function with bounded derivative on $(-\infty, \infty)$. Then there exists a unique C^1 solution of (6.7), (6.12) defined on a maximal time interval $[0, T_\infty)$. Furthermore, if $T_\infty < \infty$, then $\|\partial_x U(\cdot, t)\|_{L^\infty} \to \infty$ as $t \uparrow T_\infty$.*

In the presence of nonlinearity, classical solutions have typically finite lifespan, as shown by the following result of John (1974).

Theorem 6.2. *Assume (6.7) is strictly hyperbolic and genuinely nonlinear. If $U_0(\cdot)$ is a C^2 function with compact support and its derivative is sufficiently small on $(-\infty, \infty)$, then the classical solution of (6.7), (6.12) breaks down at a finite time, $T_\infty < \infty$.*

As a consequence of the above result, one may only hope to find weak solutions of the Cauchy problem in the large, i.e. L^∞ functions U that satisfy (6.7) in the sense of distributions. Beyond L^∞, which provides the broadest framework for weak solutions, the space BV of functions of bounded variation is a natural function class in which solutions should be sought. A bounded measurable function $U(x, t)$ has (locally) bounded variation when its distributional derivatives $\partial_x U$ and $\partial_t U$ are (locally) finite Radon measures. The domain of any BV function U is partitioned into three pairwise-disjoint subsets, namely, the *regular set* $\mathcal{C}$, the *jump set* $\mathcal{J}$, and the *residual set* $\mathcal{R}$, which have the following properties.

(a) At any point $(\bar{x}, \bar{t}) \in \mathcal{C}$, U attains an approximate limit (in the sense of Lebesgue) U_0, and $U(\bar{x}, \bar{t}) = U_0$

(b) $\mathcal{J}$ is the countable union of C^1 arcs. With any point $(\bar{x}, \bar{t}) \in \mathcal{J}$ is associated a slope s such that U attains distinct approximate limits U_- and U_+ at $(\bar{x}, \bar{t})$, relative to the half-planes $x < \bar{x} + s(t - \bar{t})$ and $x > \bar{x} + s(t - \bar{t})$.

(c) The one-dimensional Hausdorff measure of $\mathcal{R}$ vanishes.

When U is a BV solution of (6.7), the arcs of $\mathcal{J}$ are interpreted as *shock waves* and the slope s is the speed of propagation of these waves. The fact that U satisfies

(6.7) in the sense of distributions manifests itself in the *Rankine–Hugoniot jump conditions*,
$$F(U_+) - F(U_-) = s[U_+ - U_-], \qquad (6.13)$$
which hold at any point of the jump set. In particular, (6.13) implies that for weak shocks, $|U_+ - U_-| \ll 1$, the speed s must be close to one of the characteristic speeds, say to λ_i. In that case the shock is dubbed an *i-shock*. It is easily seen that the set of states U_+ joined to U_- by an i-shock lie on a smooth curve, which is called the i-shock curve through U_-.

When dealing with weak solutions of (6.7), one immediately encounters a nasty problem of nonuniqueness. This may be seen even at the most elementary level, say the Cauchy problem for the Burgers equation (6.9), with initial data
$$u(x, 0) = \begin{cases} -1, & x < 0, \\ 1, & x > 0, \end{cases} \qquad (6.14)$$
which admits the following two solutions:
$$u(x, t) = \begin{cases} -1, & x < 0,\ t > 0, \\ 1, & x > 0,\ t > 0, \end{cases} \qquad (6.15)$$
and
$$u(x, t) = \begin{cases} -1, & x < -t,\ t > 0, \\ x/t, & -t \leqslant x \leqslant t,\ t > 0, \\ 1, & x > t,\ t > 0. \end{cases} \qquad (6.16)$$

The need thus arises to devise admissibility criteria that will hopefully single out a unique weak solution to the initial-value problem for systems (6.7). Several such criteria have indeed been proposed, motivated by physical and/or mathematical considerations, and there exists extensive literature that compares and contrasts them. It seems that a consensus has been reached on this issue for solutions with weak shocks of strictly hyperbolic systems, while the discussion is still going on about more general situations. The most direct admissibility conditions seek to identify admissible shocks, departing from the premise that a general *BV* solution should be admissible if and only if all of its shocks are admissible. The most important shock admissibility criteria are the following.

(a) The *Lax E-condition* (Lax 1957), according to which an i-shock joining the states U_- and U_+ and propagating with speed s is admissible if and only if
$$\lambda_i(U_-) \geqslant s \geqslant \lambda_i(U_+). \qquad (6.17)$$

(b) The *Liu E-condition* (Liu 1976), according to which an i-shock joining the states U_- and U_+ and propagating with speed s is admissible if and only if the speed s_0 of any i-shock that joins U_- with any state U_0 that lies on the segment of the i-shock curve that connects U_- and U_+ exceeds s, i.e. $s_0 > s$.

The Lax E-condition suffices to identify admissible weak shocks in genuinely nonlinear systems, while the more stringent Liu E-condition should be employed when the system is not genuinely nonlinear.

As regards existence of admissible BV solutions to the Cauchy problem, the current state of affairs is summarized in the following theorem.

Theorem 6.3. *Assume the system (6.7) is strictly hyperbolic and the initial data $U_0(\cdot)$ is a function of bounded variation, with sufficiently small total variation. Then there exists a global BV solution of the Cauchy problem (6.7), (6.12), which is admissible, in the sense that the Liu E-condition holds at any point of approximate jump discontinuity. This solution depends continuously in L^1 on its initial values.*

The above result was established recently by Bianchini and Bressan (2003), who construct the solution U of (6.7), (6.12) by the *vanishing viscosity* method, namely, as the $\varepsilon \downarrow 0$ limit of solutions U_ε of the parabolic system

$$\partial_t U_\varepsilon(x,t) + \partial_x F(U_\varepsilon(x,t)) = \varepsilon \partial_x^2 U_\varepsilon(x,t), \qquad (6.18)$$

with initial condition $U_\varepsilon(x, 0) = U_0(x)$, $-\infty < x < \infty$. This approach is also motivated by continuum physics. Earlier constructions of the solutions, under certain restrictions on the system (6.7), were given in important papers of Glimm (1965), by the random choice method, Liu (1981), by the wave tracing method, and Bressan (1992), by the wave front tracking method. Solutions for general strictly hyperbolic systems have also been constructed by the front tracking method in Iguchi and LeFloch (2003).

6.4 Entropy

A function $\eta(U)$ is called an *entropy* for the system of conservation laws (6.7), associated with *entropy flux* $q(U)$, if

$$Dq(U) = D\eta(U)DF(U). \qquad (6.19)$$

The meaning of (6.19) is that any classical (i.e. at least Lipschitz continuous) solution U of (6.7) satisfies automatically the additional conservation law:

$$\partial_t \eta(U(x,t)) + \partial_x q(U(x,t)) = 0. \qquad (6.20)$$

The integrability condition for (6.19) reads

$$D^2\eta(U)DF(U) = DF(U)^T D^2\eta(U). \qquad (6.21)$$

We are only interested in nonlinear entropies, because any linear function $\eta(U) = C^T U$ does satisfy (6.21) but just yields a trivial linear combination of equations (6.7) as the 'new' conservation law (6.20). For scalar conservation laws, $n = 1$, (6.21) is an obvious identity, and thus any function $\eta(U)$ may qualify as an entropy. For $n = 2$, (6.21) reduces to a scalar, linear, second-order, hyperbolic partial differential equation, whose solutions yield a rich family of entropies $\eta(U)$. For $n \geq 3$, however, (6.21) becomes a formally overdetermined system of $\frac{1}{2}n(n-1)$ equations for the single function $\eta(U)$. There is an interesting, albeit very special, class of hyperbolic systems of conservation laws, namely, those endowed with a *coordinate system of Riemann invariants*, for which the $\frac{1}{2}n(n-1)$ conditions of (6.21) are internally compatible and thus a rich family of entropies still exists. The reader may find information on the remarkable properties of these special systems in the book by Serre (1999, 2000). It should be emphasized, however, that there is a much broader class of systems that are endowed with exceptional, nontrivial entropies. As noted by Friedrichs and Lax (1971), when (6.7) is *symmetric*, that is $DF(U)^T = DF(U)$, then (6.21) is satisfied by $\eta(U) = \frac{1}{2}|U|^2$. Conversely, if (6.21) holds for some $\eta(U)$ which is nontrivial in the sense that $\det D^2\eta(U) \neq 0$, then the change of state vector from U to $V = D\eta(U)$ renders the system (6.7) symmetric. Thus, it is precisely the *symmetrizable systems* that are endowed with nontrivial entropies.

The hyperbolic systems of conservation laws arising in continuum physics are all symmetrizable. This is not a mere coincidence but is a result of the policy of enforcing the second law of thermodynamics by judicial selection of the constitutive equations. We have already encountered this process in Section 6.2, where we saw that by imposing the constitutive assumption (6.6) we secure that the system (6.10) of conservation laws of isothermal thermoelasticity is endowed with the entropy–entropy flux pair:

$$\eta(u, v) = \psi(u) + \tfrac{1}{2}v^2, \qquad q(u, v) = -v\sigma(u). \tag{6.22}$$

It should be noted that (6.19) does not generally guarantee that (6.20) will also hold even for weak solutions of (6.7). Motivated by the experience from continuum physics, briefly discussed in Section 6.2, it is natural to attempt to characterize admissible weak solutions of (6.7) by requiring that, in addition to (6.7), they satisfy, in the sense of distributions, the entropy inequality

$$\partial_t \eta(U(x, t)) + \partial_x q(U(x, t)) \leq 0, \tag{6.23}$$

for a judiciously selected entropy–entropy flux pair. In physical applications, the selection of (η, q) shall be dictated by thermodynamical considerations. In all these cases, the experience is that the relevant $\eta(U)$ is a strictly convex function. Consequently, the requirement that (6.23) should be satisfied for an entropy $\eta(U)$ which is strictly convex has been incorporated in the general theory of weak solutions of hyperbolic systems of conservation laws, even in the absence of any connection with a particular physical application.

Another motivation for requiring convexity of η in (6.23) is provided by the following argument of Lax (1971). Suppose a solution U of (6.7) is constructed by the vanishing viscosity method, that is $U = \lim_{\varepsilon \to 0} U_\varepsilon$, where U_ε is a solution of the parabolic system (6.18), and the convergence is boundedly almost everywhere. Upon multiplying (6.18) by $D\eta(U_\varepsilon)$ and making use of (6.19), we arrive at

$$\partial_t \eta(U_\varepsilon) + \partial_x q(U_\varepsilon) = \varepsilon \partial_x^2 \eta(U_\varepsilon) - \varepsilon (\partial_x U_\varepsilon)^T D^2 \eta(U_\varepsilon)(\partial_x U_\varepsilon). \quad (6.24)$$

We examine the right-hand side of (6.24), as $\varepsilon \downarrow 0$. The first term tends to zero, in the sense of distributions. When $\eta(U)$ is convex, the second term is nonnegative. Therefore, the limit U will satisfy the entropy inequality (6.23).

There is general belief among physicists that the second law of thermodynamics is essentially a statement of stability. In the remainder of this paper, the aim is to demonstrate that an entropy inequality (6.23), with $\eta(U)$ strictly convex, has a number of implications on the stability of solutions of (6.7). The first step in that direction is to show that, whenever they exist, classical solutions of (6.7), (6.12) are stable within the class of weak solutions that satisfy (6.23).

Theorem 6.4. *Let $\bar{U}$ be a classical (i.e. Lipschitz continuous) solution of (6.7) on a strip $(-\infty, \infty) \times [0, T)$, with initial values $\bar{U}(x, 0) = \bar{U}_0(x)$, $-\infty < x < \infty$. Suppose now that $U \in L^\infty((-\infty, \infty) \times [0, T))$ is a weak solution of (6.7), (6.12) which satisfies (6.23). Then there are positive constants s, a, b such that*

$$\int_{-R}^{R} |U(x, \tau) - \bar{U}(x, \tau)|^2 \, dx \leq a e^{b\tau} \int_{-R-s\tau}^{R+s\tau} |U_0(x) - \bar{U}_0(x)|^2 \, dx \quad (6.25)$$

holds for any $R > 0$ and $\tau \in [0, T)$.

Sketch of Proof. On $\mathbb{R}^n \times \mathbb{R}^n$ define the functions

$$h(U, \bar{U}) = \eta(U) - \eta(\bar{U}) - D\eta(\bar{U})[U - \bar{U}], \quad (6.26)$$
$$f(U, \bar{U}) = q(U) - q(\bar{U}) - D\eta(\bar{U})[F(U) - F(\bar{U})], \quad (6.27)$$
$$G(U, \bar{U}) = F(U) - F(\bar{U}) - DF(\bar{U})[U - \bar{U}], \quad (6.28)$$

which are all of quadratic order in $U - \bar{U}$. Furthermore, since η is strictly convex, (6.26) is positive definite,

$$h(U, \bar{U}) \geq c|U - \bar{U}|^2, \quad (6.29)$$

and there is a positive constant s such that

$$|f(U, \bar{U})| \leq s h(U, \bar{U}). \quad (6.30)$$

Now U satisfies (6.7) and (6.23), while $\bar{U}$ satisfies

$$\partial_t \bar{U}(x, t) + \partial_x F(\bar{U}(x, t)) = 0, \quad (6.31)$$
$$\partial_t \eta(\bar{U}(x, t)) + \partial_x q(\bar{U}(x, t)) = 0. \quad (6.32)$$

Therefore, upon evaluating h, f and G along the two solutions U and $\bar{U}$, we obtain

$$\partial_t h(U, \bar{U}) + \partial_x f(U, \bar{U})$$
$$\leqslant -(\partial_t \bar{U})^{\mathrm{T}} D^2 \eta(\bar{U})[U - \bar{U}] - (\partial_x \bar{U})^{\mathrm{T}} D^2 \eta(\bar{U})[F(U) - F(\bar{U})]$$
$$= (\partial_x \bar{U})^{\mathrm{T}} \{DF(\bar{U})^{\mathrm{T}} D^2 \eta(\bar{U})[U - \bar{U}] - D^2 \eta(\bar{U})[F(U) - F(\bar{U})]\}$$
$$= -(\partial_x \bar{U})^{\mathrm{T}} D^2 \eta(\bar{U}) G(U, \bar{U}), \tag{6.33}$$

where use has been made of (6.21). Finally, we integrate (6.33) over the trapezoidal region $\{(x,t) : 0 \leqslant t \leqslant \tau, \ -R - s(\tau - t) \leqslant x \leqslant R + s(\tau - t)\}$ and combine (6.29) and (6.30) with Gronwall's inequality to obtain (6.25). $\square$

The above proof combines ideas of Dafermos (1979) and DiPerna (1979). For the special, albeit very important, system of conservation laws of gas dynamics, Chen et al. (2002) show that a similar argument applies even when $\bar{U}$ is the weak solution of the special, but again very important, Riemann problem. In general, however, an L^2 stability estimate like (6.25) cannot hold when both U and $\bar{U}$ are weak solutions. For systems that are genuinely nonlinear, an L^1 stability estimate has been established by Bressan et al. (1999) and Bianchini and Bressan (2003) for pairs of weak solutions with small total variation.

When (η, q) is any entropy–entropy flux pair for (6.7) and U is a BV weak solution, then $\partial_t \eta(U) + \partial_x q(U)$ is a measure. Because of (6.19), this measure vanishes on the regular set $\mathcal{C}$ of U, and is thus concentrated on the jump set $\mathcal{J}$. In that case, the entropy inequality (6.23) reduces to

$$q(U_+) - q(U_-) - s[\eta(U_+) - \eta(U_-)] \leqslant 0, \tag{6.34}$$

for every point of jump discontinuity of U. Thus, in the context of BV solutions, the entropy inequality is in effect a shock admissibility condition.

When the system (6.7) is genuinely nonlinear, a weak shock will satisfy (6.34), with $\eta(U)$ strictly convex, if and only if the Lax E-condition (6.17) holds. This provides additional justification for the policy of employing convex entropies in the inequality (6.34). It also suggests that, for genuinely nonlinear systems, a single entropy inequality (6.34), with $\eta(U)$ strictly convex, would suffice for singling out a unique admissible solution of the Cauchy problem.

When (6.7) is not genuinely nonlinear, it has been shown that the Liu E-condition, discussed in Section 6.3, implies the entropy inequality (6.34), but the converse is generally false. Thus, in that case a single entropy inequality is no longer capable of singling out a unique admissible solution. A possible remedy is to postulate (6.34) for all convex entropies of the system. This may indeed be effective for systems that are endowed with very rich families of convex entropies, such as the scalar conservation laws (see Kruzkov 1970), but fails in general.

An alternative idea for singling out a unique solution in the absence of genuine nonlinearity is to employ a single entropy–entropy flux pair (η, q) but to strengthen the entropy inequality. Note that (6.23) implies in particular that the total entropy $\int_{-\infty}^{\infty} \eta(U(x,t))\,dx$ is time decreasing. It is tempting to experiment with the conjecture that admissible solutions should dissipate the total entropy at maximal rate. When U is a BV solution, it is easy to see that

$$\frac{d^+}{dt} \int_{-\infty}^{\infty} \eta(U(x,t))\,dx = \sum \{q(U_+) - q(U_-) - s[\eta(U_+) - \eta(U_-)]\}, \quad (6.35)$$

where the summation on the right-hand side runs over all shocks that intersect the t-time line. The *entropy rate condition* stipulates that admissible solutions U should minimize, at any t, the right-hand side of (6.35), over all solutions $\bar{U}$ that coincide with U along the t-time line: $\bar{U}(\cdot, t) = U(\cdot, t)$. It has been shown (Dafermos 1989) that when the entropy rate condition holds then all weak shocks satisfy the Liu E-condition.

6.5 Quenching of Oscillations

A general feature of nonlinear partial differential equations is that the weak limit of weakly convergent sequences of solutions, or approximate solutions, is not necessarily a solution. This creates major difficulties in the program of constructing solutions by numerical or abstract procedures, but even beyond that it is a manifestation of instability. This issue is particularly germane to hyperbolic conservation laws, because hyperbolicity induces the existence of sequences of highly oscillatory solutions which may converge weakly but not strongly. One may construct such sequences, for the system (6.7), by the following procedure. Consider any two constant states V and W that may be connected by a shock of speed s, i.e. the Rankine–Hugoniot condition (6.13) holds:

$$F(V) - F(W) = s[V - W]. \quad (6.36)$$

On the (x, t)-plane, start with any finite family of parallel straight lines with slope s and define a function which is constant on each strip confined between two adjacent straight lines, taking the values V and W in alternating order. It is clear that this function is a weak solution of (6.7). Moreover, by judiciously selecting the number and location of the straight lines, it is possible to construct a sequence of such solutions which converges in L^∞ weak* to a function $U(x,t) = \rho(x - st)V + [1 - \rho(x - st)]W$, where $\rho(\xi)$ is an arbitrary measurable function from $(-\infty, \infty)$ to $[0, 1]$. Clearly, such a U will not be a weak solution of (6.7), unless F happens to be affine along the straight line segment that joins V and W. Notice now that this instability may be averted if one deals exclusively with solutions that satisfy an entropy inequality (6.23), for some nontrivial entropy–entropy flux pair (η, q). Indeed in that case

$$q(V) - q(W) - s[\eta(V) - \eta(W)] \neq 0 \quad (6.37)$$

and hence condition (6.34) would not allow jumps both from V to W and from W to V. In other words, the entropy inequality rules out solutions with rapid planar oscillations.

The question is whether an entropy inequality may quench any rapidly oscillating wave pattern. The following proposition demonstrates that this is indeed the case, at least in a very special setting. We shall work with the Burgers equation (6.9), equipped with the particular entropy–entropy flux pair $(\frac{1}{2}u^2, \frac{1}{3}u^3)$, so that the entropy inequality (6.23) reads

$$\partial_t(\tfrac{1}{2}u^2) + \partial_x(\tfrac{1}{3}u^3) \leq 0. \tag{6.38}$$

Proposition 6.5. *Assume $\{u_j\}$ is a uniformly bounded sequence of L^∞ weak solutions of Burgers' equation,*

$$\partial_t u_j + \partial_x(\tfrac{1}{2}u_j^2) = 0, \tag{6.39}$$

which satisfy the entropy inequality

$$\partial_t(\tfrac{1}{2}u_j^2) + \partial_x(\tfrac{1}{3}u_j^3) \leq 0, \tag{6.40}$$

in the sense of distributions. Then there is a subsequence which converges boundedly almost everywhere to some function u which satisfies the Burgers equation (6.9) together with the entropy inequality (6.38).

Sketch of Proof. By virtue of (6.39), there is a bounded sequence $\{\phi_j\}$ in $W^{1,\infty}$ such that

$$u_j = \partial_x \phi_j, \qquad \tfrac{1}{2}u_j^2 = -\partial_t \phi_j. \tag{6.41}$$

There is a subsequence of $\{u_j\}$, still denoted by $\{u_j\}$, such that

$$u_j \to u, \qquad \tfrac{1}{2}u_j^2 \to \tfrac{1}{2}v^2, \qquad \tfrac{1}{3}u_j^3 \to \tfrac{1}{3}w^3, \tag{6.42}$$

in L^∞ weak*, and at the same time

$$\phi_j \to \phi, \tag{6.43}$$

uniformly on compact sets, where $\partial_x \phi = u$, $-\partial_t \phi = \tfrac{1}{2}v^2$. To prove the assertion of the proposition, it will suffice to show $u^2 = v^2$.

Notice the identity

$$\tfrac{1}{3}u_j^3(u_j - u) - \tfrac{1}{2}u_j^2(\tfrac{1}{2}u_j^2 - \tfrac{1}{2}v^2)$$
$$= \tfrac{1}{3}u_j^3 \partial_x(\phi_j - \phi) + \tfrac{1}{2}u_j^2 \partial_t(\phi_j - \phi)$$
$$= \partial_x[\tfrac{1}{3}u_j^3(\phi_j - \phi)] + \partial_t[\tfrac{1}{2}u_j^2(\phi_j - \phi)] - [\partial_t(\tfrac{1}{2}u_j^2) + \partial_x(\tfrac{1}{3}u_j^3)](\phi_j - \phi). \tag{6.44}$$

We now let $j \to \infty$ in (6.44). By virtue of (6.40), the bracket on the third term of the right-hand side is a locally bounded measure, uniformly in j. Therefore,

(6.43) implies that the right-hand side of (6.44) tends to zero, in the sense of distributions. Thus, by virtue of (6.42), (6.44) yields

$$\tfrac{1}{12}u_j^4 - \tfrac{1}{3}w^3 u + \tfrac{1}{4}v^4 \to 0, \quad \text{as } j \to \infty, \tag{6.45}$$

in the sense of distributions. Next we note the elementary inequality

$$(\tfrac{1}{2}u_j^2 - \tfrac{1}{2}u^2)^2 \leqslant (u_j - u)(\tfrac{1}{3}u_j^3 - \tfrac{1}{3}u^3). \tag{6.46}$$

Letting $j \to \infty$ in (6.46) and using (6.42) and (6.45), we conclude

$$(\tfrac{1}{2}u^2 - \tfrac{1}{2}v^2)^2 \leqslant 0, \tag{6.47}$$

whence $\tfrac{1}{2}u^2 = \tfrac{1}{2}v^2$, and hence $\{u_j\}$ converges strongly to u.

The above proof, which employs a trick due to Tartar, will give the reader a taste of the application of the method of *compensated compactness* to the theory of hyperbolic conservation laws. The idea that entropy inequalities induce quenching of oscillations was conceived by Tartar (1983), who applied it to scalar conservation laws, and was developed by DiPerna (1983) for systems of two conservation laws. There is an extensive literature on the subject (the book by Serre (1999, 2000) is a good source), but the requirement of existence of a rich family of entropies limits the applicability of the method.

References

Bianchini, S. and Bressan, A. 2003 Vanishing viscosity solutions of nonlinear hyperbolic systems. *Ann. Math.* (In the press.)

Bressan, A. 1992 Global solutions of systems of conservation laws by wave-front tracking. *J. Math. Analysis Appl.* **170**, 414–432.

Bressan, A. 2000 *Hyperbolic Systems of Conservation Laws*. Oxford University Press.

Bressan, A., Liu, T.-P. and Yang, T. 1999 L^1 stability estimates for $n \times n$ conservation laws. *Arch. Ration. Mech. Analysis* **149**, 1–22.

Chen, G.-Q., Frid, H. and Li, Y. 2002 Uniqueness and stability of Riemann solutions with large oscillation in gas dynamics. *Commun. Math. Phys.* **228**, 201–217.

Dafermos, C. M. 1979 The second law of thermodynamics and stability. *Arch. Ration. Mech. Analysis* **70**, 167–179.

Dafermos, C. M. 1989 Admissible wave fans in nonlinear hyperbolic systems. *Arch. Ration. Mech. Analysis* **106**, 243–260.

Dafermos, C. M. 2000 *Hyperbolic Conservation Laws in Continuum Physics*. Springer.

DiPerna, R. J. 1979 Uniqueness of solutions to hyperbolic conservation laws. *Indiana Univ. Math. J.* **28**, 137–188.

DiPerna, R. J. 1983 Convergence of approximate solutions to conservation laws. *Arch. Ration. Mech. Analysis* **82**, 27–70.

Friedrichs, K. O. and Lax, P. D. 1971 Systems of conservation equations with a convex extension. *Proc. Natl Acad. Sci. USA* **68**, 1686–1688.

Glimm, J. 1965 Solutions in the large for nonlinear hyperbolic systems of equations. *Commun. Pure Appl. Math.* **18**, 697–715.

Holden, H. and Risebro, N. H. 2002 *Front Tracking for Hyperbolic Conservation Laws*. Springer.

Iguchi, T. and LeFloch, P. G. 2003 Existence theory for hyperbolic systems of conservation laws with general flux functions. *Arch. Ration. Mech. Analysis* **168**, 165–244.

John, F. 1974 Formation of singularities in one-dimensional nonlinear wave propagation. *Commun. Pure Appl. Math.* **27**, 377–405.

Kruzkov, S. 1970 First-order quasilinear equations with several space variables. *Mat. USSR Sbornik* **10**, 217–273.

Lax, P. D. 1957 Hyperbolic systems of conservation laws. *Commun. Pure Appl. Math.* **10**, 537–566.

Lax, P. D. 1971 Shock waves and entropy. In *Contributions to Functional Analysis* (ed. E. A. Zarantonello), pp. 603–634. Academic Press.

LeFloch, P. G. 2002 *Hyperbolic Systems of Conservation Laws*. Birkhäuser.

Liu, T.-P. 1976 The entropy condition and the admissibility of shocks. *J. Math. Analysis Appl.* **53**, 78–88.

Liu, T.-P. 1981 Admissible solutions of hyperbolic conservation laws. *Mem. AMS* **30**, No. 240.

Serre, D. 1999, 2000 *Systems of Conservation Laws*, Vols 1 and 2. Cambridge University Press.

Smoller, J. 1994 *Shock Waves and Reaction–Diffusion Equations*, 2nd edn. Springer.

Tartar, L. C. 1983 The compensated compactness method applied to systems of conservation laws. In *Systems of Nonlinear Partial Differential Equations* (ed. J. M. Ball), pp. 263–285. Reidel, Dordrecht.

Chapter Seven

Irreversibility and the Second Law of Thermodynamics

Jos Uffink
Institute for History and Foundations of Science

The second law of thermodynamics has a curious status. Many modern physicists regard it as an obsolete relic from a bygone age, while many others, even today, consider it one of the most firmly established and secure achievements of science ever accomplished.

From the perspective of the foundations of physics, a particularly interesting question is its relationship with the notion of irreversibility. It has often been argued, in particular by Planck, that the second law expresses, and characterizes, the irreversibility of all natural processes. This has led to much-debated issues, such as whether the distinction between past and future can be grounded in the second law, or how to reconcile the second law with an underlying microscopic mechanical (and hence reversible) theory. But it is not easy to make sense of these debates, since many authors mean different things by the same terms.

The purpose of this contribution is to provide some clarification by distinguishing three different meanings of the notion of (ir)reversibility and to study how they relate to different versions of the second law. A more extensive discussion is given in Uffink (2001).

7.1 Three Concepts of (Ir)reversibility

Many physical theories employ a state space Γ consisting of all possible states of a system. An instantaneous state is thus represented as a point s in Γ and a process as a parametrized curve:

$$\mathcal{P} = \{s_t \in \Gamma : t_i \leq t \leq t_f\}.$$

The laws of a theory usually allow only a definite class of such processes (e.g. the solutions of the equations of motion). Call this class $\mathcal{W}$, the set of all possible worlds (according to this theory). Now let R be an involution (i.e. $R^2 s = s$) that turns a state s into its 'time reversal' Rs. In classical mechanics, for example, R is the transformation which reverses the sign of all momenta and magnetic fields. In a theory like classical thermodynamics, in which the state does not contain velocity-like parameters, one may take R to be the identity transformation.

Further, define the time reversal $\mathcal{P}^*$ of a process $\mathcal{P}$ by

$$\mathcal{P}^* = \{(Rs)_{-t} : -t_f \leqslant t \leqslant -t_i\}.$$

The theory (or a law) is called *time-reversal invariant* (TRI) if the class $\mathcal{W}$ is closed under time reversal, i.e. if and only if

$$\mathcal{P} \in \mathcal{W} \implies \mathcal{P}^* \in \mathcal{W}. \tag{7.1}$$

According to this definition[1] the form of the laws themselves (and a given choice for R) determines whether the theory is TRI or not. And it is straightforward to show that classical mechanics is indeed TRI. Note also that the term 'time reversal' is not meant literally. That is to say, we consider *processes* whose reversal is or is not allowed by a physical law, not a reversal of time itself. The prefix is only intended to distinguish the term from a spatial reversal. Furthermore, note that it is not relevant here whether the reversed processes $\mathcal{P}^*$ occur in the actual world. It is sufficient that the theory allows them. Thus, the fact that the Sun never rises in the west is no obstacle to celestial mechanics qualifying as TRI.

Is this theme of time-reversal (non)invariance related to the second law? Even though the criterion is unambiguous, its application to thermodynamics is not a matter of routine. In contrast to mechanics, thermodynamics does not possess equations of motion. This, in turn, is due to the fact that thermodynamical processes only take place after an external intervention on the system (e.g. removing a partition, establishing thermal contact with a heat bath, pushing a piston, etc.). They do not refer to the autonomous behaviour of a free system. This is not to say that time plays no role. Classical thermodynamics, in the formulation of Clausius, Kelvin, or Planck, is concerned with processes occurring in the course of time, and its second law is clearly not TRI. However, in other formulations, such as those by Gibbs, Carathéodory, or Lieb and Yngvason, this is less clear.

My main theme, however, is the notion of *(ir)reversibility*. This term is attributed to processes rather than theories or laws. But in the philosophy of physics literature it is intimately connected with time-reversal invariance. More precisely, one calls a process $\mathcal{P}$ allowed by a given theory irreversible if and only if the reversed process $\mathcal{P}^*$ is excluded by this theory. Obviously, such a process $\mathcal{P}$ exists only if the theory in question is not TRI. Conversely, every nonTRI theory admits irreversible processes in this sense. They constitute the hallmark of time-reversal variance and, therefore, discussions about (ir)reversibility and

---

[1] Illner and Neunzert (1987) proposed an alternative definition of time-reversal invariance. Supposing the theory is deterministic, its laws specify evolution operators $U(t_1, t_0)$ such that $s_t = U(t, t_0)s_0$. In that case, one can define the TRI of the theory by the requirement $U^{-1}(t, t_0)RU(t, t_0) = R$. This definition has the advantage that it does not rely on the possible worlds' semantics. However, it applies only for deterministic theories, and not to thermodynamics.

(non)TRI in philosophy of physics coincide for the most part. However, in thermodynamics, the term is commonly employed with other meanings.

The thermodynamics literature often uses the term 'irreversibility' to denote an aspect of our experience which, for want of a better word, one might also call *irrecoverability*. In many processes, the transition from an initial state s_i to a final state s_f, cannot be fully 'undone,' once the process has taken place. In other words, there is no process which starts off from state s_f and restores the initial state s_i completely. Ageing and dying, wear and tear, erosion and corruption are the obvious examples. This is the sense of irreversibility that Planck intended, when he called it the essence of the second law.

Many writers have emphasized this theme of irrecoverability in connection with the second law. Indeed, Eddington introduced his famous phrase, 'the arrow of time,' in a general discussion of the 'running-down of the Universe,' and illustrated it with many examples of processes involving 'irrevocable changes,' including the example of Humpty Dumpty, who, allegedly, could not be put together again after his great fall. In retrospect, one might perhaps say that a better expression for this theme is the *ravages* of time rather than its arrow.

(Ir)recoverability differs from (non)TRI in at least two respects. First, the only thing that matters here is the retrieval of the original state s_i. It is not necessary that one can find a process $\mathcal{P}^*$ which retraces all the intermediate stages of the original process in reverse order. A second difference is that we are dealing with a *complete* recovery. Planck repeatedly emphasized that this condition includes the demand that all auxiliary systems that may have been employed in the original process are brought back to their initial state. Now, although one might argue that a similar demand should also be included in the definition of TRI, the problem is here that the auxiliary systems are often not characterizable as thermodynamical systems.

Schematically, the idea can be expressed as follows. Let s be a state of the system and Z a (formal) state of its environment. Let $\mathcal{P}$ be some process that brings about the following transition:

$$\langle s_i, Z_i \rangle \xrightarrow{\mathcal{P}} \langle s_f, Z_f \rangle. \tag{7.2}$$

Then $\mathcal{P}$ is reversible in Planck's sense if and only if there exists[2] another process $\mathcal{P}'$ that produces

$$\langle s_f, Z_f \rangle \xrightarrow{\mathcal{P}'} \langle s_i, Z_i \rangle. \tag{7.3}$$

However, the term 'reversible' is also used in yet a third sense, which has no straightforward connection with the arrow of time at all. It is often used to denote processes which proceed so delicately and slowly that the system can

---

[2] One might read 'exists' here as 'allowed by the theory,' i.e. as $\mathcal{P}' \in \mathcal{W}$. But this is not Planck's view. He emphasized that $\mathcal{P}'$ might employ any appliances *available in nature*, rather than allowed by a theory. This is a third respect in which his sense of reversibility differs from that of TRI.

be regarded as remaining in equilibrium 'up to a negligible error' during the entire process. We shall see that this is the meaning embraced by Clausius. Actually, it appears to be the most common usage of the term in the physical and chemical literature (see, for example, Denbigh 1989; Hollinger and Zenzen 1985). A more apt name for this kind of process is *quasistatic*. Of course, the above way of speaking is vague, and has to be amended by criteria specifying what kind of 'errors' are intended and when they are 'small.' These criteria take the form of a limiting procedure so that, strictly speaking, reversibility is here not an attribute of a particular process but of a series of processes.

Again, quasistatic processes are not necessarily the same as those called reversible in the previous two senses. For example, the motion of an ideal harmonic oscillator is reversible in the sense of Planck, but it is not quasistatic. Conversely, the discharge of a charged condenser through a very high resistance can be made to proceed quasistatically, but even then it remains irreversible in Planck's sense.

Comparison with the notion of TRI is hampered by the fact that 'quasistatic' is not strictly a property of a process. Perhaps the following example might be helpful. Consider a process $\mathcal{P}_N$ in which a system, originally at temperature θ_1, is consecutively placed in thermal contact with a sequence of N heat baths, each at a slightly higher temperature than the previous one, until it reaches a temperature θ_2. By making N large, and the temperature steps small, such a process becomes quasistatic, and we can approximate it by a curve in the space of equilibrium states. However, for any N, the time-reversal of the process is impossible.

The reason why so many authors nevertheless call such a curve 'reversible' is that one can consider a second process $\mathcal{Q}_N$, in which the system, originally at temperature θ_2, is placed into contact with a series of heat baths, each slightly *colder* than the previous one. Again, each process $\mathcal{Q}_N$ is nonTRI. A fortiori, no $\mathcal{Q}_N$ is the time reversal of any $\mathcal{P}_N$. Yet, if we now take the quasistatic limit, the state change of the system will follow the same curve in equilibrium space as in the previous case, traversed in the opposite direction. The point is, of course, that precisely because this curve is not itself a process, the notion of time reversal does not apply to it.

7.2 Early Formulations of the Second Law

The work of the founding fathers Carnot, Clausius, and Kelvin (William Thomson) can be divided into two lines: one main line dealing exclusively with cyclic processes; and another addressing (also) noncyclic processes. In this section I will discuss both.

The first line starts with the work of Carnot (1824). Carnot studied cyclic processes performed by an arbitrary system in interaction with two heat reservoirs (the furnace and the refrigerator) at temperatures θ^+ and θ^-, while doing work on some third body. Let $Q^+(\mathcal{C})$, $Q^-(\mathcal{C})$ and $W(\mathcal{C})$ denote, respectively, the heat absorbed from the furnace, the heat given off to the refrigerator, and the work done by the system during the cycle $\mathcal{C}$. He assumed that the heat reservoirs remain unchanged while they exchange heat with the system.

Carnot's main assumptions were that (i) heat is a conserved substance, i.e. $Q^+(\mathcal{C}) = Q^-(\mathcal{C})$, and (ii) the impossibility of a *perpetuum mobile* of the first kind. This second assumption is Carnot's Principle.

Carnot's Principle. *It is impossible to perform a repeatable cycle in which the only result is the performance of (positive) work.*

Note that Carnot did not object to the performance of (positive) work in a cycle as such. Rather, his point was that, due to the assumption that the heat reservoirs act as invariable buffers, the cycle could be repeated arbitrarily often. Thus, violation of the above principle would provide *unlimited* production of work at no cost whatsoever. This he regards as inadmissible.

By a well-known reductio ad absurdum argument, he obtained the following theorem.

Carnot's Theorem.

(a) *The efficiency $\eta(\mathcal{C}) := W(\mathcal{C})/Q^+(\mathcal{C})$ is bounded by a universal maximum C, which depends only on the temperatures θ^+ and θ^-:*

$$\eta(\mathcal{C}) \leqslant C(\theta^+, \theta^-). \tag{7.4}$$

(b) *This maximum is attained if the cycle $\mathcal{C}$ is 'reversible.'*

In fact, Carnot did not use the term 'reversible.' So one might ask how he conceived of the condition in (b). Actually, he discusses the issue twice. He starts his argument with an example: the Carnot cycle for steam. In passing, he notes its relevant feature: 'The operations we have just described might have been performed in an inverse direction and order' (Mendoza 1960, p. 11). This feature, of course, turns out to be crucial for the claim that this cycle has maximal efficiency. Later, he realized that a more precise formulation of this claim was desirable, and he formulates a necessary and sufficient criterion for maximum efficiency (Mendoza 1960, p. 13): it should be avoided that bodies of different temperature come into direct thermal contact. He notes that this criterion cannot be met exactly, but can be approximated as closely as we wish. In modern terms, the criterion is that the process should be quasistatic at all stages which involve heat exchange.

Accordingly, even at this early stage, there are two plausible options for a definition of a 'reversible' cycle. Either we focus on the crucial property of the Carnot cycle that it can also be run backwards. This is the option chosen by Kelvin in 1851. Of course, this is a natural choice, since this property is essential to the proof of the theorem. Or else one can focus on Carnot's necessary and sufficient condition and use this as a definition of a reversible cycle. This is more or less the option followed by Clausius in 1864. He called a cyclic process reversible (*umkehrbar*) if and only if it proceeds quasistatically.

Carnot's work proved very valuable, a quarter of a century later, when Kelvin showed in 1848 that it could be used to devise an absolute temperature scale. But in the meantime, serious doubts had appeared about the conservation of heat. Thus, when the importance of his theorem was recognized, the adequacy of Carnot's original derivation had already become suspect. Therefore, Clausius (1850) and Kelvin (1851) sought to obtain Carnot's Theorem on a different footing. They replaced Carnot's assumption (i) by the Joule–Mayer principle stating the equivalence of heat and work, i.e. $Q^+(\mathcal{C}) = Q^-(\mathcal{C}) + JQ(\mathcal{C})$, where J is Joule's constant. Instead of Carnot's assumption (ii), they adopted the impossibility of a *perpetuum mobile* of the second kind.

Clausius–Kelvin Principle. *It is impossible to perform a cycle[3] in which the only effect is*

> *to let heat pass from a cooler to a hotter body (Clausius),*
> *to perform work and cool a single heat reservoir (Kelvin).*

They showed that Carnot's Theorem, by that time called the 'second thermodynamic law' or the '*Zweite Hauptsatz*,' can be recovered.

In a series of papers, Clausius and Kelvin extended and reformulated the result. In 1854 Kelvin showed that the absolute temperature scale $T(\theta)$ can be chosen such that $C(T^+, T^-) = J(1 - T^-/T^+)$, or equivalently

$$\frac{Q^+(\mathcal{C})}{T^+} = \frac{Q^-(\mathcal{C})}{T^-}. \tag{7.5}$$

Generalizing the approach to cycles involving an arbitrary number of heat reservoirs, they obtained the formulation:[4]

$$\oint_\mathcal{C} \frac{dQ}{T} = 0 \quad \text{if } \mathcal{C} \text{ is reversible} \tag{7.6}$$

and

$$\oint_\mathcal{C} \frac{dQ}{T} \leq 0 \quad \text{if } \mathcal{C} \text{ is not reversible}. \tag{7.7}$$

---

[3] In the usual formulation of these principles, the unlimited repeatability of the cycle is not stressed as much as it was by Carnot. However, one may infer that it was at least intended by Kelvin when he wrote in his introduction, 'Whenever in what follows, *the work done or the mechanical effect produced* by a thermodynamic engine is mentioned without qualification, it must be understood that the mechanical effect produced, either in a non-varying machine, or in a complete cycle, or any number of complete cycles of a periodical engine, is meant' (Kestin 1976, p. 177).

[4] From here on, Q is regarded as positive when absorbed by and negative when given off by the system.

Note that here T stands for the absolute temperature of the heat reservoirs; it is only in the case of (7.6) that T can be equated with the temperature of the system.

Let us now investigate the connections with the themes of Section 7.1. All three authors adopt a principle which is manifestly nonTRI: it forbids the occurrence of certain cyclic processes while allowing their reversal. However, the main object in the work of Clausius and Kelvin considered above was to obtain part (a) of Carnot's Theorem, or its generalization (7.6). These results are TRI, and accordingly the nonTRI element did not receive much attention. Indeed, Kelvin never mentions relation (7.7) at all, and indeed calls (7.6) 'the full expression of the second thermodynamic law.' Clausius (1854) discusses (7.7) only very briefly.

It is much harder to find a connection with irrecoverability. All the papers considered here are only concerned with cyclic processes. There can therefore be no question of irrecoverable changes in that system, or of a monotonically changing quantity. If one insists on finding such a connection, the only option is to take the environment into account, in particular the heat reservoirs. Indeed, nowadays one would argue that if a system performs an irreversible cycle, the total entropy of the heat reservoirs increases. But such a view would be problematic here. First, we are at a stage in which the very existence of an entropy function is yet to be established. One cannot assume that the heat reservoirs already possess an entropy, without running the risk of circularity. Moreover, the heat reservoirs are conceived of as buffers with infinite heat capacity, and it is not straightforward to include them in an entropy balance. The connection with irrecoverability therefore remains dubious.

The second line announced at the beginning of this section consists mainly of three papers: Kelvin (1852) and Clausius (1864, 1865). They differ from earlier and later works by the same authors because they explicitly address noncyclic processes. Kelvin (1852) is a very brief note on the 'universal tendency towards the dissipation of energy.' He argued that natural processes in general bring about 'unreversible' changes, so that a full restoration of the initial state is impossible. Clearly, Kelvin uses the term 'unreversible' here in the sense of 'irrecoverable.' He claims that this tendency is a necessary consequence of his (1851) principle mentioned above. Moreover, he draws an eschatological conclusion: in the distant future, life on Earth must perish. It is here that we first encounter the 'terroristic nimbus' of the second law: the heat death of the Universe.[5]

Starting in 1862, Clausius also addresses noncyclic processes, and some years later, reaches a similar conclusion. He notes in 1865 that the validity of (7.6)

---

[5] The lack of any argument for Kelvin's bold claims has puzzled many commentators. It has been suggested (Smith and Wise 1989) that the source for these claims is perhaps to be found in his religious beliefs rather than in thermodynamics.

implies that the integral

$$\int_{s_1}^{s_2} \frac{\mathrm{d}Q}{T}$$

is independent of the integration path, and can be used to define a new function of state, called entropy S, such that

$$S(s_2) - S(s_1) = \int_{s_1}^{s_2} \frac{\mathrm{d}Q}{T}, \qquad (7.8)$$

where the integral is performed for an *umkehrbar* (i.e. quasistatic) process. For an *unumkehrbar* process he uses relation (7.7) to obtain

$$\int_{s_1}^{s_2} \frac{\mathrm{d}Q}{T} \leqslant S(s_2) - S(s_1). \qquad (7.9)$$

If this latter process is adiabatic, i.e. if there is no heat exchange with the environment, one may put $\mathrm{d}Q = 0$ and it follows that

$$S(s_2) \geqslant S(s_1). \qquad (7.10)$$

Hence we obtain Clausius' version of the second law.

The Second Law. *For every nonquasistatic process in an adiabatically isolated system which begins and ends in an equilibrium state, the entropy of the final state is greater than or equal to that of the initial state. For every quasistatic process in an adiabatic system, the entropy of the final state is equal to that of the initial state.*

This is the first instance of a formulation of the second law as a statement about entropy increase. Note that only the '$\geqslant$' sign is established. One often reads that for irreversible processes the strict inequality holds in (7.10), holds but this does not follow from Clausius' version. Note also that, in contrast to the common view that the entropy principle holds for isolated systems only, Clausius' result applies to *adiabatically* isolated systems.

Clausius also draws a bold inference about all natural processes and the fate of the Universe.

> The second law in the form I have given it says that all transformations taking place in nature go by themselves in a certain direction, which I have denominated the positive direction. ... The application of this law to the Universe leads to a conclusion to which W. Thomson first called attention ... namely ... that the total state of the Universe will change continually in that direction and hence will inevitably approach a limiting state.
>
> Clausius (1867, p. 42)

Noting that his theory is still not capable of treating the phenomenon of heat radiation, he 'restricts himself'—as he puts it—to an application of the theory to the Universe.

> One can express the fundamental laws of the Universe that correspond to the two main laws of thermodynamics in the following simple form:
>
> 1. The energy of the Universe is constant.
> 2. The entropy of the Universe tends to a maximum.
>
> Clausius (1867, p. 44)

These words of Clausius are probably the most often quoted, and the most controversial, in the history of thermodynamics. Even Planck admitted that the entropy of the Universe is an undefined concept (Planck 1897, Section 135). Ironically, Clausius could have avoided such criticism if he had not 'restricted' himself to the Universe but generalized his formulation to an arbitrary adiabatically isolated system (beginning and ending in equilibrium).

Another objection is that this version of the law presupposes that the initial and final states can also be connected by a quasistatic process, in order to define their entropy difference by means of (14.2). This is not trivial for transformations other than exchanges of heat and work.

To conclude, this second line of development focuses on arbitrary noncyclic processes of completely general systems. The main claim is that, apart from the quasistatic case, all such processes are irrecoverable. However, the arguments given for those grand claims are rather fragile. Kelvin provides no argument at all, and Clausius' attempts depends on rather special assumptions. A curious point is that when Clausius (1876) reworked his previous papers into a textbook he completely dropped his famous claim that the entropy of the Universe tends to a maximum. The most general statement of the second law presented in this book is again given as (7.6) and (7.7), i.e. restricted to cyclic processes.

7.3 Planck

The importance of Planck's *Vorlesungen über Thermodynamik* (Planck 1897) can hardly be underestimated. The book has gone through 11 editions, from 1897 until 1964, and still remains the most authoritative exposition of classical thermodynamics. Planck's position has always been that the second law expresses irrecoverability of all processes in nature. However, it is not easy to analyse Planck's arguments for this claim. His text differs in many small but decisive details in the various editions. I also warn that the English translation of the *Vorlesungen* is unreliable. Particularly confusing is that it uses the translation 'reversible' indiscriminately, where Planck distinguishes between the terms *umkehrbar*, which he uses in Clausius' sense, i.e. meaning 'quasistatic,' and *reversibel*, in the sense of Kelvin (1852) meaning 'recoverable.' Moreover, after Planck learned about Carathéodory's work through a review by Born in 1921, he presented a completely different argument from the eighth edition onwards.

In spite of the many intricacies in Planck's book, I shall limit myself to a brief exposition of Planck's latter argument, published first in Planck (1926). He starts from the statement that 'friction is an *irreversibel* process,' which he considers to be an expression of Kelvin's Principle. This may need some explanation, because, at first sight, this statement does not concern cyclic processes or the *perpetuum mobile* at all. But for Planck, the statement means that there exists no process which 'undoes' the consequences of friction, i.e. a process which produces no other effect than cooling a reservoir and doing work. The condition 'no other effect' here allows for the operation of any type of auxiliary system that operates in a cycle.

He then considers an adiabatically isolated fluid[6] capable of exchanging energy with its environment by means of a weight at height h. Planck asks whether it is possible to bring about a transition from an initial state s of this system to a final state s', in a process which brings about no changes in the environment other than the displacement of the weight. If Z denotes the state of the environment and h the height of the weight, the desired transition can be represented as

$$(s, Z, h) \stackrel{?}{\to} (s', Z, h').$$

He argues that, by means of '*reversibel*-adiabatic'[7] processes, one can always achieve a transition from the initial state s to an intermediary state s^* in which the volume equals that of state s' and the entropy equals that of s. That is, one can realize a transition

$$(s, Z, h) \to (s^*, Z, h^*), \quad \text{with } V(s^*) = V(s') \text{ and } S(s^*) = S(s).$$

Whether the desired final state s' can now be reached from the intermediate state s^* depends on the value of the only independent variable in which s^* and s' differ. For this variable one can choose the energy U.

There are three cases.

(1) $h^* = h'$. In this case, energy conservation implies $U(s^*) = U(s')$. Because the coordinates U and V determine the state of the fluid completely, s^* and s' must coincide.

(2) $h^* > h'$. In this case, $U(s^*) < U(s)$, and the state s' can be reached from s^* by letting the weight perform work on the system, e.g. by means of

---

[6] A fluid has, by definition, a state completely characterized by two independent variables.

[7] Apparently, Planck's pen slipped here. He means '*umkehrbar*-adiabatic.'

friction, until the weight has dropped to height h'. According to the above formulation of Kelvin's Principle, this process is irreversible (i.e. irrecoverable).

(3) $h^* < h'$ and $U(s^*) > U(s)$. In this case the desired transition is impossible. It would be the reversal of the irreversible process just mentioned in (2), i.e. produce work by cooling the system and thus realize a *perpetuum mobile* of the second kind.

Now, Planck argues that in all three cases, one can also achieve a transition from s^* to s' by means of heat exchange in an *umkehrbar* (i.e. quasistatic) process in which the volume remains fixed. For such a process he writes

$$dU = T\,dS. \tag{7.11}$$

Using the assumption that $T > 0$, it follows that, in the three cases above, U must vary in the same sense as S. That is, the cases $U(s^*) < U(s')$, $U(s^*) = U(s')$ or $U(s^*) > U(s')$, can also be characterized as $S(s^*) < S(s')$, $S(s^*) = S(s')$ and $S(s^*) > S(s')$, respectively.

An analogous argument can be constructed for a system consisting of several fluids. Just as in earlier editions of his book, Planck generalizes the conclusion (without a shred of proof) to arbitrary systems and arbitrary physical and chemical processes.

> Every process occurring in nature proceeds in the sense in which the sum of the entropies of all bodies taking part in the process is increased. In the limiting case, for reversible processes this sum remains unchanged. ... This provides an exhaustive formulation of the content of the second law of thermodynamics.
>
> Planck (1926, p. 463)

Note how much Planck's construal of the *perpetuum mobile* differs from that of Carnot and Kelvin. The latter authors considered the device which performs the cycle, as the system of interest and the reservoir as part of the environment. In contrast, for Planck, the *reservoir* is the thermodynamical system, and the engine performing the cyclic process belongs to the environment. Related to this switch of perspective is the point that the reservoir is now assumed to have a finite energy content. Thus, the state of the reservoir can change under the action of the hypothetical *perpetuum mobile* device. As a consequence, the cycle need not be repeatable, in sharp contrast to Carnot's original formulation of the idea.

Secondly, Planck's argument can hardly be regarded as satisfactory for the bold and universal formulation of the second law. It applies only to systems consisting of fluids, and relies on several implicit assumptions which can be questioned outside of this context. In particular, this holds for the assumption that there always exist functions S and T (with $T > 0$) such that $đQ = T\,dS$;

and the assumption of a rather generous supply of quasistatic processes. As we shall see in Section 7.5, Carathéodory's treatment is much more explicit on just these issues.

7.4 Gibbs

The work of Gibbs (1906) is very different from that of his European colleagues. Where they were primarily concerned with processes, Gibbs concentrates his efforts on a description of equilibrium states. He assumes that these states are completely characterized by a finite number of state variables like temperature, energy, pressure, volume, entropy, etc., but he makes no effort to prove the existence or uniqueness of these quantities from empirical principles. He proposes the following principle (Gibbs 1906, p. 56).

Gibbs's Principle. *For the equilibrium of any isolated system it is necessary and sufficient that in all possible variations of the state of the system which do not alter its energy, the variation of its entropy shall either vanish or be negative.*

Actually, Gibbs presented this statement only as 'an inference naturally suggested by the general increase of entropy which accompanies the changes occurring in any isolated material system.' But many later authors have regarded the Gibbs Principle as a formulation of the second law (see, for example, Buchdahl 1966; Callen 1960; van der Waals and Kohnstamm 1927). We can follow their lead and that of Truesdell (1986) about how the principle is to be understood.

The first point to note is then that the Gibbs Principle is not literally to be seen as a criterion for equilibrium. Indeed, this would make no sense because all states considered here are already equilibrium states. Rather, it is to be understood as a criterion for *stable* equilibrium. Second, the principle is interpreted analogous to other well-known variational principles in physics like the principle of least action, etc. Here, a 'variation' is *virtual*, i.e. it represents a comparison between two conceivable models or 'possible worlds,' and one should not think of them as (part of) a process that proceeds in the course of time in one particular world. Instead, a variational principle serves to decide which of these possible worlds is physically admissible, or, in the present case, stable.

According to this view, the Gibbs Principle tells us when a conceivable equilibrium state is stable. Such a proposition obviously has a modest scope. First, not all equilibrium states found in nature are necessarily stable. Secondly, Gibbs's Principle is more restricted than Clausius' statement of the second law in the sense that it applies to *isolated* (i.e. no energy exchange is allowed), and not merely adiabatically isolated, systems. More importantly, it provides no information about evolutions in the course of time; and a direction of natural processes, or a tendency towards increasing entropy, cannot be obtained from it. Hence, the second law as formulated by Gibbs has no connection with the arrow of time.

Of course, the view sketched above does not completely coincide with Gibbs's own statements. In some passages he clearly connects virtual variations to actual processes, e.g. when writing, 'it must be regarded as generally possible to produce that variation by some process' (Gibbs 1906, p. 61). Some sort of connection between variations and processes is of course indispensable if one wants to maintain the idea that this principle has implications for processes.

Probably the most elaborate attempt to provide such a connection is the presentation by Callen (1960). Here, it is assumed that, apart from its actual state, a thermodynamic system is characterized by a number of *constraints*, determined by a macroscopic experimental context. These constraints single out a particular subset $\mathcal{C}$ of Γ, consisting of states which are consistent with the constraints. It is postulated that in stable equilibrium the entropy is maximal over all states in $\mathcal{C}$.

A process is then conceived of as being triggered by the cancellation of one or more of these constraints (e.g. mixing or expansion of gases after the removal of a partition, loosening a previously fixed piston, etc.). It is assumed that such a process sets in spontaneously, after the removal of a constraint.

Now, clearly, the removal of a constraint implies an enlargement of the set $\mathcal{C}$. Hence, if we assume that the final state of this process is again a stable equilibrium state, it follows immediately that every process ends in a state of higher (or at best equal) entropy.

I will not attempt to dissect the problems of Callen's approach, except to make three remarks. First, the idea of extending the description of a thermodynamical system in such a way that apart from its state, it is also characterized by a constraint, brings along some conceptual difficulties. For if the actual state is s, it is hard to see how the class of other states contained in the same constraint set $\mathcal{C}$ is relevant to the system. It seems that on this approach the state of a system does not provide a complete description of its thermodynamical properties.

Second, the picture emerging from Callen's approach is somewhat anthropomorphic. For example, he writes, for the case in which there are no constraints, i.e. $\mathcal{C} = \Gamma$, that 'the system is free to select any one of a number of states' (Callen 1960, p. 27). This sounds as if the system is somehow able to 'probe' the set $\mathcal{C}$ and chooses its own state from the options allowed by the constraints.

Third, the result that entropy increases in a process from one equilibrium state to another depends rather crucially on the assumption that processes can be successfully modelled as the removal of constraints. But, clearly, this assumption does not apply to all natural processes. For instance, one can also trigger a process by imposing additional constraints. Hence, this approach does not attain the universal validity of the second law that Planck argued for.

7.5 Carathéodory

Carathéodory (1909) was the first mathematician to pursue a rigorous formalization of the second law. Like Gibbs, he construed thermodynamics as a theory of equilibrium states rather than (cyclic) processes. Again, a thermodynamical system is described by a state space Γ, represented as a (subset of an) n-dimensional manifold in which the thermodynamic state variables serve as coordinates. He assumes that Γ is equipped with the standard Euclidean topology. But metrical properties of the space do not play a role in the theory, and there is no preference for a particular system of coordinates.

However, the coordinates are not completely arbitrary. Carathéodory distinguishes between 'thermal coordinates' and 'deformation coordinates.' (In typical applications, temperature or energy are thermal coordinates, whereas volumes are deformation coordinates.) The *state* of a system is specified by both types of coordinates; the *shape* (*Gestalt*) of the system by the deformation coordinates alone. It seems to be assumed that the deformation coordinates remain meaningful in the description of the system when the system is not in equilibrium, whereas the thermal coordinates are defined only for equilibrium states. In any case, it is assumed that one can obtain every desired final shape from every initial state by means of an adiabatic process.

The idea is now to develop the theory in such a way that the second law provides a characteristic mathematical structure of state space. The fundamental concept is a relation that represents whether state t can be reached from state s in an adiabatic process.[8] This relation is called *adiabatic accessibility*, and I will denote it, following Lieb and Yngvason, by $s \prec t$. This notation may suggest that the relation has the properties of an ordering relation. And, indeed, given its intended interpretation, this would be very natural. But Carathéodory does not state or rely on these properties anywhere in his paper.

In order to introduce the second law, Carathéodory proposes an empirical claim: from an arbitrary given initial state it is not possible to reach every final state by means of adiabatic processes. Moreover, such inaccessible final states can be found in every neighborhood of the initial state. However, he immediately rejects this preliminary formulation, because it fails to take into account the finite precision of physical experiments. Therefore, he strengthens the claim by the idea that there must be a small region surrounding the inaccessible state, consisting of points which are also inaccessible.

The second law thus receives the following formulation.

Carathéodory's Principle. *In every open neighborhood $U_s \subset \Gamma$ of an arbitrary chosen state s there are states t such that for some open neighborhood*

---

[8] Carathéodory's definition of an 'adiabatic process' is as follows. He calls a container adiabatic if the system contained in it remains in equilibrium, regardless of what occurs in the environment, as long as the container is not moved nor changes its shape. Thus, the only way of inducing a process of a system in an adiabatic container is by deformation of the walls of the vessel (e.g. a change of volume or stirring). A process is said to be adiabatic if it takes place while the system remains in an adiabatic container.

U_t within U_s all states r within U_t cannot be reached adiabatically from s. Formally,

$$\forall s \in \Gamma \; \forall U_s \; (\exists t \in U_s) \wedge (\exists U_t \subset U_s) \; \forall r \in U_t : s \not\prec r. \qquad (7.12)$$

He then restricts his discussion to so-called 'simple systems,' defined by four additional conditions.

(i) The system has only a single independent thermal coordinate. Physically, this means that the system has no internal adiabatic walls since in that case it would have parts with several independent temperatures. By convention, the state of a simple system is written as $s = (x_0, \ldots, x_{n-1})$, where x_0 is the thermal coordinate.

(ii) For any given pair of an initial state and final shape of the system there is more than one adiabatic process $\mathcal{P}$ that connects them, requiring different amounts of work. For example, for a gas initially in any given state one can obtain an arbitrary final value for its volume by adiabatic expansion or compression. This change of volume can proceed slowly or fast, and these procedures indeed require different amounts of work.

(iii) The amounts of work done in the processes just mentioned form a connected interval. In other words, if for a given initial state and a given final shape there are adiabatic processes $\mathcal{P}_1$, $\mathcal{P}_2$ connecting them, which require the work $W(\mathcal{P}_1)$ and $W(\mathcal{P}_2)$, respectively, then there are also adiabatic processes $\mathcal{P}$ with any value of $W(\mathcal{P})$, for $W(\mathcal{P}_1) \leqslant W(\mathcal{P}) \leqslant W(\mathcal{P}_2)$.

In order to formulate the fourth demand, Carathéodory considers a special kind of adiabatic process. He argues that one can perform an adiabatic process from any given initial state to any given final shape, in such a way that the deformation coordinates follows some prescribed continuous functions of time:

$$x_1(t), \ldots, x_{n-1}(t). \qquad (7.13)$$

Note that the system will generally not remain in equilibrium in such a process, and therefore the behaviour of the thermal coordinate x_0 remains unspecified.

Consider a series of such adiabatic processes in which the velocity of the deformation becomes 'infinitely slow,' i.e. a series in which the derivatives $\dot{x}_1(t), \ldots, \dot{x}_{n-1}(t)$ converge uniformly towards zero. He calls this limit a *quasistatic change of state*. The final demand is now as follows.

(iv) In a quasistatic change of state, the work done on the system converges to a value W, depending only on the given initial state and final shape, which can be expressed as a path integral along the locus of the functions (7.13),

$$W = \int dW = \int p_1 \, dx_1 + \cdots + p_n \, dx_{n-1},$$

where $p_1, \ldots, p_n$ denote some functions on Γ. Physically, this demand says that for adiabatic processes, in the quasistatic limit, there is neither internal friction nor hysteresis.

Carathéodory's version of the first law (which I have not discussed here) can then be invoked to show that

$$W = U(s_\mathrm{f}) - U(s_\mathrm{i}), \tag{7.14}$$

or, in other words, the work done in the quasistatic limit equals the energy difference between the final and initial states. He argues that by this additional condition the value of the thermal coordinate of the final state is also uniquely fixed. Since the choice of a final shape is arbitrary, this also holds for all intermediate stages of the process. Thus, a quasistatic adiabatic change of state corresponds to a unique curve in Γ.

With this concept of a 'simple system,' Carathéodory obtains the following theorem.

Carathéodory's Theorem. *For simple systems, Carathéodory's Principle is equivalent to the proposition that the differential form $đQ := dU - đW$ possesses an integrable divisor, i.e. there exist functions S and T on the state space Γ such that*

$$đQ = T\, dS. \tag{7.15}$$

Thus, for simple systems, every equilibrium state can be assigned a value for entropy and absolute temperature. Curves representing quasistatic adiabatic changes of state are characterized by the differential equation $đQ := dU - đW = 0$, and by virtue of (7.15) one can conclude that (if $T \neq 0$) entropy remains constant. Obviously, the functions S and T are not uniquely determined by the relation (7.15). Carathéodory discusses further conditions to determine the choice of T and S up to a constant of proportionality, and extends the discussion to composite simple systems. However, I will not discuss this issue.

Before we proceed to the discussion of the relation of this formulation to the arrow of time, I want to mention a number of strong and weak points of the approach. A major advantage of Carathéodory's approach is that it provides a suitable mathematical formalism for the theory, and brings it in line with other modern physical theories. The way this is done is comparable with the (contemporary) development of relativity theory. There, Einstein's original approach, which starts from empirical principles such as the light postulate and the relativity principle, was replaced by an abstract geometrical structure, Minkowski spacetime, where these empirical principles are incorporated in the local properties of the metric. Similarly, Carathéodory constructs an abstract state space where an empirical statement of the second law is converted into a local topological property. Furthermore, all coordinate systems are treated on the same footing (as long as there is only one thermal coordinate, and they generate the same topology).[9] Note further that the environment of the system is never mentioned explicitly in his treatment of the theory. This too is a big conceptual advantage.

But Carathéodory's work has also provoked objections, in particular because of its high level of abstraction. Many complain that the absence of an explicit reference to a *perpetuum mobile* obscures the physical content of the second law. The question has been raised (e.g. by Planck (1926)) of whether the principle of Carathéodory has any empirical content at all. However, Landsberg (1964) has shown that for simple systems Kelvin's Principle implies Carathéodory's Principle, so that any violation of the latter would also be a violation of the former.

Other problems in Carathéodory's approach concern the additional assumptions needed to obtain the result (7.15). In the first place, we have seen that the result is restricted to simple systems, involving four additional auxiliary conditions. Falk and Jung (1959) objected that the division of five assumptions into four pertaining to simple systems and one 'principle,' expressing a general law of nature, seems ad hoc. Indeed, the question of whether Carathéodory's Principle can claim empirical support for nonsimple systems still seems to be open.

Secondly, Bernstein (1960) has pointed out defects in the proof of Carathéodory's Theorem. What his proof actually establishes is merely the *local* existence of functions S and T obeying (7.15). But this does not mean there exists a single pair of functions, defined globally on Γ, that obey (7.15). In fact, a purely local proposition like Carathéodory's Principle is too weak to guarantee the existence of a global entropy function.

For the purpose of this contribution, of course, we need to investigate whether and how this work relates to the arrow of time. We have seen that Carathéodory, like Gibbs, conceives of thermodynamics as a theory of equilibrium states, rather than processes. But his concept of 'adiabatic accessibility' does refer to processes between equilibrium states. The connection with the arrow of time is therefore more subtle than in the case of Gibbs.

In order to judge the time-reversal invariance of the theory of Carathéodory according to the criterion (7.1) on p. 122 it is necessary to specify a time-reversal transformation R. It seems natural to choose this in such a way that $Rs = s$ and $R(\prec) = \succ$ (since the time reversal of an adiabatic process from s to t is an adiabatic process from t to s). Then Carathéodory's Principle is *not* TRI. Indeed, the principle forbids that Γ contains a 'minimal state' (i.e. a state s from which one can reach all states in some neighborhood of s). It allows models where a 'maximal' state exists (i.e. a state s from which one can reach no

---

[9] Indeed, the analogy with relativity theory can be stretched even further. Lieb and Yngvason call the set $\mathcal{F}_s = \{t : s \prec t\}$ the 'forward cone' of s. This is analogous to the future light cone of a point p in Minkowski spacetime (i.e. the set of all points q which are 'causally accessible' from p). Thus, Carathéodory's Principle implies that s is always on the boundary of its own forward cone.

other state in some neighborhood of s). Time reversal of such a model violates Carathéodory's Principle. However, this noninvariance manifests itself only in rather pathological cases. (For a fluid, a 'maximal' state would be one for which temperature and volume cannot be increased.) If we exclude the existence of such maxima, Carathéodory's theory becomes TRI.

Carathéodory also discusses the notorious notion of irreversibility. Consider, for a simple system, the class of all final states s' with a given shape $(x'_1, \ldots, x'_{n-1})$ that are adiabatically accessible from a given initial state $s = (x_0, \ldots, x_{n-1})$. For example, an adiabatically isolated gas is expanded from some initial state (T, V) to some desired final volume V'. The expansion may take place by moving a piston, slowly or more or less suddenly. The set of final states that can be reached in this fashion differ only in the value of their thermal coordinate x'_0. Due to demand (iii) above, the class of accessible final states constitute a connected curve. Carathéodory argues that, for reasons of continuity, the values of S attained on this curve will also constitute a connected interval. Now, among the states of the curve there is the final state, say t, of a quasistatic adiabatic change of state starting from s. And we know that $S(t) = S(s)$. Carathéodory argues that this entropy value $S(s)$ cannot be an internal point of this interval. Indeed, if it were an internal point, then there would exist a small interval $(S(s) - \varepsilon, S(s) + \varepsilon)$ such that the corresponding states on the curve would all be accessible from s. Moreover, it is always assumed that we can change the deformation coordinates in an arbitrary fashion by means of adiabatic state changes. By quasistatic adiabatic changes of state we can even do this with constant entropy. But then all states in a neighborhood of s would be adiabatically accessible, which violates Carathéodory's Principle.

Therefore, all final states with the final shape $(x'_1, \ldots, x'_{n-1})$ that can be reached from the given point s must have an entropy in an interval of which $S(s)$ is a boundary point. Or in other words, they all lie on one and the same side of the hypersurface $S = \text{const}$. By reasons of continuity he argues that this must be the same side for all initial states. Whether this is the side where entropy is higher or lower than that of the initial state remains an open question. According to Carathéodory, a further appeal to empirical experience is necessary to decide this issue.

He concludes (Carathéodory 1909, p. 378):

> [It] follows from our conclusions that, when for any change of state the value of the entropy has not remained constant, one can find no adiabatic change of state, which is capable of returning the considered system from its final state back to its initial state. *Every change of state, for which the entropy varies is 'irreversible.'*

Without doubt, this conclusion sounds pleasing to the ears of anyone who believes that irreversibility is the genuine trademark of the second law. But a few remarks are in order.

First, Carathéodory's conclusion is neutral with respect to time reversal: both increase and decrease of entropy are irreversible! Planck objected that the approach is not strong enough to characterize the direction of irreversible processes. In fact, Carathéodory (1925) admitted this point. He stressed that an additional appeal to experience is necessary to conclude that changes of entropy in adiabatic processes are always positive (if $T > 0$). In other words, in Carathéodory's approach this is not a consequence of the second law.

A second remark is that 'irreversible' here means that the change of state cannot be undone in an *adiabatic* process. This is yet another meaning for the term, different from those we have discussed before. The question is then, of course, whether changes of states that cannot be undone by an adiabatic process might perhaps be undone by some other process. Indeed, it is not hard to find examples of this possibility: consider a cylinder of ideal gas in thermal contact with a heat reservoir. When the piston is pulled out quasistatically,[10] the gas does work, while it takes in heat from the reservoir. Its entropy increases, and the process would thus qualify as irreversible in Carathéodory's sense. But Planck's book discusses this case as an example of a reversible process! Indeed, when the gas is quasistatically recompressed, the heat is restored to the reservoir and the initial state is recovered for both system and reservoir. Thus, Carathéodory's concept of 'irreversibility' does not coincide with Planck's.

There is also another way to investigate whether Carathéodory's approach captures the content of the second law *à la* Clausius, Kelvin or Planck, namely, by asking whether this approach allows models in which these formulations of the second law are invalid. An example is obtained by applying the formalism to a fluid while swapping the meaning of terms in each of the three pairs: heat/work, thermal/deformation coordinate, and adiabatic/without any exchange of work. The validity of Carathéodory's formalism is invariant under this operation, and a fluid remains a simple system. Indeed, we obtain, as a direct analog of (7.15), $đW = p\,dV$ for all quasistatic processes of a fluid. This shows that, in the present interpretation, pressure and volume play the role of temperature and entropy, respectively. Furthermore, irreversibility makes sense here too. For fluids with positive pressure, one can increase the volume of a fluid without doing work, but one cannot decrease volume without doing work. But still, the analog of the principles of Clausius of Kelvin are false: a fluid with low pressure can very well do positive work on another fluid with high pressure by means of a lever or hydraulic mechanism. And, thus, the sum of all volumes of a composite system can very well decrease, even when no external work is provided.

---

[10] Carathéodory's precise definition of the term 'quasistatic' is, of course, applicable to adiabatic processes only. I use the term here in the looser sense of Section 7.1.

7.6 Lieb and Yngvason

Lieb and Yngvason (1999) have recently provided a major contribution, by means of an elaborate and rigorous approach to the second law. In this context, I cannot do justice to their work, and will only sketch the main ideas, as far as they are relevant to my topic.

On a formal level, this work builds upon the approaches of Carathéodory (1909) and Giles (1964). (In its physical interpretation, however, it is more closely related to Planck, as we will see below.) A system is represented by a state space Γ on which a relation $\prec$ of adiabatic accessibility is defined. All axioms mentioned below are concerned with this relation. Further, Lieb and Yngvason introduce a formal operation of combining two systems in states s and t into a composite system in state (s, t), and the operation of 'scaling,' i.e. the construction of a copy in which all its extensive quantities are multiplied by a positive factor α. This is denoted by a multiplication of the state with α. These scaled states αs belong to a scaled state space $\Gamma_{(\alpha)}$. The main axioms of Lieb and Yngvason apply to all states $s \in \bigcup_\alpha \Gamma_{(\alpha)}$ (and compositions of such states). They read as follows.

A1. Reflexivity. $s \prec s$.

A2. Transitivity. $s \prec t$ and $t \prec r$ imply $s \prec r$.

A3. Consistency. $s \prec s'$ and $t \prec t'$ imply $(s, t) \prec (s', t')$.

A4. Scale invariance. If $s \prec t$, then $\alpha s \prec \alpha t$ for all $\alpha > 0$.

A5. Splitting and recombination. For all $0 < \alpha < 1$, $s \prec (\alpha s, (1-\alpha)s)$ and $(\alpha s, (1-\alpha)s) \prec s$.

A6. Stability. If there are states t_0 and t_1 such that $(s, \varepsilon t_0) \prec (r, \varepsilon t_1)$ holds for a sequence of ε's converging to zero, then $s \prec r$.

7. Comparability Hypothesis.[11]   For all states s, t in the same space Γ, $s \prec t$ or $t \prec s$.[12]

The Comparability Hypothesis has, as its name already indicates, a lower status than the axioms. It is intended as a characterization of a particular type of thermodynamical systems, namely, of 'simple' systems and systems composed of such 'simple' systems.[13] A substantial part of their paper is devoted to an

---

[11] Actually, Lieb and Yngvason call this the 'Comparison Hypothesis'. I believe, however, that the present name is more apt.

[12] The clause 'in the same space Γ' means that the hypothesis is not intended for the comparison of states of scaled systems. Thus, it is not demanded that we can either adiabatically transform a state of 1 mole of oxygen into a state of 2 moles of oxygen or conversely.

[13] Beware that the present meaning of the term does not coincide with that of Carathéodory. For simple systems in Carathéodory's sense the Comparability Hypothesis need not hold.

attempt to derive this hypothesis from further axioms. I will, however, not go into this.

The aim of the work is to derive the following result.

Entropy Principle (Lieb and Yngvason version). *There exists a function*[14] *S defined on all states of all systems such that when s and t are comparable, then*

$$s \prec t \quad \text{if and only if } S(s) \leqslant S(t). \tag{7.16}$$

Lieb and Yngvason (1999, pp. 19, 20) interpret the result (7.16) as an expression of the second law.

> It says that entropy must increase in an irreversible process ... [and] ... the physical content of [(7.16)] ... [is that]... adiabatic processes not only increase entropy but an increase in entropy also dictates which adiabatic processes are possible (between comparable states, of course).

The question of whether this result actually follows from their assumptions is somewhat involved. They show that the entropy principle follows from Axioms A1–A6 and the Comparability Hypothesis under some special conditions which, physically speaking, exclude mixing and chemical reactions. To extend the result, an additional ten axioms are needed (three of which serve to derive the Comparability Hypothesis). And even then, only a weak form of the above entropy principle is actually obtained, where 'iff' in (7.16) is replaced by 'implies.'

Before considering the interpretation of this result more closely, a few general remarks are in order. This approach combines mathematical precision with clear and plausible axioms and achieves a powerful theorem. This is true progress in the formulation of the second law. Note that the theorem is obtained without appealing to anything remotely resembling Carathéodory's Principle. This is undoubtedly an advantage for those who judge that principle too abstract. In fact the axioms and hypothesis mentioned above allow models which violate the principle of Carathéodory (Lieb and Yngvason 1999, p. 91). For example, it may be that all states are mutually accessible, in which case the entropy function S is simply a constant on Γ.

For the purpose of this paper, the question is whether there is a connection with the arrow of time in this formulation of the second law. As before, there are two aspects to this question: irreversibility and time-reversal (in)variance. We have seen that Lieb and Yngvason interpret the relation (7.16) as saying that entropy must increase in irreversible processes. At first sight, this interpretation is curious. Adiabatic accessibility is not the same thing as irreversibility. So how can the above axioms have implications for irreversible processes?

---

[14] Actually, the Lieb–Yngvason entropy principle also states the additivity and extensivity of the entropy function.

This puzzle is resolved when we consider the physical interpretation which Lieb and Yngvason (1999, p. 17) propose for the relation $\prec$.

Adiabatic Accessibility. *A state t is adiabatically accessible from a state s, in symbols $s \prec t$, if it is possible to change the state from s to t by means of an interaction with some device (which may comprise mechanical and electric parts as well as auxiliary thermodynamic systems) and a weight, in such a way that the auxiliary system returns to its initial state at the end of the process, while the weight may have changed its position in a gravitational field.*

This view is rather different from Carathéodory's, or, indeed, anyone else's: clearly, this term is not intended to refer to processes occurring in a thermos flask. As the authors explicitly emphasize, even processes in which the system is *heated* are adiabatic, in the present sense, when this heat is generated by an electrical current from a dynamo driven by descending weight. Actually, the condition that the auxiliary systems return to their initial state in the present concept is strongly reminiscent of Planck's concept of 'reversible'!

This is not to say, of course, that they are identical. As we have seen before, a process $\mathcal{P}$ involving a system, an environment and a weight at height h, which produces the transition

$$\langle s, Z, h \rangle \xrightarrow{\mathcal{P}} \langle s', Z', h' \rangle,$$

is reversible for Planck if and only if there exists a 'recovery' process $\mathcal{P}'$ which produces

$$\langle s', Z', h' \rangle \xrightarrow{\mathcal{P}'} \langle s, Z, h \rangle.$$

Here, the states Z and Z' may differ from each other. For Lieb and Yngvason, a process

$$\langle s, Z, h \rangle \xrightarrow{\mathcal{P}} \langle s', Z', h' \rangle$$

is adiabatic if and only if $Z = Z'$. But in all his discussions, Planck always restricted himself to such reversible processes 'which leave no changes in other bodies,' i.e. obeying the additional requirement $Z = Z'$. These processes are adiabatic in the present sense.

A crucial consequence of this is that, in the present sense, it follows that if a process $\mathcal{P}$ as considered above is adiabatic, any recovery process $\mathcal{P}'$ is automatically adiabatic too. Thus, we can now conclude that if an adiabatic process is accompanied by an entropy increase, it cannot be undone, i.e. it is irreversible in Planck's sense. This explains why the result (7.16) is seen as a formulation of a principle of entropy increase. In fact, we can reason as follows: assume s and t are states which are mutually comparable, and that $S(s) < S(t)$. According to (7.16), we then have $s \prec t$ and $t \not\prec s$. This means that there exists a process from s to t which proceeds without producing any change in auxiliary systems except, possibly, a displacement of a single weight.

At the same time there exists no such process from t to s. The first-mentioned process is therefore irreversible in Planck's sense. Thus we have at last achieved a conclusion implying the existence of irrecoverable processes by means of a satisfactory argument!

However, it must be noted that this conclusion is obtained only for systems obeying the Comparability Hypothesis and under the exclusion of mixing and chemical processes. The weak version of the entropy principle, which is derived when we drop the latter restriction, does not justify this conclusion. Moreover, note that it would be incorrect to construe (7.16) as a characterization of *processes*. The relation $\prec$ is interpreted in terms of the *possibility* of processes. As remarked in Section 7.5, one and the same change of state can very well be obtained (or undone) by means of different processes, some of which are adiabatic and others not. Thus, when $S(s) < S(t)$ for comparable states, this does not mean that *all* processes from s to t are irreversible, but only that there exists an adiabatic irreversible process between these states. So the entropy principle here is not the universal proposition of Planck.

The next question concerns the time-reversal (in)variance of this approach. As before, we can look upon the axioms as singling out a class of possible worlds $\mathcal{W}$. It is easy to show, using the implementation of time reversal used earlier, i.e. replacing $\prec$ by $\succ$, the six general axioms, and the Comparability Hypothesis, are TRI![15] The fact that it is not necessary to introduce time-reversal noninvariance into the formalism to obtain the second law is very remarkable.

However, there remains one problematic aspect of the proposed physical interpretation. It refers to the state of auxiliary systems in the environment of the system. Thus, we are again confronted by the old and ugly questions, when shall we say that the state of such auxiliary systems has changed, and when are we fully satisfied that their initial state is restored? These questions remain rather intractable from the point of view of thermodynamics, when one allows arbitrary auxiliary systems (e.g. living beings) whose states are not represented by the thermodynamical formalism. Thus, the question of when the relation $\prec$ holds cannot be decided in thermodynamical terms.

7.7 Discussion

We have seen that there is a large variety in the connections between irreversibility and the second law. At one end of the spectrum, there is Planck's view that the second law expresses the irreversibility of all processes in nature. A convincing derivation of this bold claim has, however, never been given. At the other

---

[15] This conclusion cannot be extended to the complete set of axioms proposed by Lieb and Yngvason. In particular, their Axioms A7 and T1, which address mixing and equilibration processes, are explicitly nonTRI. (I thank Jakob Yngvason for pointing this out to me.) However, these axioms are needed only in the derivation of the (TRI) Comparability Hypothesis, and not in the derivation of the entropy principle.

extreme, we find Gibbs's approach, which completely avoids any connection with time.

But even for approaches belonging to the middle ground, the term 'irreversible' is used with various meanings: time-reversal noninvariant, irrecoverable, and quasistatic. In the long-standing debate on the question how the second law relates to statistical mechanics, however, most authors have taken irreversibility in the sense of time-reversal noninvariance. The point that the term usually means something very different in thermodynamics has been almost completely overlooked.

The more careful and formal approaches by Carathéodory and, in particular, Lieb and Yngvason rather yield a surprising conclusion. It is possible to build up a precise formulation of the second law without introducing a nonTRI element in the discussion. The resulting formalism, therefore, remains strictly neutral to the question of whether entropy increases or decreases. It implies only that an entropy function can be constructed consistently, i.e. as either increasing between adiabatically accessible states of *all* simple systems, or decreasing. At the same time, the Lieb–Yngvason approach does imply that entropy-increasing processes between comparable states are irreversible in Planck's sense. This, of course, once more shows the independence of the two notions.

Finally, I would like to point out an analogy between the axiomatization of thermodynamics in the Carathéodory and Lieb–Yngvason approach and that of special relativity in the approach of Robb (1921). In both cases, we start out with a particular relationship $\prec$ which is assumed to exist between points of a certain space. In relativity, this is the relation of connectability by a causal signal. In both cases, it is postulated that this relation forms a pre-order. In both cases, important partial results show that the forward sectors $\mathcal{C}_s = \{t : s \prec t\}$ are convex and nested and that s is on the boundary of $\mathcal{C}_s$. And in both cases the aim is to show that the space is 'orientable' (Earman 1974) and admits a global function which increases in the forward sector. If this analogy is taken seriously, the Lieb–Yngvason entropy principle has just as much to do with TRI as the fact that Minkowski space-time admits a global time coordinate.

There is, however, also an important disanalogy. In thermodynamics, the space Γ represents the states of a system, and an important feature is that we can combine systems into a composite, or divide one into subsystems. In relativity, the space represent the whole of space-time, and there is no question of combining several of these.

Due to the possibility of combining systems, the Lieb–Yngvason entropy principle does yield an additional result: entropy can be defined consistently, in the sense just mentioned. For some, this result may be sufficient to conclude that the principle does express some form of irreversibility. However, as has been emphasized by Schrödinger (1950), a formulation of the second law which states that all systems can only change their entropy in the same sense is not in contradiction to the time-reversal invariance of an underlying microscopic theory.

References

Bernstein, B. 1960 Proof of Carathéodory's local theorem and its global application to thermostatics. *J. Math. Phys.* **1**, 222–224.

Buchdahl, H. A. 1966 *The Concepts of Classical Thermodynamics*. Cambridge University Press.

Callen, H. B. 1960 *Thermodynamics*. Wiley, New York.

Carathéodory, C. 1909 Untersuchungen über die Grundlagen der Thermodynamik. *Math. Annalen* **67**, 355–386.

Carathéodory, C. 1925 Über die Bestimmung der Energie und der absoluten Temperatur mit Hilfe von reversiblen Prozessen. *Sitzungsberichte der preußischen Akademie der Wissenschaften, Math. Phys. Klasse*, pp. 39–47.

Clausius, R. 1854 Über eine veränderte Form des zweiten Hauptsatzes der mechanischen Wärmetheorie. *Poggendroff's Annln Phys.* **93**, 481–506.

Clausius, R. 1864 Ueber die Concentration von Wärme- und Lichtstrahlen und die Gränze Ihre Wirkung. In *Abhandlungen über die Mechanischen Wärmetheorie*, Vol. 1, pp. 322–361. Vieweg & Sohn, Braunschweig.

Clausius, R. 1865 Ueber verschiedene für die Anwendung bequeme Formen der Haubtgleichungen der mechanischen Wärmetheorie. *Vierteljahrschrift der naturforschenden Gesellschaft (Zürich)* **10**, 1–59. (Also in Clausius (1867), pp. 1–56, and translated in Kestin (1976), pp. 162–193.)

Clausius, R. 1867 *Abhandlungungen über die Mechanische Wärmetheorie*, Vol. 2. Vieweg & Sohn, Braunschweig.

Clausius, R. 1876 *Die Mechanische Wärmetheorie*. Vieweg & Sohn, Braunschweig.

Denbigh, K. 1989 The many faces of irreversibility. *Br. J. Phil. Sci.* **40**, 501–518.

Earman, J. 1974 An attempt to add a little direction to 'the problem of the direction of time'. *Phil. Sci.* **41**, 15–47.

Falk, G. and Jung, H. 1959 Axiomatik der Thermodynamik. In *Handbuch der Physik* (ed. S. Flügge), Vol. III/2. Springer.

Gibbs, J. W. 1906 *The Scientific Papers of J. Willard Gibbs*, Vol. 1. *Thermodynamics*. Longmans, London.

Giles, R. 1964 *Mathematical Foundations of Thermodynamics*. Pergamon, Oxford.

Hollinger, H. B. and Zenzen, M. J. 1985 *The Nature of Irreversibility*. Reidel, Dordrecht.

Illner, R. and Neunzert, H. 1987 The concept of irreversibility in the kinetic theory of gases. *Transp. Th. Statist. Phys.* **16**, 89–112.

Kelvin, Lord 1852 On a universal tendency in nature to the dissipation of mechanical energy. *Trans. R. Soc. Edinb.* **20**, 139–142. (Also in Kestin (1976), pp. 194–197.)

Kestin, J. 1976 *The Second Law of Thermodynamics*. Dowden, Hutchinson and Ross, Stroudsburg, PA.

Landsberg, P. T. 1964 A deduction of Carathéodory's Principle from Kelvin's Principle. *Nature* **201**, 485–486.

Lieb, E. H. and Yngvason, J. 1999 The physics and mathematics of the second law of thermodynamics. *Phys. Rep.* **310**, 1–96. Erratum, **314** (1999), 669. (Also at http://xxx.lanl.gov/abs/cond-mat/9708200.)

Mendoza, E. (ed.) 1960 *Reflections on the Motive Power of Fire by Sadi Carnot and Other Papers on the Second Law of Thermodynamics by E. Clapeyron and R. Clausius.* Dover, New York.

Planck, M. 1897 *Vorlesungen über Thermodynamik.* Veit, Leipzig.

Planck, M. 1926 Über die Begrundung des zweiten Hauptsatzes der Thermodynamik. *Sitzungsberichte der preußischen Akademie der Wissenschaften, Math. Phys. Klasse*, pp. 453–463.

Robb, A. A. 1921 *The Absolute Relations of Time and Space.* Cambridge University Press.

Schrödinger, E. 1950 Irreversibility. *Proc. R. Irish Acad.* A **53**, 189–195.

Smith, C. and Wise, M. N. 1989 *Energy and Empire: a Biographical Study of Lord Kelvin.* Cambridge University Press.

Truesdell, C. 1986 What did Gibbs and Carathéodory leave us about thermodynamics? In *New Perspectives in Thermodynamics* (ed. J. Serrin), pp. 101–123. Springer.

Uffink, J. 2001 Bluff your way in the second law of thermodynamics. *Stud. Hist. Phil. Mod. Phys.* **32**, 305–394.

van der Waals, J. D. and Kohnstamm, Ph. 1927 *Lehrbuch der Thermostatik.* Barth, Leipzig.

Chapter Eight

The Entropy of Classical Thermodynamics

Elliott H. Lieb
Princeton University

Jakob Yngvason
Universität Wien

The essence of the second law of classical thermodynamics is the entropy principle, which asserts the existence of an additive and extensive entropy function, S, that is defined for all equilibrium states of thermodynamic systems and whose increase characterizes the possible state changes under adiabatic conditions. It is one of the few really fundamental physical laws (in the sense that no deviation, however tiny, is permitted) and its consequences are far reaching. This principle is independent of models, statistical mechanical or otherwise, and can be understood without recourse to Carnot cycles, ideal gases and other assumptions about such things as heat, temperature, reversible processes, etc., as is usually done. Also, the well-known formula of statistical mechanics, $S = -\sum p \log p$, is not needed for the derivation of the entropy principle.

This contribution is in part a summary of our joint work (see Lieb and Yngvason 1999), where the existence and uniqueness of S is proved to be a consequence of certain basic properties of the relation of adiabatic accessibility among equilibrium states. We also present some open problems and suggest directions for further study.

Foreword

At the conference 'International Symposium on Entropy,' hosted by the Max Planck Institute for Physics of Complex Systems, Dresden, 26–28 June 2000, one of us (J.Y.) contributed a talk with the above title. It was a review of our work (Lieb and Yngvason 1999) on the mathematical foundations of classical

---

E.H.L.'s work was partly supported by US National Science Foundation grant PHY98-20650.
J.Y.'s work was partly supported by the Adalsteinn Kristjansson Foundation, University of Iceland.

© 2000 by the authors. Reproduction of this article, by any means, is permitted for noncommercial purposes.

This article also appears under the title 'The Mathematical Structure of the Second law of Thermodynamics' in *Contemporary Developments in Mathematics 2001*, pp. 89–129, International Press (2002), which should be referenced as the primary publication.

thermodynamics. An extensive summary of this work was published in Lieb and Yngvason (1998). It was also published in Lieb and Yngvason (2000b) with additional sections added. A shorter summary, addressed particularly to physicists, appeared in Lieb and Yngvason (2000a). An expanded version of Lieb and Yngvason (2000b) appeared in Lieb and Yngvason (2002) and we reproduce it here. Section 8.1 is primarily from Lieb and Yngvason (1998) but is augmented by proofs of all theorems. The present version is therefore mathematically complete, but the original paper (Lieb and Yngvason 1999) is recommended for additional insights and extensive discussions. Section 8.2 is primarily from Lieb and Yngvason (2000b) and Section 8.3 is mainly from Lieb and Yngvason (2000a).

8.1 A Guide to Entropy and the Second Law of Thermodynamics

This chapter is intended for readers who, like us, were told that the second law of thermodynamics is one of the major achievements of the 19th century, that it is a logical, perfect and unbreakable law, but who were unsatisfied with the 'derivations' of the entropy principle as found in textbooks and in popular writings.

A glance at the books will inform the reader that the law has 'various formulations' (which is a bit odd for something so fundamental) but they all lead to the existence of an entropy function whose reason for existence is to tell us which processes can occur and which cannot. An interesting summary of these various points of view is in Uffink (2001). Contrary to convention, we shall refer to the existence of entropy as *the* second law. This, at least, is unambiguous. The entropy we are talking about is that defined by thermodynamics (and *not* some analytic quantity, usually involving expressions such as $-p \ln p$, that appears in information theory, probability theory, and statistical mechanical models).

Why, one might ask, should a mathematician be interested in the second law of thermodynamics, which, historically, had something to do with attempts to understand and improve the efficiency of steam engines? The answer, as we perceive it, is that the law is really an interesting mathematical theorem about orderings on sets, with profound physical implications. The axioms that constitute this ordering are somewhat peculiar from the mathematical point of view and might not arise in the ordinary ruminations of abstract thought. They are special, but important, and they are driven by considerations about the world, which is what makes them so interesting. Maybe an ingenious reader will find an application of this same logical structure to another field of science.

Classical thermodynamics, as it is usually presented, is based on three laws (plus one more, due to Nernst, which is mainly used in low-temperature physics and is not immutable like the others). In brief, these are as follows.

The zeroth law. This expresses the transitivity of equilibrium, and is often said to imply the existence of temperature as a parametrization of equilibrium

THE ENTROPY OF CLASSICAL THERMODYNAMICS

states. We use it below but formulate it without mentioning temperature. In fact, temperature makes no appearance here until almost the very end.

The first law. This is conservation of energy. It is a concept from mechanics and provides the connection between mechanics (and things like falling weights) and thermodynamics. We discuss this later on when we introduce simple systems; the crucial usage of this law is that it allows energy to be used as one of the parameters describing the states of a simple system.

The second law. Three popular formulations of this law are as follows.

- No process is possible the sole result of which is that heat is transferred from a body to a hotter one (*Clausius*).
- No process is possible the sole result of which is that a body is cooled and work is done (*Kelvin* (and *Planck*)).
- In any neighborhood of any state there are states that cannot be reached from it by an adiabatic process (*Carathéodory*).

All three are supposed to lead to the entropy principle (defined below). These steps can be found in many books and will not be trodden again here. Let us note in passing, however, that the first two use concepts such as hot, cold, heat, cool, that are intuitive but have to be made precise before the statements are truly meaningful. No one has seen 'heat,' for example. The last (which uses the term 'adiabatic process,' to be defined below) presupposes some kind of parametrization of states by points in $\mathbb{R}^n$, and the usual derivation of entropy from it assumes some sort of differentiability; such assumptions are beside the point as far as understanding the meaning of entropy goes.

The basic input in our analysis of the second law is a certain kind of ordering on a set and is denoted by

$$\prec$$

(pronounced 'precedes'). It is transitive and reflexive as in A1, A2 below, but $X \prec Y$ and $Y \prec X$ does not imply $X = Y$, so it is a 'preorder.' The big question is whether $\prec$ can be encoded in an ordinary, real-valued function on the set, denoted by S, such that if X and Y are related by $\prec$, then $S(X) \leqslant S(Y)$ if and only if $X \prec Y$. The function S is also required to be additive and extensive in a sense that will soon be made precise.

A helpful analogy is the question, when can a vector field, $V(x)$, on $\mathbb{R}^3$ be encoded in an ordinary function, $f(x)$, whose gradient is V? The well-known answer is that a necessary and sufficient condition is that curl $V = 0$. Once V is observed to have this property one thing becomes evident and important: it is necessary to measure the integral of V only along some curves—not all curves—in order to deduce the integral along *all* curves. The encoding then has enormous predictive power about the nature of future measurements of V.

In the same way, knowledge of the function S has enormous predictive power in the hands of chemists, engineers and others concerned with the ways of the physical world.

Our concern will be the existence and properties of S, starting from certain natural axioms about the relation $\prec$. We present our results with slightly abridged versions of some proofs, but full details, and a discussion of related previous work on the foundations of classical thermodynamics, are given in Lieb and Yngvason (1999). The literature on this subject is extensive and it is not possible to give even a brief account of it here, except for mentioning that the previous work closest to ours is that of Giles (1964) and Buchdahl (1966) (see also Cooper 1967; Duistermaat 1968; Roberts and Luce 1968). (The situation is summarized more completely in Lieb and Yngvason (1999).) These other approaches are also based on an investigation of the relation $\prec$, but the overlap with our work is only partial. In fact, a major part of our work is the derivation of a certain property (the 'Comparison Hypothesis' below), which is taken as an axiom in the other approaches. Realizing the full power of this property was a remarkable and largely unsung achievement of Giles (1964).

Let us begin the story with some basic concepts.

1. *Thermodynamic system.* Physically, this consists of certain specified amounts of certain kinds of matter, e.g. a gram of hydrogen in a container with a piston, or a gram of hydrogen and a gram of oxygen in two separate containers, or a gram of hydrogen and two grams of hydrogen in separate containers. The system can be in various states which, physically, are *equilibrium states*. The space of states of the system is usually denoted by a symbol such as Γ and states in Γ by X, Y, Z, etc.

Physical motivation aside, a state space, mathematically, is just a set—to begin with; later on we will be interested in embedding state spaces in some convex subset of some $\mathbb{R}^{n+1}$, i.e. we will introduce coordinates. As we said earlier, however, the entropy principle is quite independent of coordinatization, Carathéodory's Principle notwithstanding.

2. *Composition and scaling of states.* The notion of Cartesian product, $\Gamma_1 \times \Gamma_2$, corresponds simply to the two (or more) systems being side by side on the laboratory table; mathematically, it is just another system (called a *compound system*), and we regard the state space $\Gamma_1 \times \Gamma_2$ as the same as $\Gamma_2 \times \Gamma_1$. Likewise, when forming multiple compositions of state spaces, the order and the grouping of the spaces is immaterial. Thus $(\Gamma_1 \times \Gamma_2) \times \Gamma_3$, $\Gamma_1 \times (\Gamma_2 \times \Gamma_3)$ and $\Gamma_1 \times \Gamma_2 \times \Gamma_3$ are to be identified as far as composition of state spaces is concerned. Points in $\Gamma_1 \times \Gamma_2$ are denoted by pairs (X, Y), and in $\Gamma_1 \times \cdots \times \Gamma_N$ by N-tuples $(X_1, \ldots, X_N)$ as usual. The subsystems comprising a compound system are physically independent systems, but they are allowed to interact with each other for a period of time and thereby alter each other's state.

The concept of scaling is crucial. It is this concept that makes our thermodynamics inappropriate for microscopic objects like atoms or cosmic objects like stars. For each state space Γ and number $\lambda > 0$ there is another state space, denoted by $\Gamma^{(\lambda)}$ with points denoted by λX. This space is called a *scaled copy* of Γ. Of course, we identify $\Gamma^{(1)} = \Gamma$ and $1X = X$. We also require $(\Gamma^{(\lambda)})^{(\mu)} = \Gamma^{(\lambda\mu)}$ and $\mu(\lambda X) = (\mu\lambda)X$. The physical interpretation of $\Gamma^{(\lambda)}$ when Γ is the space of one gram of hydrogen is simply the state space of λ grams of hydrogen. The state λX is the state of λ grams of hydrogen with the same 'intensive' properties as X, e.g. pressure, while 'extensive' properties like energy, volume, etc., are scaled by a factor λ (by definition).

For any given Γ we can form Cartesian product state spaces of the type $\Gamma^{(\lambda_1)} \times \Gamma^{(\lambda_2)} \times \cdots \times \Gamma^{(\lambda_N)}$. These will be called *multiple scaled copies* of Γ.

The notation $\Gamma^{(\lambda)}$ should be regarded as merely a mnemonic at this point, but later on, with the embedding of Γ into $\mathbb{R}^{n+1}$, it will literally be $\lambda \Gamma = \{\lambda X : X \in \Gamma\}$ in the usual sense.

3. *Adiabatic accessibility*. Now we come to the ordering. We say $X \prec Y$ (with X and Y possibly in *different* state spaces) if Y is *adiabatically accessible* from X according to the definition below. Different state spaces can occur, e.g. if there is mixing or a chemical reaction between two states of a compound system to produce a state in a third system.

What does this mean? Mathematically, we are just given a list of pairs $X \prec Y$. There is nothing more to be said, except that later on we will assume that this list has certain properties that will lead to interesting theorems about this list, and will lead, in turn, to the existence of an *entropy function*, S, characterizing the list.

The physical interpretation is quite another matter. In textbooks a process taking X to Y is usually called adiabatic if it takes place in 'thermal isolation,' which in turn means that 'no heat is exchanged with the surroundings.' Such concepts (heat, thermal, etc.) appear insufficiently precise to us and we prefer the following version, which is in the spirit of Planck's formulation of the second law (Planck 1926) and avoids those concepts. Our definition of adiabatic accessibility might at first sight appear to be less restrictive than the usual one, but as discussed in Lieb and Yngvason (1999, pp. 29 and 54), in the end anything that we call an adiabatic process (meaning that Y is adiabatically accessible from X) can also be accomplished in 'thermal isolation' as the concept is usually understood. Our definition has the great virtue (as discovered by Planck) that it avoids having to distinguish between work and heat—or even having to define the concept of heat. We emphasize, however, that the theorems do not require agreement with our physical definition of adiabatic accessibility; other definitions are conceivably possible. We also emphasize that we do not care about the temporal development involved in the state change; we only care about the net result for the system and the rest of the Universe.

> *A state Y is adiabatically accessible from a state X, in symbols $X \prec Y$, if it is possible to change the state from X to Y by means of an interaction with some device consisting of some auxiliary system and a weight, in such a way that the auxiliary system returns to its initial state at the end of the process, whereas the weight may have risen or fallen.*

The role of the 'weight' in this definition is merely to provide a particularly simple source (or sink) of mechanical energy. Note that an adiabatic process, physically, does not have to be gentle, or 'static,' or anything of the kind. It can be arbitrarily violent and destructive, so long as the system is brought back to equilibrium! The 'device' need not be a well-defined mechanical contraption. It can be another thermodynamic system, and even a gorilla jumping up and down on the system, or a combination of these—as long as the device returns to its initial state. The device can have intelligence, e.g. it can contain a clever scientist whose strategy depends on the progress of the experiment. Only the initial state X and the final state Y matter.

An example might be useful here. Take a pound of hydrogen in a container with a piston. The states are describable by two numbers, energy and volume, the latter being determined by the position of the piston. Starting from some state, X, we can take our hand off the piston and let the volume increase explosively to a larger one. After things have calmed down, call the new equilibrium state Y. Then $X \prec Y$. Question: Is $Y \prec X$ true? Answer: No. To get from Y to X adiabatically we would have to use some machinery and a weight, with the machinery returning to its initial state, and there is no way this can be done. Using a weight we can, indeed, recompress the gas to its original volume, but we will find that the energy is then larger than its original value.

Let us write

$$X \prec\prec Y \quad \text{if } X \prec Y \text{ but not } Y \prec X \text{ (written } Y \not\prec X\text{)}.$$

In this case we say that we can go from X to Y by an *irreversible adiabatic process*. If $X \prec Y$ and $Y \prec X$, we say that X and Y are *adiabatically equivalent* and write

$$X \stackrel{A}{\sim} Y.$$

Equivalence classes under $\stackrel{A}{\sim}$ are called *adiabats*.

4. *Comparability*. Given two states X and Y in two (same or different) state spaces, we say that they are comparable if $X \prec Y$ or $Y \prec X$ (or both). This turns out to be a crucial notion. Two states are not always comparable; a necessary condition is that they have the same material composition in terms of the chemical elements. For example, since water is H_2O and the

atomic weights of hydrogen and oxygen are 1 and 16, respectively, the states in the compound system of 2 g of hydrogen and 16 g of oxygen are comparable with states in a system consisting of 18 g of water (but not with 11 g of water or 18 g of oxygen).

Actually, the classification of states into various state spaces is done mainly for conceptual convenience. The second law deals only with states, and the only thing we really have to know about any two of them is whether or not they are comparable. Given the relation $\prec$ for all possible states of all possible systems, we can ask whether this relation can be encoded in an entropy function according to the following principle.

Entropy Principle. *There is a real-valued function on all states of all systems (including compound systems) called* entropy *and denoted by S such that the following holds.*

(a) MONOTONICITY. *When X and Y are comparable states then*

$$X \prec Y \quad \text{if and only if } S(X) \leqslant S(Y). \tag{8.1}$$

(b) ADDITIVITY AND EXTENSIVITY. *If X and Y are states of some (possibly different) systems and if (X, Y) denotes the corresponding state in the compound system, then the entropy is additive for these states, i.e.*

$$S(X, Y) = S(X) + S(Y). \tag{8.2}$$

S is also extensive, i.e. for or each $\lambda > 0$ and each state X and its scaled copy $\lambda X \in \Gamma^{(\lambda)}$ (defined in concept 2 on pp. 150 and 151)

$$S(\lambda X) = \lambda S(X). \tag{8.3}$$

A formulation logically equivalent to (a), not using the word 'comparable,' is the following pair of statements:

$$X \stackrel{A}{\sim} Y \Longrightarrow S(X) = S(Y) \quad \text{and} \quad X \prec\prec Y \Longrightarrow S(X) < S(Y). \tag{8.4}$$

The right hand-side is especially noteworthy. It says that entropy must increase in an irreversible adiabatic process.

The additivity of entropy in compound systems is often just taken for granted, but it is one of the startling conclusions of thermodynamics. First of all, the content of additivity, (8.2), is considerably more far reaching than one might think from the simplicity of the notation. Consider four states X, X', Y, Y' and suppose that $X \prec Y$ and $X' \prec Y'$. One of our axioms, A3, will be that then $(X, X') \prec (Y, Y')$, and (8.2) contains nothing new or exciting. On the other hand, the compound system can well have an adiabatic process in which $(X, X') \prec (Y, Y')$ but $X \not\prec Y$. In this case, (8.2) conveys much information. Indeed, by monotonicity, there will be many cases of this kind because the inequality $S(X) + S(X') \leq S(Y) + S(Y')$ certainly does not imply that $S(X) \leq S(Y)$. The fact that the inequality $S(X) + S(X') \leq S(Y) + S(Y')$ tells us *exactly* which adiabatic processes are allowed in the compound system (among comparable states), independent of any detailed knowledge of the manner in which the two systems interact, is astonishing and is at the *heart of thermodynamics*. The second reason that (8.2) is startling is this: from (8.1) alone, restricted to one system, the function S can be replaced by $29S$ and still do its job, i.e. satisfy (8.1). However, (8.2) says that it is possible to calibrate the entropies of all systems (i.e. simultaneously adjust all the undetermined multiplicative constants) so that the entropy $S_{1,2}$ for a compound $\Gamma_1 \times \Gamma_2$ is $S_{1,2}(X, Y) = S_1(X) + S_2(Y)$, even though systems 1 and 2 are totally unrelated!

We are now ready to ask some basic questions.

Q1. Which properties of the relation $\prec$ ensure existence and (essential) uniqueness of S?

Q2. Can these properties be derived from simple physical premises?

Q3. Which convexity and smoothness properties of S follow from the premises?

Q4. Can temperature (and hence an ordering of states by 'hotness' and 'coldness') be defined from S and what are its properties?

The answer to Q1 can be given in the form of six axioms that are reasonable, simple, 'obvious,' and unexceptionable. An additional, crucial assumption is also needed, but we call it a 'hypothesis' instead of an axiom because we show later how it can be derived from some other axioms, thereby answering Q2.

A1. Reflexivity. $X \stackrel{A}{\sim} X$.

A2. Transitivity. If $X \prec Y$ and $Y \prec Z$, then $X \prec Z$.

A3. Consistency. If $X \prec X'$ and $Y \prec Y'$, then $(X, Y) \prec (X', Y')$.

A4. Scaling Invariance. If $\lambda > 0$ and $X \prec Y$, then $\lambda X \prec \lambda Y$.

A5. Splitting and Recombination. $X \stackrel{A}{\sim} ((1-\lambda)X, \lambda X)$ for all $0 < \lambda < 1$. Note that the two state spaces are different. If $X \in \Gamma$, then the state space on the right side is $\Gamma^{(1-\lambda)} \times \Gamma^{(\lambda)}$.

A6. Stability. If $(X, \varepsilon Z_0) \prec (Y, \varepsilon Z_1)$ for some Z_0, Z_1 and a sequence of ε's tending to zero, then $X \prec Y$. This axiom is a substitute for continuity, which we cannot assume because there is no topology yet. It says that 'a grain of dust cannot influence the set of adiabatic processes.'

An important lemma is that (A1)–(A6) imply the *cancellation law*, which is used in many proofs. It says that for any three states X, Y, Z

$$(X, Z) \prec (Y, Z) \Longrightarrow X \prec Y. \tag{8.5}$$

THE ENTROPY OF CLASSICAL THERMODYNAMICS   155

Proof. We show that $(X, Z) \prec (Y, Z)$ implies $(X, Z/2) \prec (Y, Z/2)$ and hence $(X, Z/2^n) \prec (Y, Z/2^n)$ for all $n = 1, 2, \ldots$. By the stability axiom, A6, this implies $X \prec Y$.

The argument for $(X, Z/2) \prec (Y, Z/2)$ is as follows:

$$
\begin{aligned}
(X, Z/2) &\stackrel{A}{\sim} (X/2, X/2, Z/2) &&\text{(by A5 and A3)} \\
&\prec (X/2, Y/2, Z/2) &&\text{(by } (X, Z) \prec (Y, Z)\text{, using A3 and A4)} \\
&\prec (Y/2, Y/2, Z/2) &&\text{(again by } (X, Z) \prec (Y, Z)\text{, using A3 and A4)} \\
&\stackrel{A}{\sim} (Y, Z/2) &&\text{(by A5 and A3).}
\end{aligned}
$$

$\square$

The next concept plays a key role in our treatment.

CH. Definition. We say that the *Comparison Hypothesis* (CH) holds for a state space Γ if all pairs of states in Γ are comparable.

Note that A3, A4 and A5 automatically extend comparability from a space Γ to certain other cases, e.g. $X \prec ((1 - \lambda)Y, \lambda Z)$ for all $0 \leqslant \lambda \leqslant 1$ if $X \prec Y$ and $X \prec Z$. On the other hand, comparability on Γ alone does not allow us to conclude that X is comparable to $((1 - \lambda)Y, \lambda Z)$ if $X \prec Y$ but $Z \prec X$. For this, one needs CH on the product space $\Gamma^{(1-\lambda)} \times \Gamma^{(\lambda)}$, which is not implied by CH on Γ.

The significance of A1–A6 and CH is borne out by the following theorem.

Theorem 8.1 (equivalence of entropy and A1–A6, given CH). *The following are equivalent for a state space Γ.*

(i) *The relation $\prec$ between states in (possibly different) multiple scaled copies of Γ, e.g. $\Gamma^{(\lambda_1)} \times \Gamma^{(\lambda_2)} \times \cdots \times \Gamma^{(\lambda_N)}$, is characterized by an entropy function, S, on Γ in the sense that*

$$(\lambda_1 X_1, \lambda_2 X_2, \ldots) \prec (\lambda'_1 X'_1, \lambda'_2 X'_2, \ldots) \tag{8.6}$$

is equivalent to the condition that

$$\sum_i \lambda_i S(X_i) \leqslant \sum_j \lambda'_j S(X'_j) \tag{8.7}$$

whenever

$$\sum_i \lambda_i = \sum_j \lambda'_j. \tag{8.8}$$

(ii) *The relation $\prec$ satisfies conditions A1–A6, and CH holds for every multiple scaled copy of Γ.*

This entropy function on Γ is unique up to affine equivalence, i.e. $S(X) \to aS(X) + B$, with $a > 0$.

Proof. The implication (i) ⇒ (ii) is obvious. To prove the converse and also the uniqueness of entropy, pick two reference points $X_0 \prec\prec X_1$ in Γ. (If there are no such points, then entropy is simply constant and there is nothing more to prove.) To begin with, we focus attention on the 'strip' $\{X : X_0 \prec X \prec X_1\}$. (See Figure 8.1.) In the following it is important to keep in mind that, by axiom A5, $X \stackrel{A}{\sim} ((1-\lambda)X, \lambda X)$, so X can be thought of as a point in $\Gamma^{(1-\lambda)} \times \Gamma^{(\lambda)}$, for any $0 \leqslant \lambda \leqslant 1$.

Consider uniqueness first. If S is any entropy function satisfying (i), then necessarily $S(X_0) < S(X_1)$, and $S(X) \in [S(X_0), S(X_1)]$. Hence there is a unique $\lambda \in [0, 1]$ such that

$$S(X) = (1-\lambda)S(X_0) + \lambda S(X_1). \tag{8.9}$$

By (i), in particular additivity and extensivity of S and the fact that $X \stackrel{A}{\sim} ((1-\lambda)X, \lambda X)$, this is *equivalent* to

$$X \stackrel{A}{\sim} ((1-\lambda)X_0, \lambda X_1). \tag{8.10}$$

Because (8.10) is a property of X_0, X_1, X which is independent of S, and because of the equivalence of (8.10) and (8.9) for any entropy function, any other entropy function, S' say, must satisfy (8.9) with the *same* λ but with $S(X_0)$ and $S(X_1)$ replaced by $S'(X_0)$ and $S'(X_1)$, respectively. This proves that entropy is uniquely determined up to the choice of the entropy for the two reference points. A change of this choice clearly amounts to an affine transformation of the entropy function.

The equivalence of (8.10) and (8.9) provides also a clue for constructing entropy. Using only the properties of the relation $\prec$ one must produce a unique λ satisfying (8.10). The uniqueness of such a λ, if it exists, follows from the more general fact that

$$((1-\lambda)X_0, \lambda X_1) \prec ((1-\lambda')X_0, \lambda' X_1) \tag{8.11}$$

is equivalent to

$$\lambda \leqslant \lambda'. \tag{8.12}$$

This equivalence follows from $X_0 \prec\prec X_1$, using A4, A5 and the cancellation law, (8.5).

To find λ we consider

$$\lambda_{\max} = \sup\{\lambda : ((1-\lambda)X_0, \lambda X_1) \prec X\} \tag{8.13}$$

and

$$\lambda_{\min} = \inf\{\lambda : X \prec ((1-\lambda)X_0, \lambda X_1)\}. \tag{8.14}$$

Making use of the stability axiom, A6, one readily shows that the sup and inf are achieved, and hence

$$((1-\lambda_{\max})X_0, \lambda_{\max} X_1) \prec X \tag{8.15}$$

and
$$X \prec ((1 - \lambda_{\min})X_0, \lambda_{\min} X_1). \tag{8.16}$$
Hence, by A2,
$$((1 - \lambda_{\max})X_0, \lambda_{\max} X_1) \prec ((1 - \lambda_{\min})X_0, \lambda_{\min} X_1) \tag{8.17}$$
and thus (contrary to what the notation might suggest)
$$\lambda_{\max} \leqslant \lambda_{\min}. \tag{8.18}$$
That $\lambda_{\max}$ cannot be strictly smaller than $\lambda_{\min}$ follows from the Comparison Hypothesis for the state spaces $\Gamma^{(1-\lambda)} \times \Gamma^{(\lambda)}$: If $\lambda > \lambda_{\max}$, then $((1 - \lambda)X_0, \lambda X_1) \prec X$ cannot hold, and hence, by (CH), the alternative, i.e.
$$X \prec ((1 - \lambda)X_0, \lambda X_1), \tag{8.19}$$
must hold. Likewise, $\lambda < \lambda_{\min}$ implies
$$((1 - \lambda)X_0, \lambda X_1) \prec X. \tag{8.20}$$
Hence, if $\lambda_{\max} < \lambda_{\min}$, we have produced a whole interval of λ's satisfying (8.10). This contradicts the statement made earlier that (8.10) specifies at most one λ. At the same time we have shown that $\lambda = \lambda_{\min} = \lambda_{\max}$ satisfies (8.10). Hence we can define the entropy by (8.9), assigning some fixed, but arbitrarily, chosen values $S(X_0) < S(X_1)$ to the reference points. For the special choice $S(X_0) = 0$ and $S(X_1) = 1$ we have the *basic formula for S* (see Figure 8.1):
$$\boxed{S(X) = \sup\{\lambda : ((1 - \lambda)X_0, \lambda X_1) \prec X\},} \tag{8.21}$$
or, equivalently,
$$\boxed{S(X) = \inf\{\lambda : X \prec ((1 - \lambda)X_0, \lambda X_1)\}.} \tag{8.22}$$

The existence of λ satisfying (8.9) can also be shown for X outside the 'strip,' i.e. for $X \prec X_0$ or $X_1 \prec X$, by simply interchanging the roles of X, X_0 and X_1 in the considerations above. For these cases we use the convention that $(X, -Y) \prec Z$ means $X \prec (Y, Z)$, and $(Y, 0Z) = Y$. If $X \prec X_0$, λ in equation (8.9) will be $\leqslant 0$, and if $X_1 \prec X$, it will be $\geqslant 1$.

Our conclusion is that every $X \in \Gamma$ is equivalent, in the sense of $\overset{A}{\sim}$, to a scaled composition of the reference points X_0 and X_1. By A5 this also holds for all points in multiple scaled copies of Γ, where by A4 we can assume that the total 'mass' in (8.8) is equal to 1. Moreover, by the definition of S, the left and right sides of (8.7) are just the corresponding compositions of $S(X_0)$ and $S(X_1)$. To see that S characterizes the relation on multiple scaled copies, it is thus sufficient to show that (8.11) holds if and only if
$$(1 - \lambda)S(X_0) + \lambda S(X_1) \leqslant (1 - \lambda')S(X_0) + \lambda' S(X_1). \tag{8.23}$$
Since $S(X_0) < S(X_1)$ this is just the equivalence of (8.11) and (8.12) that was already mentioned. □

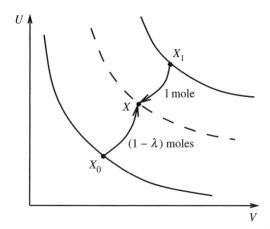

Figure 8.1. The entropy of X is, according to equation (8.21), determined by the largest amount of X_1 that can be transformed adiabatically into X, with the help of a complementary amount of X_0. The coordinates U (energy) and V (work coordinates) are irrelevant for equation (8.21), but are important in the context of the *simple systems* to be discussed later.

Remark 8.2. (i) The formula (8.21) for entropy is reminiscent of an old definition of heat by Laplace and Lavoisier in terms of the amount of ice that a body can melt: $S(X)$ is the maximal amount of substance in the state X_1 that can be transformed into the state X with the help of a complementary amount in the state X_0. According to (8.22) this is also the minimal amount of substance in the state X_1 that is needed to transfer a complementary amount in the state X into the state X_0. Note also that any λ satisfying (8.19) is an upper bound and any λ satisfying (8.20) is a lower bound to $S(X)$.

(ii) The construction of entropy in the above proof requires CH to hold for the twofold scaled products $\Gamma^{(1-\lambda)} \times \Gamma^{\lambda}$. It is not sufficient that CH holds for Γ alone, but in virtue of the other axioms it necessarily holds for all multiple scaled products of Γ if it holds for the twofold scaled products.

(iii) Theorem 8.1 states the properties a binary relation on a set must have in order to be characterized by a function satisfying our additivity and extensivity requirements. The set here is the union of all multiple scaled products of Γ. In a quite different context, this mathematical problem has already been discussed by Herstein and Milnor (1953) for what they call a 'mixture set,' which in our terminology corresponds to the union of all twofold scaled products $\Gamma^{(1-\lambda)} \times \Gamma^{\lambda}$. The main result of their paper is very similar to Theorem 8.1 for this special case.

Theorem 8.1 extends to products of multiple scaled copies of different systems, i.e. to general *compound* systems. This extension is an immediate consequence of the following theorem, which is proved by applying Theorem 8.1

THE ENTROPY OF CLASSICAL THERMODYNAMICS

to the product of the system under consideration with some standard reference system.

Theorem 8.3 (consistent entropy scales). *Assume that CH holds for all compound systems. For each system Γ let S_Γ be some definite entropy function on Γ in the sense of Theorem 8.1. Then there are constants a_Γ and $B(\Gamma)$ such that the function S, defined for all states of all systems by*

$$S(X) = a_\Gamma S_\Gamma(X) + B(\Gamma) \qquad (8.24)$$

for $X \in \Gamma$, satisfies additivity (8.2), extensivity (8.3), and monotonicity (8.1) in the sense that whenever X and Y are in the same state space then

$$X \prec Y \quad \text{if and only if } S(X) \leqslant S(Y). \qquad (8.25)$$

Proof. The entropy function defined by (8.21) will in general not satisfy additivity and extensivity, because the reference points can be quite unrelated for the different state spaces Γ. Note that (8.21) both fixes the states where the entropy is defined to be 0 (those that are $\overset{A}{\sim} X_0$) and also an arbitrary entropy unit for Γ by assigning the value 1 to the states $\overset{A}{\sim} X_1$. To obtain an additive and extensive entropy, it is first necessary to choose the points where the entropy is 0 in a way that is compatible with these requirements.

This can be achieved by considering the formal vector space spanned by all systems and choosing a Hamel basis of systems $\{\Gamma_\alpha\}$ in this space such that every system can be written uniquely as a scaled product of a finite number of the Γ_α's. Pick an arbitrary point in each state space in the basis, and define for each state space Γ a corresponding point $X_\Gamma \in \Gamma$ as a composition of these basis points. Then

$$X_{\Gamma_1 \times \Gamma_2} = (X_{\Gamma_1}, X_{\Gamma_2}) \quad \text{and} \quad X_{t\Gamma} = tX_\Gamma. \qquad (8.26)$$

Assigning the entropy 0 to these points X_Γ is clearly compatible with additivity and extensivity.

To ensure that the entropy unit is the same for all state spaces, choose some fixed space Γ_0 with fixed reference points $Z_0 \prec\prec Z_1$. For any Γ consider the product space $\Gamma \times \Gamma_0$ and the entropy function $S_{\Gamma \times \Gamma_0}$ defined by (8.21) in this space with reference points (X_Γ, Z_0) and (X_Γ, Z_1). Then $X \mapsto S_{\Gamma \times \Gamma_0}(X, Z_0)$ defines an entropy function on Γ by the cancellation law (8.5). It it is additive and extensive by the properties (8.26) of X_Γ, and by Theorem 8.1 it is related to any other entropy function on Γ by an affine transformation.

An *explicit formula* for this additive and extensive entropy is

$$S(X) = \sup\{\lambda : (X_\Gamma, \lambda Z_1)\} \prec (X, \lambda Z_0) \qquad (8.27)$$
$$= \inf\{\lambda : (X, \lambda Z_0) \prec (X_\Gamma, \lambda Z_1)\}, \qquad (8.28)$$

because

$$(X, Z_0) \overset{A}{\sim} ((1-\lambda)(X_\Gamma, Z_0), \lambda(X_\Gamma, Z_1)) \qquad (8.29)$$

is equivalent to
$$(X, \lambda Z_0) \stackrel{A}{\sim} (X_\Gamma, \lambda Z_1) \tag{8.30}$$
by the cancellation law. □

Theorem 8.3 is what we need, except for the question of mixing and chemical reactions, which is treated at the end and which can be put aside at a first reading. In other words, as long as we do not consider adiabatic processes in which systems are converted into each other (e.g. a compound system consisting of a vessel of hydrogen and a vessel of oxygen is converted into a vessel of water), the entropy principle has been verified. If that is so, the reader may justifiably ask what remains to be done. The answer is twofold. First, Theorem 8.3 requires that CH holds for *all* systems, including compound ones, and we are not content to take this as an axiom. Second, important notions of thermodynamics such as 'thermal equilibrium' (which will eventually lead to a precise definition of 'temperature') have not so far appeared. We shall see that these two points (i.e. thermal equilibrium and CH) are not unrelated.

As for CH, other authors (see Buchdahl 1966; Cooper 1967; Giles 1964; Roberts and Luce 1968) essentially *postulate* that it holds for all systems by making it axiomatic that comparable states fall into equivalence classes. (This means that the conditions $X \prec Z$ and $Y \prec Z$ always imply that X and Y are comparable; likewise, they must be comparable if $Z \prec X$ and $Z \prec Y$.) Replacing the concept of a 'state space' by that of an equivalence class, the Comparison Hypothesis then holds in these other approaches *by assumption* for all state spaces. We, in contrast, would like to derive CH from something that we consider more basic. Two ingredients will be needed: the analysis of certain special but commonplace systems, called 'simple systems,' and some assumptions about thermal contact (the 'zeroth law') that will act as a kind of glue holding the parts of a compound systems in harmony with each other.

A *simple system* is one whose state space can be identified with some open convex subset of some $\mathbb{R}^{n+1}$ with a distinguished coordinate denoted by U, called the *energy*, and additional coordinates $V \in \mathbb{R}^n$, called *work coordinates*. The energy coordinate is the way in which thermodynamics makes contact with mechanics, where the concept of energy arises and is precisely defined. The fact that the amount of energy in a state is independent of the manner in which the state was arrived at is, in reality, the first law of thermodynamics. A typical (and often the only) work coordinate is the volume of a fluid or gas (controlled by a piston); other examples are deformation coordinates of a solid or magnetization of a paramagnetic substance.

Our goal is to show, with the addition of a few more axioms, that CH holds for simple systems and their scaled products. In the process, we will introduce more structure, which will capture the intuitive notions of thermodynamics; thermal equilibrium is one.

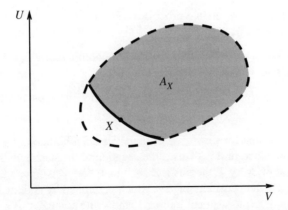

Figure 8.2. The coordinates U and V of a simple system. The state space (bounded by dashed line) and the forward sector A_X (shaded) of a state X are convex, by Axiom A7. The boundary of A_X (full line) is an adiabat, cf. Theorem 8.5.

Here, for the first time in our theory, coordinates are introduced. Up to now state spaces were fairly abstract things; there was no topology. Calculus, for example, played no role—contrary to the usual presentation of classical thermodynamics. For simple systems we are talking about points in $\mathbb{R}^n$ and we can thus talk about 'open sets,' 'convexity,' etc. In particular, if we take a point X and scale it to tX, then this scaling now has the usual concrete meaning it always has in $\mathbb{R}^n$, namely, all coordinates of X are multiplied by the positive number t. The notion of $+$, as in $X + Y$, had no meaning heretofore, but now it has the usual one of addition of vectors in $\mathbb{R}^n$.

First, there is an axiom about convexity.

A7. Convex combination. If X and Y are states of a simple system and $t \in [0, 1]$, then

$$(tX, (1-t)Y) \prec tX + (1-t)Y, \tag{8.31}$$

in the sense of ordinary convex addition of points in $\mathbb{R}^{n+1}$. A straightforward consequence of this axiom (and A5) is that the *forward sectors*

$$A_X := \{Y \subset \Gamma : X \prec Y\} \tag{8.32}$$

of states X in a simple system Γ are *convex* sets (see Figure 8.2).

Another consequence is a connection between the existence of irreversible processes and Carathéodory's Principle (Boyling 1972; Carathéodory 1909) mentioned above.

Lemma 8.4. *Assume (A1)–(A7) for $\Gamma \subset \mathbb{R}^N$ and consider the following statements.*

(a) **Existence of irreversible processes.** *For every $X \in \Gamma$ there is a $Y \in \Gamma$ with $X \prec\prec Y$.*

(b) **Carathéodory's Principle.** *In every neighborhood of every $X \in \Gamma$ there is a $Z \in \Gamma$ with $X \not\sim^A Z$.*

Then (a) $\Rightarrow$ (b) *always. If the forward sectors in Γ have interior points, then* (b) $\Rightarrow$ (a).

Proof. Suppose that for some $X \in \Gamma$ there is a neighborhood, $\mathcal{N}_X$, of X such that $\mathcal{N}_X$ is contained in A_X. (This is the negation of the statement that in every neighborhood of every X there is a Z such that $X \prec Z$ is false.) Let $Y \in A_X$ be arbitrary. By the convexity of A_X, which is implied by Axiom A7, X is an interior point of a line segment joining Y and some point $Z \in \mathcal{N}_X$, and, again by A7,

$$(tZ, (1-t)Y) \prec X \stackrel{A}{\sim} (tX, (1-t)X) \tag{8.33}$$

for some $t \in (0, 1)$. But we also have that $(tX, (1-t)X) \prec (tZ, (1-t)X)$ since $Z \in A_X$. This implies, by the cancellation law and A4, that $Y \prec X$. Thus we conclude that for some X, we have that $X \prec Y$ implies $X \stackrel{A}{\sim} Y$. This contradicts (a). In particular, we have shown that (a) $\Rightarrow$ (b).

Conversely, assuming that (a) is false, there is a point X_0 whose forward sector is given by $A_{X_0} = \{Y : Y \stackrel{A}{\sim} X_0\}$. Let X be an interior point of A_{X_0}, i.e. there is a neighborhood of X, $\mathcal{N}_X$, which is entirely contained in A_{X_0}. All points in $\mathcal{N}_X$ are adiabatically equivalent to X_0, however, and hence to X, since $X \in \mathcal{N}_X$. Thus (b) is false. □

We need three more axioms for simple systems, which will take us into an analytic detour. The first of these establishes (a) above.

A8. Irreversibility. For each $X \in \Gamma$ there is a point $Y \in \Gamma$ such that $X \prec\prec Y$. (This axiom is implied by A14 below, but is stated here separately because important conclusions can be drawn from it alone.)

A9. Lipschitz tangent planes. For each $X \in \Gamma$ the *forward sector* $A_X = \{Y \in \Gamma : X \prec Y\}$ has a *unique* support plane at X (i.e. A_X has a *tangent plane* at X). The tangent plane is assumed to be a *locally Lipschitz continuous* function of X, in the sense explained below.

A10. Connectedness of the boundary. The boundary ∂A_X (relative to the open set Γ) of every forward sector $A_X \subset \Gamma$ is connected. (This is technical and conceivably can be replaced by something else.)

Axiom A8 plus Lemma 8.4 asserts that every X lies on the boundary ∂A_X of its forward sector. Although Axiom A9 asserts that the convex set, A_X, has a true tangent at X only, it is an easy consequence of Axiom A2 that A_X has a

THE ENTROPY OF CLASSICAL THERMODYNAMICS

true tangent everywhere on its boundary. To say that this tangent plane is locally Lipschitz continuous means that if $X = (U^0, V^0)$, then this plane is given by

$$U - U^0 + \sum_1^n P_i(X)(V_i - V_i^0) = 0, \tag{8.34}$$

with locally Lipschitz continuous functions P_i. The function P_i is called the generalized *pressure* conjugate to the work coordinate V_i (where V_i is the volume, and P_i is the ordinary pressure).

Lipschitz continuity and connectedness is a well-known guarantee of uniqueness of the solution to the coupled differential equations

$$\frac{\partial u}{\partial V_j}(V) = -P_j(u(V), V) \quad \text{for } j = 1, \ldots, n, \tag{8.35}$$

which describes the boundary ∂A_X of A_X.

With these axioms one can now prove that the Comparison Hypothesis holds for the state space Γ of a simple system.

Theorem 8.5 (CH for simple systems). *If X and Y are states of the same simple system, then either $X \prec Y$ or $Y \prec X$. Moreover,*

$$X \stackrel{A}{\sim} Y \iff Y \in \partial A_X \iff X \in \partial A_Y.$$

Proof. The proof is carried out in several steps, which also provide further information about the forward sectors.

Step 1: A_X is closed. We have to prove that if $Y \in \Gamma$ is on the boundary of A_X, then Y is in A_X. For this purpose we can assume that the set A_X has full dimension, i.e. the interior of A_X is not empty. If, on the contrary, A_X lay in some lower-dimensional hyperplane, then the following proof would work, without any changes, simply by replacing Γ by the intersection of Γ with this hyperplane.

Let W be any point in the interior of A_X. Since A_X is convex, and Y is on the boundary of A_X, the half-open line segment joining W to Y (call it $[W, Y)$, bearing in mind that $Y \notin [W, Y)$) lies in A_X. The prolongation of this line beyond Y lies in the complement of A_X and has at least one point (call it Z) in Γ. (This follows from the fact that Γ is open and $Y \in \Gamma$.) For all sufficiently large integers n, the point Y_n defined by

$$\frac{n}{n+1} Y_n + \frac{1}{n+1} Z = Y \tag{8.36}$$

belongs to $[W, Y)$. We claim that $(X, (1/n)Z) \prec (Y, (1/n)Y)$. If this is so, then we are done because, by the stability axiom, A6, $X \prec Y$.

To prove the last claim, first note that $(X, (1/n)Z) \prec (Y_n, (1/n)Z)$ because $X \prec Y_n$ and by Axiom A3. By scaling, A4, the convex combination Axiom A7, and (8.36)

$$\left(Y_n, \frac{1}{n}Z\right) = \frac{n+1}{n}\left(\frac{n}{n+1}Y_n, \frac{1}{n+1}Z\right) \prec \frac{n+1}{n}Y. \quad (8.37)$$

But this last equals $(Y, (1/n)Y)$ by the splitting axiom, A5. Hence

$$\left(X, \frac{1}{n}Z\right) \prec \left(Y, \frac{1}{n}Y\right).$$

Step 2: A_X has a nonempty interior. A_X is a convex set by Axiom A7. Hence, if A_X had an empty interior, it would necessarily be contained in a hyperplane. (An illustrative picture to keep in mind here is that A_X is a closed (two-dimensional) disc in $\mathbb{R}^3$ and X is some point inside this disc and not on its perimeter. This disc is a closed subset of $\mathbb{R}^3$ and X is on its boundary (when the disc is viewed as a subset of $\mathbb{R}^3$). The hyperplane is the plane in $\mathbb{R}^3$ that contains the disc.)

Any hyperplane containing A_X is a support plane to A_X at X, and by Axiom A9 the support plane, Π_X, is unique, so $A_X \subset \Pi_X$. If $Y \in A_X$, then $A_Y \subset A_X \subset \Pi_X$ by transitivity, A2. By the irreversibility Axiom A8, there exists a $Y \in A_X$ such that $A_Y \neq A_X$, which implies that the convex set $A_Y \subset \Pi_X$, regarded as a subset of Π_X, has a boundary point in Π_X. If $Z \in \Pi_X$ is such a boundary point of A_Y, then $Z \in A_Y$ because A_Y is closed. By transitivity, $A_Z \subset A_Y \subset \Pi_X$, and $A_Z \neq \Pi_X$ because $A_Y \neq A_X$.

Now A_Y, considered as a subset of Π_X, has an $(n-1)$-dimensional supporting hyperplane at Z (because Z is a boundary point). Call this hyperplane Π'_Z. Since $A_Z \subset A_Y$, Π'_Z is a supporting hyperplane for A_Z, regarded as a subset of Π_X. Any n-dimensional hyperplane in $\mathbb{R}^{n+1}$ that contains the $(n-1)$-dimensional hyperplane $\Pi'_Z \subset \Pi_X$ clearly supports A_Z at Z, where A_Z is now considered as a convex subset of $\mathbb{R}^{n+1}$. Since there are infinitely many such n-dimensional hyperplanes in $\mathbb{R}^{n+1}$, we have a contradiction to the uniqueness Axiom A9.

Step 3: $Y \in \partial A_X \Rightarrow X \in \partial A_Y$ and hence $A_X = A_Y$. We present here only a sketch of the proof (for details see Lieb and Yngvason 1999, Theorems 3.5 and 3.6). First, using the convexity axiom, A7, and the existence of a tangent plane of A_X at X, one shows that the boundary points of A_X can be written as $(u_X(V), V)$, where u_X is a solution to the equation system (8.35). Here V runs through the set

$$\rho_X = \{V : (U, V) \in \partial A_X \text{ for some } U\}. \quad (8.38)$$

Secondly, the solution of (8.35) that passes through X is unique by the Lipschitz condition for the pressure. In particular, if $Y \in \partial A_X$, then u_X must coincide on $\rho_Y \subset \rho_X$ with the solution u_Y through Y. The proof is completed by showing that $\rho_X = \rho_Y$; this uses the fact that ρ_X is connected by Axiom A10 and also that ρ_X is open. For the latter it is important that no tangent plane of A_X can be parallel to the U-axis, because of Axiom A9.

Step 4: $X \notin A_Y \Rightarrow Y \in A_X$. Let Z be some point in the interior of A_Y and consider the line segment L joining X to Z. If we assume $X \notin A_Y$, then part of L lies outside A_Y, and therefore L intersects ∂A_Y at some point $W \in \partial A_Y$. By Step 3, A_Y and A_W are the same set, so $W \prec Z$ (because $Y \prec Z$). We claim that this implies $X \prec Z$ also. This can be seen as follows.

We have $W = tX + (1-t)Z$ for some $t \in (0, 1)$. By A7, A5, $W \prec Z$, and A3

$$(tX, (1-t)Z) \prec W \stackrel{A}{\sim} (tW, (1-t)W) \prec (tW, (1-t)Z). \qquad (8.39)$$

By transitivity, A2, and the cancellation law, (8.5), $tX \prec tW$. By scaling, A4, $X \prec W$ and hence, by A2, $X \prec Z$.

Since Z was arbitrary, we learn that Interior(A_Y) $\subset A_X$. Since A_X and A_Y are both closed by Step 1, this implies $A_Y \subset A_X$ and hence, by A1, $Y \in A_X$.

Step 5: $X \stackrel{A}{\sim} Y \Longleftrightarrow Y \in \partial A_X$. By Step 1, A_X is closed, so $\partial A_X \subset A_X$. Hence, if $Y \in \partial A_X$, then $X \prec Y$. By Step 3, $Y \in \partial A_X$ is equivalent to $X \in \partial A_Y$, so we can also conclude that $Y \prec X$. The implication $\Longleftarrow$ is thus clear. On the other hand, $X \stackrel{A}{\sim} Y$ implies $A_X = A_Y$ by Axiom A2 and thus $\partial A_X = \partial A_Y$. But $Y \in \partial A_Y$ by Axioms A1, A8 and Lemma 8.4. Thus the adiabats, i.e. the $\stackrel{A}{\sim}$ equivalence classes, are exactly the boundaries of the forward sectors. $\square$

Remark 8.6. It can also be shown from our axioms that the orientation of forward sectors w.r.t. the energy axis is the same for *all* simple systems (cf. Lieb and Yngvason 1999, Theorems 3.3 and 4.2). By convention we choose the direction of the energy axis so that the energy always *increases* in adiabatic processes at fixed work coordinates. When temperature is defined later, this will imply that temperature is always positive. Since spin systems in magnetic fields are sometimes regarded as capable of having 'negative temperatures,' it is natural to ask what in our axioms excludes such situations. The answer is, Convexity, A7, together with Axiom A8. The first would imply that if the energy can both increase and decrease in adiabatic processes, then a state of maximal energy is also in the state space. But such a state would also have maximal entropy and thus violate A8. From our point of view, 'negative temperature' states should not be regarded as true equilibrium states.

Before leaving the subject of simple systems let us remark on the connection with Carathéodory's development. The point of contact is the fact that $X \in \partial A_X$. We assume that A_X is convex and use transitivity and Lipschitz continuity to arrive, eventually, at Theorem 8.5. Carathéodory uses Frobenius' Theorem, plus assumptions about differentiability to conclude the existence, locally, of a surface containing X. Important *global* information, such as Theorem 8.5, is then not easy to obtain without further assumptions, as discussed, for example, in Boyling (1972).

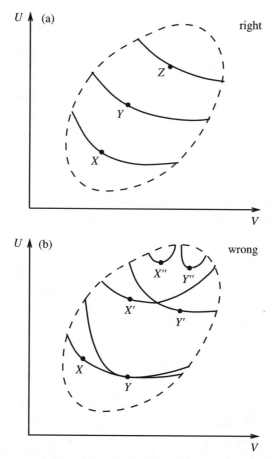

Figure 8.3. This figure illustrates Theorem 8.5, i.e. the fact that forward sectors of a simple system are nested. (b) What could, in principle, go wrong—but does not.

The next topic is *thermal contact* and the zeroth law, which entails the very special assumptions about $\prec$ that we mentioned earlier. It will enable us to establish CH for products of several systems, and thereby show, via Theorem 8.3, that entropy exists and is additive. Although we have established CH for a simple system, Γ, we have not yet established CH even for a product of two copies of Γ. This is needed in the definition of S given in (8.21). The S in (8.21) is determined up to an affine shift and we want to be able to calibrate the entropies (i.e. adjust the multiplicative and additive constants) of all systems so that they work together to form a global S satisfying the entropy principle. We need five more axioms. They might look a bit abstract, so a few words of introduction might be helpful.

THE ENTROPY OF CLASSICAL THERMODYNAMICS

In order to relate systems to each other, in the hope of establishing CH for compounds, and thereby an additive entropy function, some way must be found to put them into contact with each other. Heuristically, we imagine two simple systems (the same or different) side by side, and fix the work coordinates (e.g. the volume) of each. Bring them into 'thermal contact' (e.g. by linking them to each other with a copper thread) and wait for equilibrium to be established. The total energy U will not change but the individual energies, U_1 and U_2 will adjust to values that depend on U and the work coordinates. This new system (with the thread permanently connected) then behaves like a simple system (with one energy coordinate) but with several work coordinates (the union of the two work coordinates). Thus, if we start initially with $X = (U_1, V_1)$ for system 1 and $Y = (U_2, V_2)$ for system 2, and if we end up with $Z = (U, V_1, V_2)$ for the new system, we can say that $(X, Y) \prec Z$. This holds for every choice of U_1 and U_2 whose sum is U. Moreover, after thermal equilibrium is reached, the two systems can be disconnected, if we wish, and once more form a compound system, whose component parts we say are in thermal equilibrium. That this is transitive is the zeroth law.

Thus, not only can we make compound systems consisting of independent subsystems (which can interact, but separate again), we can also make a new simple system out of two simple systems. To do this an energy coordinate has to disappear, and thermal contact does this for us. This is formalized in the following two axioms.

A11. Thermal join. For any two simple systems with state spaces Γ_1 and Γ_2, there is another *simple* system, called the *thermal join* of Γ_1 and Γ_2, with state space

$$\Delta_{12} = \{(U, V_1, V_2) : U = U_1 + U_2 \text{ with } (U_1, V_1) \in \Gamma_1, (U_2, V_2) \in \Gamma_2\}. \tag{8.40}$$

If $X = (U_1, V_1) \in \Gamma_1$ and $Y = (U_2, V_2) \in \Gamma_2$, we define

$$\theta(X, Y) := (U_1 + U_2, V_1, V_2) \in \Delta_{12}. \tag{8.41}$$

It is assumed that the formation of a thermal join is an adiabatic operation for the compound system, i.e.

$$(X, Y) \prec \theta(X, Y). \tag{8.42}$$

A12. Thermal splitting. For any point $Z \in \Delta_{12}$ there is at least one pair of states, $X \in \Gamma_1, Y \in \Gamma_2$, such that

$$Z = \theta(X, Y) \stackrel{A}{\sim} (X, Y). \tag{8.43}$$

Definition 8.7. If $\theta(X, Y) \stackrel{A}{\sim} (X, Y)$, we say that the states X and Y are in *thermal equilibrium* and write

$$X \stackrel{T}{\sim} Y. \tag{8.44}$$

All and A12 together say that for each choice of the individual work coordinates there is a way to divide up the energy U between the two systems in a stable manner. A12 is the stability statement, for it says that joining is reversible, i.e. once the equilibrium has been established, one can cut the copper thread and retrieve the two systems back again, but with a special partition of the energies. This reversibility allows us to think of the thermal join, which is a simple system in its own right, as a special subset of the product system, $\Gamma_1 \times \Gamma_2$, which we call the *thermal diagonal*.

Axioms A11 and A12, together with the general Axioms A4, A5, A7, and our assumption that a compound state (X, Y) is identical to (Y, X), imply that the relation $\overset{T}{\sim}$ is reflexive and symmetric:

Lemma 8.8.

(i) $X \overset{T}{\sim} X$.

(ii) If $X \overset{T}{\sim} Y$, then $Y \overset{T}{\sim} X$.

Proof. (i) Let $X = (U, V)$. Then, by A11,

$$(X, X) \prec \theta(X, X) = (2U, V, V). \tag{8.45}$$

By A12, this is, for some U' and U'' with $U' + U'' = 2U$,

$$\overset{A}{\sim} ((U', V), (U'', V)), \tag{8.46}$$

which, using A4, A7, and finally A5, is

$$\overset{A}{\sim} 2(\tfrac{1}{2}(U', V), \tfrac{1}{2}(U'', V)) \prec 2(\tfrac{1}{2}(U' + U''), V) = 2X \overset{A}{\sim} (X, X). \tag{8.47}$$

Hence $(X, X) \overset{A}{\sim} \theta(X, X)$, i.e. $X \overset{T}{\sim} X$.

(ii) By A11 and A12 we have quite generally

$$\theta(X, Y) = (U_X + U_Y, V_X, V_Y) \overset{A}{\sim} ((U', V_X), (U'', V_Y)) \tag{8.48}$$

for some U', U'' with $U' + U'' = U_X + U_Y = U_Y + U_X$. Since composition of states is commutative (i.e. $(A, B) = (B, A)$, as we stated when explaining the basic concepts) we obtain, using A11 again,

$$((U', V_X), (U'', V_Y)) = ((U'', V_Y), (U', V_X)) \prec (U'' + U', V_Y, V_X)$$
$$= \theta(Y, X). \tag{8.49}$$

Interchanging X and Y we thus have $\theta(X, Y) \overset{A}{\sim} \theta(Y, X)$. Hence, if $X \overset{T}{\sim} Y$, i.e. $(X, Y) \overset{A}{\sim} \theta(X, Y)$, then $(Y, X) = (X, Y) \overset{A}{\sim} \theta(X, Y) \overset{A}{\sim} \theta(Y, X)$, i.e. $Y \overset{T}{\sim} X$. □

Remark 8.9. Instead of presenting a formal proof of the symmetry of $\overset{T}{\sim}$, it might seem more natural to simply identify $\theta(X, Y)$ with $\theta(Y, X)$ by an additional axiom. This could be justified both from the physical interpretation of the thermal join, which is symmetric in the states (connect the systems with a copper thread), and also because both joins have the same energy and the same work coordinates, only written in different order. But since we really only need that $\theta(X, Y) \overset{A}{\sim} \theta(Y, X)$ and this follows from the axioms as they stand, it is not necessary to postulate such an identification. Likewise, it is not necessary to identify $\theta(\theta(X, Y), Z)$ with $\theta(X, \theta(Y, Z))$ formally by an axiom, because we do not need it. It is possible, using the present axioms, to prove $\theta(\theta(X, Y), Z) \overset{A}{\sim} \theta(X, \theta(Y, Z))$, but we do not do so since we do not need this either.

We now come to the famous zeroth law, which says that the thermal equilibrium is transitive, and hence (by Lemma 8.8) an equivalence relation.

A13. Zeroth law of thermodynamics. If $X \overset{T}{\sim} Z$ and $Z \overset{T}{\sim} Y$, then $X \overset{T}{\sim} Y$.

The zeroth law is often taken to mean that the equivalence classes can be labeled by an 'empirical' temperature, but we do not want to mention temperature at all at this point. It will appear later.

There are two more axioms about thermal contact, but before we state them we draw two simple conclusions from A11, A12, and A13.

Lemma 8.10.

(i) *If* $X \overset{T}{\sim} Y$, *then* $\lambda X \overset{T}{\sim} \mu Y$ *for all* $\lambda, \mu > 0$.

(ii) *If* $X \overset{T}{\sim} Y$ *and* $Z \overset{T}{\sim} X$, *then* $Z \overset{T}{\sim} \theta(X, Y)$.

Proof. (i) By the zeroth law, A13, it suffices to show that $\lambda X \overset{T}{\sim} X$ and $\mu Y \overset{T}{\sim} Y$. For this we use similar arguments as in Lemma 8.8(i):

$$(\lambda X, X) \prec \theta(\lambda X, X) = ((1+\lambda)U, \lambda V, V) \overset{A}{\sim} ((\lambda U', \lambda V), (U'', V)) \quad (8.50)$$

with $\lambda U' + U'' = (1+\lambda)U$. By A4, A7, and A5 this is

$$\overset{A}{\sim} (1+\lambda)\left(\frac{\lambda}{1+\lambda}(U', V), \frac{1}{1+\lambda}(U'', V)\right)$$

$$\prec (1+\lambda)\left(\frac{1}{1+\lambda}(\lambda U' + U''), V\right)$$

$$= (1+\lambda)X \overset{A}{\sim} (\lambda X, X). \quad (8.51)$$

(ii) By the zeroth law, it suffices to show that $X \overset{T}{\sim} \theta(X, Y)$, i.e.

$$(X, \theta(X, Y)) \overset{A}{\sim} \theta(X, \theta(X, Y)). \quad (8.52)$$

The left side of this equation is $\prec$ the right side by A11, so we need only show

$$\theta(X, \theta(X, Y)) \prec (X, \theta(X, Y)). \tag{8.53}$$

Now, since $X \stackrel{T}{\sim} Y$ and hence also $2X \stackrel{T}{\sim} Y$ by (i), the right side of (8.53) is

$$(X, \theta(X, Y)) \stackrel{A}{\sim} (X, (X, Y)) = ((X, X), Y) \stackrel{A}{\sim} (2X, Y) \stackrel{A}{\sim} \theta(2X, Y)$$
$$= (2U_X + U_Y, 2V_X, V_Y). \tag{8.54}$$

(Here A5, A4, and A3 have been used, besides (i).) On the other hand, using A12 twice as well as A3, we have for some $X' = (U', V_X)$, $X'' = (U'', V_X)$ and $Y' = (U''', V_Y)$ with $U' + U'' + U''' = 2U_X + U_Y$:

$$\theta(X, \theta(X, Y)) \stackrel{A}{\sim} (X', (X'', Y')) = ((X', X''), Y'). \tag{8.55}$$

By convexity A7 and scaling A4, as above, this is

$$\prec (X' + X'', Y') \prec \theta(X' + X'', Y') = (2U_X + U_Y, 2V_X, V_Y). \tag{8.56}$$

But this is $\stackrel{A}{\sim} (X, \theta(X, Y))$ by (8.54). □

We now turn to the remaining two axioms about thermal contact.

A14 requires that for every adiabat (i.e. an equivalence class w.r.t. $\stackrel{A}{\sim}$) there exists at least one isotherm (i.e. an equivalence class w.r.t. $\stackrel{T}{\sim}$), containing points on both sides of the adiabat. Note that, for each given X, only two points in the entire state space Γ are required to have the stated property. This assumption essentially prevents a state space from breaking up into two pieces that do not communicate with each other. Without it, counterexamples to CH for compound systems can be constructed (cf. Lieb and Yngvason 1999, Section 4.3). A14 implies A8, but we listed A8 separately in order not to confuse the discussion of simple systems with thermal equilibrium.

A15 is a technical and can perhaps be eliminated. Its physical motivation is that a sufficiently large copy of a system can act as a heat bath for other systems. When temperature is introduced later, A15 will have the meaning that all systems have the same temperature range. This postulate is needed if we want to be able to bring every system into thermal equilibrium with every other system.

A14. Transversality. If Γ is the state space of a simple system and if $X \in \Gamma$, then there exist states $X_0 \stackrel{T}{\sim} X_1$ with $X_0 \prec\prec X \prec\prec X_1$.

A15. Universal temperature range. If Γ_1 and Γ_2 are state spaces of simple systems, then, for every $X \in \Gamma_1$ and every V belonging to the projection of Γ_2 onto the space of its work coordinates, there is a $Y \in \Gamma_2$ with work coordinates V such that $X \stackrel{T}{\sim} Y$.

THE ENTROPY OF CLASSICAL THERMODYNAMICS

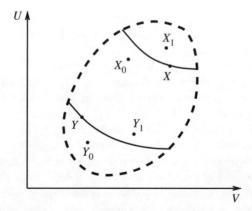

Figure 8.4. Transversality, A14, requires that on each side of the adiabat through any point X there are points, X_0 and X_1, that are in thermal equilibrium with each other.

The reader should note that the concept 'thermal contact' has appeared, but not temperature or hot and cold or anything resembling the Clausius or Kelvin–Planck formulations of the second law. Nevertheless, we come to the main achievement of our approach. *With these axioms we can establish CH for products of simple systems* (each of which satisfies CH, as we already know). The proof has two parts. In the first, we consider multiple scaled copies of the *same* simple system and use the thermal join and in particular transversality to reduce the problem to comparability within a single simple system, which is already known to hold by Theorem 8.5. The basic idea here is that with X, X_0, X_1 as in A14, the states $((1-\lambda)X_0, \lambda X_1)$ and $((1-\lambda)X, \lambda X)$ can be regarded as states of the *same* simple system and are, therefore, comparable. *This is the key point needed for the construction of S, according to (8.21)*. The importance of transversality is thus brought into focus. In the second part we consider products of *different* simple systems. This case is more complicated and requires all the Axioms A1–A14, in particular the zeroth law, A13.

Lemma 8.11 (CH in multiple scaled copies of a simple system). *For any simple system Γ, all states of the form $(\lambda_1 Y_1, \lambda_2 Y_2, \dots)$ with $Y_i \in \Gamma$ and $\sum_i \lambda_i$ fixed are comparable.*

Proof. By scaling invariance of the relation $\prec$ (Axiom A4) we may assume that $\sum_i \lambda_i = 1$. Now suppose $Y_1, \dots, Y_N, Y_1', \dots, Y_M' \in \Gamma$, and $\lambda_1, \dots, \lambda_N$, $\lambda_1', \dots, \lambda_M' \in \mathbb{R}$ with $\sum_i \lambda_i = \sum_j \lambda_j' = 1$. We shall show that for some $\bar{X}_0$, $\bar{X}_1 \in \Gamma$ with $\bar{X}_0 \prec\prec \bar{X}_1$ and some $\lambda, \lambda' \in \mathbb{R}$

$$(\lambda_1 Y_1, \dots, \lambda_N Y_N) \stackrel{A}{\sim} ((1-\lambda)\bar{X}_0, \lambda \bar{X}_1), \tag{8.57}$$

$$(\lambda_1' Y_1', \dots, \lambda_M' Y_M') \stackrel{A}{\sim} ((1-\lambda')\bar{X}_0, \lambda' \bar{X}_1). \tag{8.58}$$

This will prove the lemma, since we already know from the equivalence of (8.11) and (8.12) that the right sides of (8.57) and (8.58) are comparable.

It was already noted that if $X_0 \prec\prec X_1$ and $X_0 \stackrel{T}{\sim} X_1$, then every X in the 'strip' $\Sigma(X_0, X_1) := \{X : X_0 \prec X \prec X_1\}$ is comparable to $((1-\lambda)X_0, \lambda X_1)$ for any $0 \leqslant \lambda \leqslant 1$, due to the Axioms A5, A11, A12, and Theorem 8.5. This implies in the same way as in the proof of Theorem 8.1 that X is in fact *adiabatically equivalent* to $((1-\lambda)X_0, \lambda X_1)$ for some λ. (Namely, $\lambda = \lambda_{\max}$, defined by (8.13).) Moreover, if each of the points $Y_1, Y_2 \ldots, Y'_{M-1}, Y'_M$ is adiabatically equivalent to such a combination of a *common* pair of points $\bar{X}_0 \prec\prec \bar{X}_1$ (which need not be in thermal equilibrium), then equations (8.57) and (8.58) follow easily from the recombination Axiom A5. The existence of such a common pair of reference points is proved by the following stepwise extension of 'local' strips defined by points in thermal equilibrium.

By the transversality property, A4, the whole state space Γ can be covered by strips $\Sigma(X_0^{(i)}, X_1^{(i)})$ with $X_0^{(i)} \prec\prec X_0^{(i)}$ and $X_0^{(i)} \stackrel{T}{\sim} X_1^{(i)}$. Here i belongs to some index set. Since all adiabats ∂A_X with $X \in \Gamma$ are relatively closed in Γ we can even cover each X (and hence Γ) with the *open* strips

$$\overset{o}{\Sigma}_i := \overset{o}{\Sigma}(X_0^{(i)}, X_1^{(i)}) = \{X : X_0^{(i)} \prec\prec X \prec\prec X_0^{(i)}\}.$$

Moreover, any compact subset, C, of Γ is covered by a finite number of such strips $\overset{o}{\Sigma}_i, i = 1, \ldots, K$, and if C is connected we may assume that

$$\overset{o}{\Sigma}_i \cap \overset{o}{\Sigma}_{i+1} \neq \emptyset.$$

In particular, this holds if C is some polygonal path connecting the points $Y_1, \ldots, Y_N, Y'_1, \ldots, Y'_M$.

By Theorem 8.5, the points, $X_0^{(1)}, X_1^{(1)}, \ldots, X_0^{(K)}, X_1^{(K)}$, can be ordered according to the relation $\prec$, and there is no restriction to assume that

$$X_0^{(i)} \prec\prec X_0^{(i+1)} \prec\prec X_1^{(i)} \prec\prec X_1^{(i+1)}. \tag{8.59}$$

Let $\bar{X}_0 = X_0^{(1)}$ denote the 'smallest' and $\bar{X}_1 = X_1^{(K)}$ the 'largest' of these points. We claim that every one of the points $Y_1, \ldots, Y_N, Y'_1, \ldots, Y'_M$ is adiabatically equivalent to a combination of $\bar{X}_0$ and $\bar{X}_1$. This is based on the following general fact.

Suppose $X_0 \prec\prec X_1$, $X'_0 \prec\prec X'_1$ and

$$X_1 \stackrel{A}{\sim} ((1-\lambda_1)X'_0, \lambda_1 X'_1), \qquad X'_0 \stackrel{A}{\sim} ((1-\lambda_0)X_0, \lambda_0 X_1). \tag{8.60}$$

If

$$X \stackrel{A}{\sim} ((1-\lambda)X_0, \lambda X_1), \tag{8.61}$$

then

$$X \stackrel{A}{\sim} ((1-\mu)X_0, \mu X'_1) \tag{8.62}$$

with

$$\mu = \frac{\lambda \lambda_1}{1 - \lambda_0 + \lambda_0 \lambda_1}, \tag{8.63}$$

and if

$$X \stackrel{A}{\sim} ((1 - \lambda')X_0', \lambda'X_1'), \tag{8.64}$$

then

$$X \stackrel{A}{\sim} ((1 - \mu')X_0, \mu'X_1') \tag{8.65}$$

with

$$\mu' = \frac{\lambda'(1 - \lambda_0) + \lambda_0 \lambda_1}{1 - \lambda_0 + \lambda_0 \lambda_1}. \tag{8.66}$$

The proof of (8.62) and (8.65) is simple arithmetic, using the splitting and recombination axiom, A5, and the cancellation law, equation (8.5). Applying this successively for $i = 1, \ldots, K$ with

$$X_0 = X_0^{(1)}, \quad X_1 = X_1^{(i)}, \quad X_0' = X_0^{(i+1)}, \quad X_1' = X_1^{(i+1)},$$

proves that any $X \in \Sigma(X_0^{(1)}, X_1^{(K)})$ is adiabatically equivalent to a combination of $\bar{X}_0 = X_0^{(1)}$ and $\bar{X}_1 = X_1^{(K)}$. As already noted, this is precisely what is needed for (8.57) and (8.58). □

By Theorem 8.1, the last lemma establishes the existence of an entropy function S within the context of one simple system Γ and its scaled copies. Axiom A7 implies that S is a *concave* function of $X = (U, V) \in \Gamma$, i.e.

$$(1 - \lambda)S(U, V) + \lambda S(U', V') \leqslant S((1 - \lambda)U + \lambda U', (1 - \lambda)V + \lambda V'). \tag{8.67}$$

Moreover, by A11 and A12 and the properties of entropy described in Theorem 8.1(i),

$$(U, V) \stackrel{T}{\sim} (U', V')$$
$$\iff S(U, V) + S(U', V') = \max_W [S(W, V) + S(U + U' - W, V')]. \tag{8.68}$$

For a given Γ the entropy function is unique up to a multiplicative and an additive constant which are undetermined as long as we stay within the group of scaled copies of Γ. The next task is to show that the multiplicative constants can be adjusted to give a universal entropy valid for copies of *different* systems, i.e. to establish the hypothesis of Theorem 8.3. This is based on the following.

Lemma 8.12 (Existence of calibrators). *If Γ_1 and Γ_2 are simple systems, then there exist states $X_0, X_1 \in \Gamma_1$ and $Y_0, Y_1 \in \Gamma_2$ such that*

$$X_0 \prec\prec X_1 \quad \text{and} \quad Y_0 \prec\prec Y_1 \tag{8.69}$$

and

$$(X_0, Y_1) \stackrel{A}{\sim} (X_1, Y_0). \tag{8.70}$$

The significance of this lemma is that it allows us to fix the *multiplicative* constants by the condition

$$S_1(X_0) + S_2(Y_1) = S_1(X_1) + S_2(Y_0). \tag{8.71}$$

Proof of Lemma 8.12. The proof of this lemma is not entirely simple and it involves all the Axioms A1–A15. Consider the simple system Δ_{12} obtained by thermally joining Γ_1 and Γ_2. Let Z be some arbitrary point in Δ_{12} and consider the adiabat ∂A_Z. Any point in ∂A_Z is by Axiom A12 adiabatically equivalent to some pair $(X, Y) \in \Gamma_1 \times \Gamma_2$. There are now two alternatives.

- For some Z there are two such pairs, (X_0, Y_1) and (X_1, Y_0), such that $X_0 \prec\prec X_1$. Since $(X_0, Y_1) \stackrel{A}{\sim} Z \stackrel{A}{\sim} (X_1, Y_0)$, this implies $Y_0 \prec\prec Y_1$ and we are done.

- $(X, Y) \stackrel{A}{\sim} (\bar{X}, \bar{Y})$ with $X \stackrel{T}{\sim} Y$ and $\bar{X} \stackrel{T}{\sim} \bar{Y}$ always implies $X \stackrel{A}{\sim} \bar{X}$ (and hence also $Y \stackrel{A}{\sim} \bar{Y}$).

The task is thus to exclude the second alternative.

The second alternative is certainly excluded if the thermal splitting of some $Z = (U, V_1, V_2)$ in Δ_{12} is not unique. Indeed, if $U = U_1 + U_2 = \bar{U}_1 + \bar{U}_2$ with $U_1 < \bar{U}_1$ and $Z \stackrel{A}{\sim} (X, Y) \stackrel{A}{\sim} (\bar{X}, \bar{Y})$ with $X = (U_1, V_1)$, $\bar{X} = (\bar{U}_1, V_1)$, $Y = (U_2, V_2)$, and $\bar{Y} = (\bar{U}_2, V_2)$, then $X \prec\prec \bar{X}$ and $\bar{Y} \prec\prec Y$. Hence we may assume that for every $Z = (U, V_1, V_2) \in \Delta_{12}$ there are *unique* U_1 and U_2 with $U_1 + U_2 = U$ and $Z \stackrel{A}{\sim} ((U_1, V_1), (U_2, V_2))$.

Consider now some fixed $\bar{Z} \in \Delta_{12}$ with corresponding thermal splitting $(\bar{X}, \bar{Y})$, $\bar{X} \in \Gamma_1$, $\bar{Y} \in \Gamma_2$. We shall now show that the second alternative above leads to the conclusion that all points on the adiabat $\partial A_{\bar{Z}}$ are in thermal equilibrium with each other. By the zeroth law (Axiom A13) and since the domain of work coordinates corresponding to the adiabat $\partial A_{\bar{Z}}$ is connected (by Axiom A10), it is sufficient to show this for all points with a fixed work coordinate V_1 and all points with a fixed work coordinate V_2.

The second alternative means that if $Z \in \partial A_{\bar{Z}}$ has the thermal splitting (X, Y), then $X \stackrel{A}{\sim} \bar{X}$ and $Y \stackrel{A}{\sim} \bar{Y}$. For a given work coordinate V_1 there is a unique $\tilde{X} = (\tilde{U}_1, V_1) \in \partial A_{\bar{X}}$. (Its energy coordinate is uniquely determined as a solution of the partial differential equations for the adiabat, cf. Step 3 in the proof of Theorem 8.5.) Hence the thermal splitting of each point Z with fixed work coordinate V_1 has the form $(\tilde{X}, Y)$, and $Y \stackrel{T}{\sim} \tilde{X}$ for all such Y. By the zeroth law, all Y's are in thermal equilibrium with each other (since they are

THE ENTROPY OF CLASSICAL THERMODYNAMICS   175

in thermal equilibrium with a common $\tilde{X}$), and hence, by the zeroth law and Lemma 8.10, all the points $\theta(\tilde{X}, Y)$ are in thermal equilibrium with each other. In the same way one shows that all points in $\partial A_{\tilde{Z}}$ with fixed V_2 are in thermal equilibrium with each other.

To complete the proof we now show that if a simple system, in particular Δ_{12}, contains two points that are *not* in thermal equilibrium with each other, then there is at least one *adiabat* that contains such a pair. The case that all points in Δ_{12} are in thermal equilibrium with each other can be excluded, since by A15 it would imply the same for Γ_1 and Γ_2 and thus the thermal splitting would not be unique, contrary to assumption. (Note, however, that a world where all systems are in thermal equilibrium with each other is not in conflict with our axiom system. The entropy would then be an affine function of U and V for all systems. In this case, the first alternative above would always hold.)

In our proof of the existence of an adiabat with two points not in thermal equilibrium we shall make use of the already established entropy function S for the simple system $\Gamma = \Delta_{12}$ which characterizes the adiabats in Γ and moreover has the properties (8.67) and (8.68).

The fact that S characterizes the adiabats means that if $\mathcal{R} \subset \mathbb{R}$ denotes the range of S on Γ, then the sets

$$E_\sigma = \{X \in \Gamma : S(X) = \sigma\}, \quad \sigma \in \mathcal{R}, \tag{8.72}$$

are precisely the adiabats of Γ. Furthermore, the concavity of S—and hence its continuity on the connected open set Γ—implies that $\mathcal{R}$ is connected, i.e. $\mathcal{R}$ is an interval.

Let us assume now that for any adiabat, all points on that adiabat are in thermal equilibrium with each other. We have to show that this implies that all points in Γ are in thermal equilibrium with each other. By the zeroth law, A3, and Lemma 8.8(i), $\overset{T}{\sim}$ is an equivalence relation that divides Γ into disjoint equivalence classes. By our assumption, each such equivalence class must be a union of adiabats, which means that the equivalence classes are represented by a family of disjoint subsets of $\mathcal{R}$. Thus

$$\mathcal{R} = \bigcup_{\alpha \in \mathcal{I}} \mathcal{R}_\alpha, \tag{8.73}$$

where $\mathcal{I}$ is some index set, $\mathcal{R}_\alpha$ is a subset of $\mathcal{R}$, $\mathcal{R}_\alpha \cap \mathcal{R}_\beta = 0$ for $\alpha \neq \beta$, and $E_\sigma \overset{T}{\sim} E_\tau$ if and only if σ and τ are in some common $\mathcal{R}_\alpha$.

We will now prove that each $\mathcal{R}_\alpha$ is an open set. It is then an elementary topological fact (using the connectedness of Γ) that there can be only one nonempty $\mathcal{R}_\alpha$, i.e. all points in Γ are in thermal equilibrium with each other and our proof will be complete.

The concavity of $S(U, V)$ with respect to U for each fixed V implies the existence of an upper and lower U-derivative at each point, which we denote by $1/T_+$ and $1/T_-$, i.e.

$$\frac{1}{T_\pm}(U, V) = \pm \lim_{\varepsilon \searrow 0} \varepsilon^{-1}[S(U \pm \varepsilon, V) - S(U, V)]. \tag{8.74}$$

Equation (8.68) implies that $X \stackrel{T}{\sim} Y$ if and only if the closed intervals

$$[T_-(X), T_+(X)] \quad \text{and} \quad [T_-(Y), T_+(Y)]$$

are not disjoint. Suppose that some $\mathcal{R}_\alpha$ are not open, i.e. there is $\sigma \in \mathcal{R}_\alpha$ and either a sequence $\sigma_1 > \sigma_2 > \sigma_3 > \cdots$ converging to σ or a sequence $\sigma_1 < \sigma_2 < \sigma_3 < \cdots$ converging to σ with $\sigma_i \notin \mathcal{R}_\alpha$. Suppose the former (the other case is similar). Then (since $T_\pm$ are monotone increasing in U by the concavity of S) we can conclude that for *every* $Y \in E_{\sigma_i}$ and *every* $X \in E_\sigma$

$$T_-(Y) > T_+(X). \tag{8.75}$$

We also note, by the monotonicity of $T_\pm$ in U, that (8.75) necessarily holds if $Y \in E_\mu$ and $\mu \geqslant \sigma_i$; hence (8.75) holds for all $Y \in E_\mu$ for *any* $\mu > \sigma$ (because $\sigma_i \searrow \sigma$). On the other hand, if $\tau \leqslant \sigma$,

$$T_+(Z) \leqslant T_-(X) \tag{8.76}$$

for $Z \in E_\tau$ and $X \in E_\sigma$. This contradicts transversality, namely, the hypothesis that there is $\tau < \sigma < \mu$, $Z \in E_\tau$, $Y \in E_\mu$ such that $[T_-(Z), T_+(Z)] \cap [T_-(Y), T_+(Y)]$ is not empty. $\square$

With the aid of Lemma 8.12 we now arrive at our chief goal, which is CH for compound systems.

Theorem 8.13 (entropy principle in products of simple systems). *The Comparison Hypothesis CH (p. 155) is valid in arbitrary compounds of simple systems. Hence, by Theorem 8.3, the relation $\prec$ among states in such state spaces is characterized by an entropy function S. The entropy function is unique, up to an overall multiplicative constant and one additive constant for each simple system under consideration.*

Proof. Let Γ_1 and Γ_2 be simple systems and let $X_0, X_1 \in \Gamma_1$ and $Y_0, Y_1 \in \Gamma_2$ be points with the properties described in Lemma 8.12. By Theorem 8.1 we know that for every $X \in \Gamma_1$ and $Y \in \Gamma_2$

$$X \stackrel{A}{\sim} ((1 - \lambda_1)X_0, \lambda_1 X_1) \quad \text{and} \quad Y \stackrel{A}{\sim} ((1 - \lambda_2)Y_0, \lambda_2 Y_1) \tag{8.77}$$

for some λ_1 and λ_2. Define $Z_0 = (X_0, Y_0)$ and $Z_1 = (X_1, Y_1)$. It is then simple arithmetic, making use of (8.70) as well as Axioms A3–A5, to show that

$$(X, Y) \stackrel{A}{\sim} ((1 - \lambda)Z_0, \lambda Z_1) \tag{8.78}$$

with $\lambda = \frac{1}{2}(\lambda_1 + \lambda_2)$. By the equivalence of (8.11) and (8.12), we know that this is sufficient for comparability within the state space $\Gamma_1 \times \Gamma_2$.

Consider now a third simple system Γ_3 and apply Lemma 8.10 to $\Delta_{12} \times \Gamma_3$, where Δ_{12} is the thermal join of Γ_1 and Γ_2. By Axiom A12 the reference points in Δ_{12} are adiabatically equivalent to points in $\Gamma_1 \times \Gamma_2$, so we can repeat the reasoning above and conclude that all points in $(\Gamma_1 \times \Gamma_2) \times \Gamma_3$ are comparable. By induction, this extends to arbitrary products of simple systems. This includes multiple scaled products, because by Lemma 8.10 and Theorem 8.1, every state in a multiple scaled product of copies of a simple system Γ is adiabatically equivalent to a state in a single scaled copy of Γ. $\square$

Remark 8.14. It should be emphasized that Theorem 8.13 contains more than the entropy principle for compounds of simple systems. The core of the theorem is an assertion about the *comparability* of all states in any state space composed of simple systems. (Note that the entropy principle would trivially be true if no state was comparable to any other state.) Combining Lemma 8.12 and Theorem 8.13 we can even assert that certain compound states in *different* state spaces are comparable. What counts is that the total 'mass' of each simple system that enters the compound is the same for both states. For instance, if Γ_1 and Γ_2 are two simple systems, and $X_1, \ldots, X_N, X'_1, \ldots, X'_{N'} \in \Gamma_1$, $Y_1, \ldots, Y_M, Y'_1, \ldots, Y'_{M'} \in \Gamma_2$, then

$$Z = (\lambda_1 X_1, \ldots, \lambda_N X_N, \mu_1 Y_1, \ldots, \mu_M Y_M)$$

is comparable with

$$W = (\lambda'_1 X'_1, \ldots, \lambda'_{N'} X'_{N'}, \mu'_1 Y'_1, \ldots, \mu'_{M'} Y'_{M'})$$

provided $\sum \lambda_i = \sum \lambda'_j$, $\sum \mu_k = \sum \mu'_\ell$, and in this case $Z \prec W$ if and only if $S(Z) \leqslant S(W)$.

At last, we are now ready to define *temperature*. Concavity of S (implied by A7), Lipschitz continuity of the pressure, and the transversality condition, together with some real analysis, play key roles in the following, which answers questions Q3 and Q4 posed at the beginning.

Theorem 8.15 (entropy defines temperature). *The entropy, S, is a concave and continuously differentiable function on the state space of a simple system. If the function T is defined by*

$$\frac{1}{T} := \left(\frac{\partial S}{\partial U}\right)_V, \tag{8.79}$$

then $T > 0$ and T characterizes the relation $\stackrel{T}{\sim}$ in the sense that $X \stackrel{T}{\sim} Y$ if and only if $T(X) = T(Y)$. Moreover, if two systems are brought into thermal contact with fixed work coordinates, then, since the total entropy cannot decrease, the energy flows from the system with the higher T to the system with the lower T.

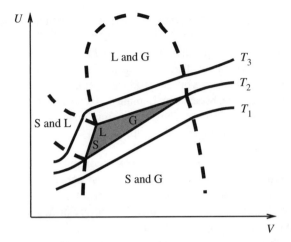

Figure 8.5. Isotherms in the (U, V)-plane near the triple point (L, liquid; G, gas; S, solid) of a simple system. (Not to scale.) In the triple point region the temperature is constant, which shows that an isotherm need not have codimension one.

Remark 8.16. The temperature need not be a strictly monotone function of U; indeed, it is not so in a 'multiphase region' (see Figure 8.5). It follows that T is not always capable of specifying a state, and this fact can cause some pain in traditional discussions of the second law—if it is recognized, which usually it is not.

Proof of Theorem 8.15. The complete proof is rather long and we shall not present all the details here. They can be found in Lieb and Yngvason (1999, Lemma 5.1 and Theorems 5.1–5.4). As in the proof of Lemma 8.12, concavity of $S(U, V)$ implies the existence of the upper and lower partial derivatives of S with respect to U and hence of the upper and lower temperatures $T_\pm$ defined by (8.74). Moreover, as also noted in the proof of Lemma 8.12,

$$X \stackrel{T}{\sim} Y \iff [T_-(X), T_+(X)] \cap [T_-(Y), T_+(Y)] \neq \emptyset. \tag{8.80}$$

The main goal is to show that $T_- = T_+ = T$, and that T is a continuous function of X.

Step 1: T_+ and T_- are locally Lipschitz continuous on adiabats. The essential input here is the local Lipschitz continuity of the pressure, i.e. for each $X \in \Gamma$ and each $r > 0$ there is a constant $C = C(X, r)$ such that

$$|P(X) - P(Y)| \leqslant C|X - Y| \tag{8.81}$$

if $|X - Y| < r$. The assertion is that

$$|T_+(X) - T_+(Y)| \leqslant c|X - Y| \tag{8.82}$$

THE ENTROPY OF CLASSICAL THERMODYNAMICS   179

for some $c = c(X, r)$, if $|X-Y| < r$ and $Y \in \partial A_X$, together with the analogous equation for T_-.

As in the proof of Theorem 8.5, Step 3, the adiabatic surface through $X = (U_0, V_0)$ is given by $(u_X(V), V)$, where u_X is the solution to the system of differential equations

$$\frac{\partial u}{\partial V_i} = -P_i, \quad i = 1, \ldots, n, \tag{8.83}$$

with the initial condition $u(V_0) = U_0$. Let us denote this solution by u_0, and consider also, for $\varepsilon > 0$, the solution u_ε with the initial condition $u(V_0) = U_0 + \varepsilon$. This latter solution determines the adiabatic surface through $X_\varepsilon = (U_0 + \varepsilon, V_0)$ (for V sufficiently close to V_0, so that $(u(V), V) \in \Gamma$).

Let S_0 denote the entropy on $(u_0(V), V)$ and S_ε the entropy on $(u_\varepsilon(V), V)$. Then, by definition,

$$T_+(U_0, V_0) = \lim_{\varepsilon \downarrow 0} \frac{\varepsilon}{S_\varepsilon - S_0} = \lim_{\varepsilon \downarrow 0} \frac{u_\varepsilon(V_0) - u_0(V_0)}{S_\varepsilon - S_0} \tag{8.84}$$

and

$$T_+(u_0(V), V) = \lim_{\varepsilon \downarrow 0} \frac{u_\varepsilon(V) - u_0(V)}{S_\varepsilon - S_0}$$
$$= T_+(U_0, V_0)\left[1 + \lim_{\varepsilon \downarrow 0} \frac{u_\varepsilon(V) - u_0(V) - \varepsilon}{\varepsilon}\right]. \tag{8.85}$$

To prove (8.82) it suffices to show that for all nonnegative ε close to 0 and V close to V_0

$$\frac{u_\varepsilon(V) - u_0(V) - \varepsilon}{\varepsilon} \leqslant D|V - V_0| \tag{8.86}$$

for some D. This estimate (with $D = 2C$) follows from (8.81), using (8.83) to write $u_\varepsilon(V) - u_\varepsilon(V_0)$ and $u_0(V) - u_0(0)$ as line integrals of the pressure (see Lieb and Yngvason 1999, p. 69).

Step 2: $T_+(X) > T_-(X) \Rightarrow T_+$ *and* T_- *are constant on* ∂A_X. This step relies on concavity of entropy, continuity of $T_\pm$ on adiabats (Step 1), and last, but not least, on the zeroth law. Without the zeroth law it is easy to give counterexamples to the assertion.

If $T_+(X) > T_-(X)$, but T_+ is not constant, then by continuity of $T_\pm$ on adiabats there exist Y, Z with $Y \stackrel{A}{\sim} Z \stackrel{A}{\sim} X$, $T_+(Y) < T_+(Z)$ but $[T_-(Y), T_+(Y)] \cap [T_-(Z), T_+(Z)] \neq \emptyset$, i.e. $Y \stackrel{T}{\sim} Z$. Now it is a general fact about a concave function, in particular $U \mapsto S(U, V)$, that the set of points where it is differentiable, i.e. where $T_+(U, V) = T_-(U, V) \equiv T(U, V)$ is dense. Moreover, if $U_1 > U_2 > \cdots$ is a sequence of such points converging to U, then $T(U_i, V)$ converges to $T_+(U, V)$. Using continuity of T_+ on the adiabat, we conclude that there exists a W such that $T_+(W) = T_-(W) = T(W)$, but

$T_+(Y) < T(W) < T_+(Z)$. This contradicts the zeroth law, because such a W would be in thermal equilibrium with Z (because $T(W) \in [T_-(Z), T_+(Z)]$) but not with Y (because $T(Y) \notin [T_-(Y), T_+(Y)]$). In the same way the assumption that T_- is not constant on the adiabat ∂A_X leads to a contradiction.

Step 3: $T_+ = T_-$. Assume $T_+(X) > T_-(X)$ for some X. Then, by Step 2, T_+ and T_- are constant on the whole adiabat ∂A_X. Now by concavity and monotonicity of S in U (cf. Remark 8.6) we have $T_+(Y) \leqslant T_-(Z)$ if $S(U_Y, V_Y) < S(U_Z, V_Z)$, $V_Y = V_Z$. Hence, if Z is such that

$$Y \prec\prec Z \text{ and } V_Y = V_Z \text{ for some } Y \overset{A}{\sim} X, \tag{8.87}$$

then $T_+(X) \leqslant T_-(Z) \leqslant T_+(Z)$. Likewise, if Z' is such that

$$Z' \prec\prec Y' \text{ and } V_{Y'} = V_{Z'} \text{ for some } Y' \overset{A}{\sim} X, \tag{8.88}$$

then

$$T_-(Z') \leqslant T_+(Z') \leqslant T_-(X) < T_+(X) \leqslant T_-(Z) \leqslant T_+(Z). \tag{8.89}$$

This means that no Z satisfying (8.87) can be in thermal equilibrium with a Z' satisfying (8.88). In the case that every point in the state space has its work coordinates in common with some point on the adiabat ∂A_X, this violates the transversality axiom, A14.

Using Axiom A5 we can also treat the case where the projection of the adiabat ∂X_X onto the work coordinates does not cover the whole range of the work coordinates for Γ, i.e. when

$$\rho_X := \{V : (U, V) \in \partial A_X \text{ for some } U\} \neq \{V : (U, V) \in \Gamma \text{ for some } U\}$$
$$=: \rho(\Gamma). \tag{8.90}$$

One considers a line of points $(U, \bar{V}) \in \Gamma$ with $\bar{V}$ fixed on the boundary of ρ_X in $\rho(\Gamma)$. One then shows that a gap between the upper and lower temperature of X, i.e. $T_-(X) < T_+(X)$, implies that points with these work coordinates $\bar{V}$ can only be in thermal equilibrium with points on one side of the adiabat ∂A_X, in contradiction to A15 and the zeroth law. See Lieb and Yngvason (1999, p. 72) for the details.

Step 4: T is continuous. By Step 3 T is uniquely defined and by Step 1 it is locally Lipschitz continuous on each adiabat, i.e.

$$|T(X) - T(X')| \leqslant c|X - X'| \tag{8.91}$$

if $X \overset{A}{\sim} X'$ and X and Y both lie in some ball of sufficiently small radius. Moreover, concavity of S and the fact that $T_+ = T_- = T$ imply that T is continuous along each line $l_V = \{(U, V) : (U, V) \in \Gamma\}$.

Let now $X_\infty, X_1, X_2, \ldots$ be points in Γ such that $X_j \to X_\infty$ as $j \to \infty$. We write $X_j = (U_j, V_j)$, we let A_j denote the adiabat ∂A_{X_j}, we let $T_j = T(X_j)$ and we set $l_j = \{(U, V_j) : (U, V_j) \in \Gamma\}$.

THE ENTROPY OF CLASSICAL THERMODYNAMICS

By Axiom A9, the slope of the tangent of A_X, i.e. the pressure $P(X)$, is locally Lipschitz continuous. Therefore, for X_j sufficiently close to X_∞ we can assume that each adiabat A_j intersects l_∞ in some point, which we denote by Y_j. Since $|X_j - X_\infty| \to 0$ as $j \to \infty$, we have that $|Y_j \to X_\infty|$ as well. In particular, we can assume that all the X_i and Y_j lie in some small ball around X_∞ so that (8.91) applies. Now

$$|T(X_j) - T(X_\infty)| \leqslant |T(X_j) - T(Y_j)| + |T(Y_j) - T(X_\infty)|, \qquad (8.92)$$

and as $j \to \infty$, $T(Y_j) - T(X_\infty) \to 0$ because Y_j and X_∞ are in l_∞. Also, $T(X_j) - T(Y_j) \to 0$ because

$$|T(X_j) - T(Y_j)| < c|X_j - Y_j| \leqslant c|X_j - X_\infty| + c|Y_j - X_\infty|.$$

Step 5: S is continuously differentiable. The adiabat through a point $X \in \Gamma$ is characterized by the once continuously differentiable function $u_X(V)$ on $\mathbb{R}^n$. Thus, $S(u_X(V), V)$ is constant, so (in the sense of distributions)

$$0 = \left(\frac{\partial S}{\partial U}\right)\left(\frac{\partial u_X}{\partial V_j}\right) + \frac{\partial S}{\partial V_j}. \qquad (8.93)$$

Since $1/T = \partial S/\partial U$ is continuous, and $\partial u_X/\partial V_j = -P_j$ is Lipschitz continuous, we see that $\partial S/\partial V_j$ is a continuous function and we have the well-known formula

$$\frac{\partial S}{\partial V_j} = \frac{P_j}{T}. \qquad (8.94)$$

Step 6: Energy flows from 'hot' to 'cold.' Let

$$X = (U_X, V_X) \quad \text{and} \quad Y = (U_Y, V_Y)$$

be two states of simple systems and assume that $T(X) > T(Y)$. By Axioms A11 and A12,

$$(X, Y) \prec \theta(X, Y) \stackrel{A}{\sim} (X', Y') \qquad (8.95)$$

with $X' = (U_{X'}, V_X)$, $Y' = (U_{Y'}, V_Y)$ and

$$U_{X'} + U_{Y'} = U_X + U_Y. \qquad (8.96)$$

Moreover, $X' \stackrel{T}{\sim} Y'$ and hence, by (8.80) and Step 3,

$$T(X') = T(Y') \equiv T^*. \qquad (8.97)$$

We claim that
$$T(X) \geqslant T^* \geqslant T(Y). \qquad (8.98)$$

(At least one of these inequalities is strict because of the uniqueness of temperature for each state.) Suppose that inequality (8.98) failed, e.g. $T^* > T(X) > T(Y)$. Then we would have that $U_{X'} > U_X$ and $U_{Y'} > U_Y$ and at least one of these would be strict (by the strict monotonicity of U with respect to T, which follows from the concavity and differentiability of S). This pair of inequalities is impossible in view of (8.96).

Since T^* satisfies (8.98), the theorem now follows from the monotonicity of U with respect to T. □

From the entropy principle and the relation

$$T = \left(\frac{\partial S}{\partial U}\right)^{-1} \tag{8.99}$$

between temperature and entropy we can now derive the usual formula for the *Carnot efficiency*

$$\eta_C := 1 - \left(\frac{T_0}{T_1}\right) \tag{8.100}$$

as an upper bound for the efficiency of a 'heat engine' that undergoes a cyclic process. Let us define a *thermal reservoir* to be a simple system whose work coordinates remains unchanged during some process. Consider a combined system consisting of a thermal reservoir and some machine, and an adiabatic process for this combined system. The entropy principle says that the total entropy change in this process is

$$\Delta S_{\text{machine}} + \Delta S_{\text{reservoir}} \geq 0. \tag{8.101}$$

Let $-Q$ be the energy change of the reservoir, i.e. if $Q \geq 0$, then the reservoir delivers energy, otherwise it absorbs energy. If T denotes the temperature of the reservoir *at the end of the process*, then, by the convexity of $S_{\text{reservoir}}$ in U, we have

$$\Delta S_{\text{reservoir}} \leq -\frac{Q}{T}. \tag{8.102}$$

Hence

$$\Delta S_{\text{machine}} - \frac{Q}{T} \geq 0. \tag{8.103}$$

Let us now couple the machine first to a 'high-temperature reservoir' which delivers energy Q_1 and reaches a final temperature T_1, and later to a 'low-temperature reservoir' which absorbs energy $-Q_0$ and reaches a final temperature T_0. The whole process is assumed to be cyclic for the machine so the entropy changes for the machine in both steps cancel. (It returns to its initial state.) Combining (8.101), (8.102), and (8.103) we obtain

$$\frac{Q_1}{T_1} + \frac{Q_0}{T_0} \leq 0, \tag{8.104}$$

which gives the usual inequality for the efficiency $\eta := (Q_1 + Q_0)/Q_1$:

$$\eta \leq 1 - \left(\frac{T_0}{T_1}\right) = \eta_C. \tag{8.105}$$

THE ENTROPY OF CLASSICAL THERMODYNAMICS   183

In textbook presentations it is usually assumed that the reservoirs are infinitely large, so that their temperature remains unchanged, but formula (8.105) remains valid for finite reservoirs, provided T_1 and T_0 are properly interpreted, as above.

Mixing and chemical reactions

The core results of our analysis have now been presented and readers satisfied with the entropy principle in the form of Theorem 8.13 may wish to stop at this point. Nevertheless, a nagging doubt will occur to some, because there are important adiabatic processes in which systems are not conserved, and these processes are not yet covered in the theory. A critical study of the usual textbook treatments should convince the reader that this subject is not easy, but in view of the manifold applications of thermodynamics to chemistry and biology, it is important to tell the whole story and not ignore such processes.

One can formulate the problem as the determination of the additive constants $B(\Gamma)$ of Theorem 8.3. As long as we consider only adiabatic processes that preserve the amount of each simple system (i.e. such that equations (8.6) and (8.8) hold), these constants are indeterminate. This is no longer the case, however, if we consider mixing processes and chemical reactions (which are not really different, as far as thermodynamics is concerned.) It then becomes a nontrivial question whether the additive constants can be chosen in such a way that the entropy principle holds. Oddly, this determination turns out to be far more complex, mathematically and physically than the determination of the multiplicative constants (Theorem 8.3). In traditional treatments one resorts to *gedanken* experiments involving idealized devices such as 'van t'Hofft boxes' which are made of idealized materials such as 'semipermeable membranes' that do not exist in the real world except in an approximate sense in a few cases. For the derivation of the entropy principle by this method, however, one needs virtually perfect 'semipermeable membranes' for all substances, and it is fair to question whether such a precise physical law should be founded on nonexistent objects. Fermi, in his famous textbook (Fermi 1956), draws attention to this problem, but, like those before and after him, chooses to ignore it and presses on. We propose a better way.

What we already know is that every system has a well-defined entropy function, e.g. for each Γ there is S_Γ, and we know from Theorems 8.3 and 8.13 that the multiplicative constants a_Γ can been determined in such a way that the sum of the entropies increases in any adiabatic process in any compound space $\Gamma_1 \times \Gamma_2 \times \cdots$. Thus, if $X_i \in \Gamma_i$ and $Y_i \in \Gamma_i$, then

$$(X_1, X_2, \ldots) \prec (Y_1, Y_2, \ldots) \quad \text{if and only if} \quad \sum_i S_i(X_i) \leqslant \sum_j S_j(Y_j),$$

(8.106)

where we have denoted S_{Γ_i} by S_i for short. The additive entropy constants do not matter here since each function S_i appears on both sides of this inequality.

It is important to note that this applies even to processes that, in intermediate steps, take one system into another, provided the total compound system is the same at the beginning and at the end of the process.

The task is to find constants $B(\Gamma)$, one for each state space Γ, in such a way that the entropy defined by

$$S(X) := S_\Gamma(X) + B(\Gamma) \quad \text{for } X \in \Gamma \tag{8.107}$$

satisfies

$$S(X) \leqslant S(Y) \tag{8.108}$$

whenever

$$X \prec Y \quad \text{with } X \in \Gamma, \ Y \in \Gamma'. \tag{8.109}$$

Additionally, we require that the newly defined entropy satisfies scaling and additivity under composition. Since the initial entropies $S_\Gamma(X)$ already satisfy them, these requirements become conditions on the additive constants $B(\Gamma)$:

$$B(\Gamma_1^{(\lambda_1)} \times \Gamma_2^{(\lambda_2)}) = \lambda_1 B(\Gamma_1) + \lambda_2 B(\Gamma_2) \tag{8.110}$$

for all state spaces Γ_1, Γ_2 under consideration and $\lambda_1, \lambda_2 > 0$. Some reflection shows us that consistency in the definition of the entropy constants $B(\Gamma)$ requires us to consider all possible chains of adiabatic processes leading from one space to another via intermediate steps. Moreover, the additivity requirement leads us to allow the use of a 'catalyst' in these processes, i.e. an auxiliary system, that is recovered at the end, although a state change *within* this system might take place. With this in mind we define quantities $F(\Gamma, \Gamma')$ that incorporate the entropy differences in all such chains leading from Γ to Γ'. These are built up from simpler quantities $D(\Gamma, \Gamma')$, which measure the entropy differences in one-step processes, and $E(\Gamma, \Gamma')$, where the 'catalyst' is absent. The precise definitions are as follows. First,

$$D(\Gamma, \Gamma') := \inf\{S_{\Gamma'}(Y) - S_\Gamma(X) : X \in \Gamma, \ Y \in \Gamma', \ X \prec Y\}. \tag{8.111}$$

If there is no adiabatic process leading from Γ to Γ', we put $D(\Gamma, \Gamma') = \infty$. Next, for any given Γ and Γ' we consider all finite chains of state spaces, $\Gamma = \Gamma_1, \Gamma_2, \ldots, \Gamma_N = \Gamma'$ such that $D(\Gamma_i, \Gamma_{i+1}) < \infty$ for all i, and we define

$$E(\Gamma, \Gamma') := \inf\{D(\Gamma_1, \Gamma_2) + \cdots + D(\Gamma_{N-1}, \Gamma_N)\}, \tag{8.112}$$

where the infimum is taken over all such chains linking Γ with Γ'. Finally, we define

$$F(\Gamma, \Gamma') := \inf\{E(\Gamma \times \Gamma_0, \Gamma' \times \Gamma_0)\}, \tag{8.113}$$

where the infimum is taken over all state spaces Γ_0. (These are the 'catalysts.')

The definition of the constants $F(\Gamma, \Gamma')$ involves a threefold infimum and may look somewhat complicated at this point. The F's, however, possess subadditivity and invariance properties that need not hold for the D's and E's, but

THE ENTROPY OF CLASSICAL THERMODYNAMICS

are essential for an application of the Hahn–Banach Theorem in the proof of Theorem 8.20 below. The importance of the F's for the problem of the additive constants is made clear by the following theorem.

Theorem 8.17 (constant entropy differences). *If Γ and Γ' are two state spaces, then for any two states $X \in \Gamma$ and $Y \in \Gamma'$*

$$X \prec Y \quad \text{if and only if } S_\Gamma(X) + F(\Gamma, \Gamma') \leqslant S_{\Gamma'}(Y). \tag{8.114}$$

Remark 8.18. Since $F(\Gamma, \Gamma') \leqslant D(\Gamma, \Gamma')$ the theorem is trivially true when $F(\Gamma, \Gamma') = +\infty$, in the sense that there is then no adiabatic process from Γ to Γ'. The reason for the title 'constant entropy differences' is that the minimum jump between the entropies $S_\Gamma(X)$ and $S_{\Gamma'}(Y)$ for $X \prec Y$ to be possible is independent of X. An essential ingredient for the proof of this theorem is equation (8.106).

Proof of Theorem 8.17. The 'only if' part is obvious because $F(\Gamma, \Gamma') \leqslant D(\Gamma, \Gamma')$. For the proof of the 'if' part we shall for simplicity assume that the infima in (8.111), (8.112), and (8.113) are minima, i.e. that they are obtained for some chain of spaces and some states in these spaces. The general case can be treated very similarly by approximation, using the stability axiom, A6.

We thus assume that

$$F(\Gamma, \Gamma') = D(\Gamma \times \Gamma_0, \Gamma_1) + D(\Gamma_1, \Gamma_2) + \cdots + D(\Gamma_N, \Gamma' \times \Gamma_0) \tag{8.115}$$

for some state spaces $\Gamma_0, \Gamma_1, \Gamma_2, \ldots, \Gamma_N$, and that

$$D(\Gamma \times \Gamma_0, \Gamma_1) = S(Y_1) - S_{\Gamma \times \Gamma_0}(\tilde{X}, X_0)$$
$$= S(Y_1) - S_\Gamma(\tilde{X}) - S_{\Gamma_0}(X_0), \tag{8.116}$$
$$D(\Gamma_i, \Gamma_{i+1}) = S_{\Gamma_{i+1}}(Y_{i+1}) - S_{\Gamma_i}(X_i), \tag{8.117}$$
$$D(\Gamma_N, \Gamma' \times \Gamma_0) = S_{\Gamma' \times \Gamma_0}(\tilde{Y}, Y_0) - S_{\Gamma_N}(X_N)$$
$$= S_{\Gamma'}(\tilde{Y}) + S_{\Gamma_0}(Y_0) - S_{\Gamma_N}(X_N) \tag{8.118}$$

for states $X_i \in \Gamma_i$ and $Y_i \in \Gamma_i$, for $i = 0, \ldots, N$, $\tilde{X} \in \Gamma$ and $\tilde{Y} \in \Gamma'$ with

$$(\tilde{X}, X_0) \prec Y_1, \quad X_i \prec Y_{i+1} \quad \text{for } i = 1, \ldots, N-1, \ X_N \prec (\tilde{Y}, Y_0). \tag{8.119}$$

Hence

$$F(\Gamma, \Gamma') = S_{\Gamma'}(\tilde{Y}) + \sum_{j=0}^{N} S_{\Gamma_j}(Y_j) - S_\Gamma(\tilde{X}) - \sum_{j=0}^{N} S_{\Gamma_j}(X_j). \tag{8.120}$$

From the assumed inequality $S_\Gamma(X) + F(\Gamma, \Gamma') \leqslant S_{\Gamma'}(Y)$ and (8.120) we conclude that

$$S_\Gamma(X) + S_{\Gamma'}(\tilde{Y}) + \sum_{j=0}^{N} S_{\Gamma_j}(Y_j) \leqslant S_\Gamma(\tilde{X}) + S_{\Gamma'}(Y) + \sum_{j=0}^{N} S_{\Gamma_j}(X_j). \tag{8.121}$$

However, both sides of this inequality can be thought of as the entropy of a state in the compound space $\hat{\Gamma} := \Gamma \times \Gamma' \times \Gamma_0 \times \Gamma_1 \times \cdots \times \Gamma_N$. The entropy principle (8.106) for $\hat{\Gamma}$ then tells us that

$$(X, \tilde{Y}, Y_0, \ldots, Y_N) \prec (\tilde{X}, Y, X_0, \ldots, X_N). \tag{8.122}$$

On the other hand, using (8.119) and Axiom A3, we have that

$$(\tilde{X}, X_0, X_1, \ldots, X_N) \prec (\tilde{Y}, Y_0, Y_1, \ldots, Y_N). \tag{8.123}$$

(The left side is here in $\Gamma \times \Gamma_0 \times \Gamma_1 \times \cdots \times \Gamma_N$ and the right side in $\Gamma' \times \Gamma_0 \times \Gamma_1 \times \cdots \times \Gamma_N$.) By A3 again, we have from (8.123) that

$$(\tilde{X}, Y, X_0, \ldots, X_N) \prec (Y, \tilde{Y}, Y_0, Y_1, \ldots, Y_N). \tag{8.124}$$

(The left side in $\Gamma \times \Gamma' \times \Gamma_0 \times \Gamma_1 \times \cdots \times \Gamma_N$, the right side in $\Gamma' \times \Gamma' \times \Gamma_0 \times \Gamma_1 \times \cdots \times \Gamma_N$.) From (8.122) and transitivity of the relation $\prec$ we then have

$$(X, \tilde{Y}, Y_0, Y_1, \ldots, Y_N) \prec (Y, \tilde{Y}, Y_0, Y_1, \ldots, Y_N), \tag{8.125}$$

and the desired conclusion, $X \prec Y$, follows from the cancellation law (8.5). $\square$

According to Theorem 8.17 the determination of the entropy constants $B(\Gamma)$ amounts to satisfying the inequalities

$$-F(\Gamma', \Gamma) \leqslant B(\Gamma) - B(\Gamma') \leqslant F(\Gamma, \Gamma') \tag{8.126}$$

together with the linearity condition (8.110). It is clear that (8.126) can only be satisfied with finite constants $B(\Gamma)$ and $B(\Gamma')$, if $F(\Gamma, \Gamma') > -\infty$. To exclude the pathological case $F(\Gamma, \Gamma') = -\infty$ we introduce our last Axiom A16, whose statement requires the following definition.

Definition 8.19. A state space Γ is said to be *connected* to another state space Γ' if there are states $X \in \Gamma$ and $Y \in \Gamma'$, and state spaces $\Gamma_1, \ldots, \Gamma_N$ with states $X_i, Y_i \in \Gamma_i, i = 1, \ldots, N$, and a state space Γ_0 with states $X_0, Y_0 \in \Gamma_0$, such that

$$(X, X_0) \prec Y_1, \quad X_i \prec Y_{i+1}, \quad i = 1, \ldots, N-1, \quad X_N \prec (Y, Y_0). \tag{8.127}$$

A16. Absence of sinks. If Γ is connected to Γ', then Γ' is connected to Γ.

This axiom excludes $F(\Gamma, \Gamma') = -\infty$ because, on general grounds, one always has

$$-F(\Gamma', \Gamma) \leqslant F(\Gamma, \Gamma'). \tag{8.128}$$

(See below.) Hence $F(\Gamma, \Gamma') = -\infty$ (which means, in particular, that Γ is connected to Γ') would imply $F(\Gamma', \Gamma) = \infty$, i.e. that there is no way back from Γ' to Γ. This is excluded by Axiom 16.

The quantities $F(\Gamma, \Gamma')$ have certain properties that allow us to use the Hahn–Banach Theorem to satisfy the inequalities (8.126), with constants $B(\Gamma)$ that

THE ENTROPY OF CLASSICAL THERMODYNAMICS 187

depend linearly on Γ, in the sense of (8.110). These properties, which follows immediately from the definition, are

$$F(\Gamma, \Gamma) = 0, \tag{8.129}$$

$$F(t\Gamma, t\Gamma') = tF(\Gamma, \Gamma') \quad \text{for } t > 0, \tag{8.130}$$

$$F(\Gamma_1 \times \Gamma_2, \Gamma_1' \times \Gamma_2') \leqslant F(\Gamma_1, \Gamma_1') + F(\Gamma_2, \Gamma_2'), \tag{8.131}$$

$$F(\Gamma \times \Gamma_0, \Gamma' \times \Gamma_0) = F(\Gamma, \Gamma') \quad \text{for all } \Gamma_0. \tag{8.132}$$

In fact, (8.129) and (8.130) are also shared by the D's and the E's. The 'subadditivity' (8.131) also holds for the E's, but the 'translational invariance' (8.132) might only hold for the F's. Equation (8.128) and, more generally, the 'triangle inequality'

$$F(\Gamma, \Gamma'') \leqslant F(\Gamma, \Gamma') + F(\Gamma', \Gamma'') \tag{8.133}$$

are simple consequences of (8.131) and (8.132). Using these properties we can now derive the following theorem.

Theorem 8.20 (universal entropy). *The additive entropy constants of all systems can be calibrated in such a way that the entropy is additive and extensive, and $X \prec Y$ implies $S(X) \leqslant S(Y)$, even when X and Y do not belong to the same state space.*

Proof. The proof is a simple application of the Hahn–Banach Theorem. Consider the set $\mathcal{S}$ of all pairs of state spaces (Γ, Γ'). On $\mathcal{S}$ we define an equivalence relation by declaring (Γ, Γ') to be equivalent to $(\Gamma \times \Gamma_0, \Gamma' \times \Gamma_0)$ for all Γ_0. Denote by $[\Gamma, \Gamma']$ the equivalence class of (Γ, Γ') and let $\mathcal{L}$ be the set of all these equivalence classes.

On $\mathcal{L}$ we define multiplication by scalars and addition in the following way:

$$t[\Gamma, \Gamma'] := [t\Gamma, t\Gamma'] \quad \text{for } t > 0, \tag{8.134}$$

$$t[\Gamma, \Gamma'] := [-t\Gamma', -t\Gamma] \quad \text{for } t < 0, \tag{8.135}$$

$$0[\Gamma, \Gamma'] := [\Gamma, \Gamma] = [\Gamma', \Gamma'], \tag{8.136}$$

$$[\Gamma_1, \Gamma_1'] + [\Gamma_2, \Gamma_2'] := [\Gamma_1 \times \Gamma_2, \Gamma_1' \times \Gamma_2']. \tag{8.137}$$

With these operations $\mathcal{L}$ becomes a vector space, which is infinite dimensional in general. The zero element is the class $[\Gamma, \Gamma]$ for any Γ, because by our definition of the equivalence relation, (Γ, Γ) is equivalent to $(\Gamma \times \Gamma', \Gamma \times \Gamma')$, which in turn is equivalent to (Γ', Γ'). Note that for the same reason $[\Gamma', \Gamma]$ is the negative of $[\Gamma, \Gamma']$.

Next, we define a function H on $\mathcal{L}$ by

$$H([\Gamma, \Gamma']) := F(\Gamma, \Gamma'). \tag{8.138}$$

Because of (8.132), this function is well defined and it takes values in $(-\infty, \infty]$. Moreover, it follows from (8.130) and (8.131) that H is homogeneous, i.e.

$$H(t[\Gamma, \Gamma']) = tH([\Gamma, \Gamma']),$$

and subadditive, i.e.

$$H([\Gamma_1, \Gamma_1'] + [\Gamma_2, \Gamma_2']) \leqslant H([\Gamma_1, \Gamma_1']) + H([\Gamma_2, \Gamma_2']).$$

Likewise,

$$G([\Gamma, \Gamma']) := -F(\Gamma', \Gamma) \qquad (8.139)$$

is homogeneous and superadditive, i.e. $G([\Gamma_1, \Gamma_1'] + [\Gamma_2, \Gamma_2']) \geqslant G([\Gamma_1, \Gamma_1']) + G([\Gamma_2, \Gamma_2'])$. By (8.128) we have $G \leqslant F$ so, by the Hahn–Banach Theorem, there exists a real-valued *linear* function L on $\mathcal{L}$ lying between G and H; that is

$$-F(\Gamma', \Gamma) \leqslant L([\Gamma, \Gamma']) \leqslant F(\Gamma, \Gamma'). \qquad (8.140)$$

Pick any fixed Γ_0 and define

$$B(\Gamma) := L([\Gamma_0 \times \Gamma, \Gamma_0]). \qquad (8.141)$$

By linearity, L satisfies $L([\Gamma, \Gamma']) = -L(-[\Gamma, \Gamma']) = -L([\Gamma', \Gamma])$. We then have

$$B(\Gamma) - B(\Gamma') = L([\Gamma_0 \times \Gamma, \Gamma_0]) + L([\Gamma_0, \Gamma_0 \times \Gamma']) = L([\Gamma, \Gamma']) \quad (8.142)$$

and hence (8.126) is satisfied. $\square$

Our final remark concerns the remaining nonuniqueness of the constants $B(\Gamma)$. This indeterminacy can be traced back to the nonuniqueness of a linear functional lying between $-F(\Gamma', \Gamma)$ and $F(\Gamma, \Gamma')$ and has two possible sources. One is that some pairs of state spaces Γ and Γ' may not be connected, i.e. $F(\Gamma, \Gamma')$ may be infinite (in which case $F(\Gamma', \Gamma)$ is also infinite by axiom A16). The other is that there might be a true gap, i.e.

$$-F(\Gamma', \Gamma) < F(\Gamma, \Gamma') \qquad (8.143)$$

might hold for some state spaces, even if both sides are finite.

In nature only states containing the same amount of the chemical elements can be transformed into each other. Hence $F(\Gamma, \Gamma') = +\infty$ for many pairs of state spaces, in particular, for those that contain different amounts of some chemical element. The constants $B(\Gamma)$ are, therefore, never unique. For each equivalence class of state spaces (with respect to the relation of connectedness), we can define a constant that is arbitrary except for the proviso that the constants should be additive and extensive under composition and scaling of systems. In our world there are 92 chemical elements (or, strictly speaking, a somewhat larger number, N, since one should count different isotopes as different elements), and this leaves us with at least 92 free constants that specify the entropy of one gram of each of the chemical elements in some specific state.

The other possible source of nonuniqueness, a nontrivial gap (8.143) for systems with the same composition in terms of the chemical elements is, as far as we know, not realized in nature, although it is a logical possibility. The true situation seems rather to be the following. Every state space Γ is connected to a distinguished state space

$$\Lambda(\Gamma) = \lambda_1 \Gamma_1 \times \cdots \times \lambda_N \Gamma_N, \tag{8.144}$$

where the Γ_i are the state spaces of one mole of each of the chemical elements, and the numbers $(\lambda_1, \ldots, \lambda_N)$ specify the amount of each chemical element in Γ. We have

$$\Lambda(t\Gamma) = t\Lambda(\Gamma) \tag{8.145}$$

and

$$\Lambda(\Gamma \times \Gamma') = \Lambda(\Gamma) \times \Lambda(\Gamma'). \tag{8.146}$$

Moreover (and this is the crucial 'experimental fact'),

$$-F(\Lambda(\Gamma), \Gamma) = F(\Gamma, \Lambda(\Gamma)) \tag{8.147}$$

for all Γ. Note that (8.147) is subject to experimental verification by measuring on the one hand entropy differences for processes that synthesize chemical compounds from the elements (possibly through many intermediate steps and with the aid of catalysts), and on the other hand for processes where chemical compounds are decomposed into the elements. This procedure need not invoke any semipermeable membranes.

It follows from (8.128), (8.133), and (8.147) that

$$F(\Gamma, \Gamma') = F(\Gamma, \Lambda(\Gamma)) + F(\Lambda(\Gamma), \Gamma') \tag{8.148}$$

and

$$-F(\Gamma', \Gamma) = F(\Gamma, \Gamma') \tag{8.149}$$

for all Γ' that are connected to Γ. Moreover, an explicit formula for $B(\Gamma)$ can be given:

$$B(\Gamma) = F(\Gamma, \Lambda(\Gamma)). \tag{8.150}$$

If $F(\Gamma, \Gamma') = \infty$, then (8.126) holds trivially, while for connected state spaces Γ and Γ' we have by (8.148) and (8.149)

$$B(\Gamma) - B(\Gamma') = F(\Gamma, \Gamma') = -F(\Gamma', \Gamma), \tag{8.151}$$

i.e. the inequalities (8.126) are saturated. It is also clear that in this case $B(\Gamma)$ is unique up to the choice of arbitrary constants for the fixed systems $\Gamma_1, \ldots, \Gamma_N$. The particular choice (8.150) corresponds to putting $B(\Gamma_i) = 0$ for the chemical elements $i = 1, \ldots, N$.

In conclusion, once the entropy constants for the chemical elements have been fixed and a temperature unit has been chosen (to fix the multiplicative constants), the universal entropy is completely fixed.

Acknowledgements

We are indebted to many people for helpful discussions, including Fred Almgren, Thor Bak, Bernard Baumgartner, Pierluigi Contucci, Roy Jackson, Anthony Knapp, Martin Kruskal, Mary Beth Ruskai, and Jan Philip Solovej. We thank Daniel Goroff for drawing our attention to the work by Herstein and Milnor (1953).

8.2 Some Speculations and Open Problems

1. As we have stressed, the purpose of the entropy function is to quantify the list of equilibrium states that can evolve from other equilibrium states. The evolution can be arbitrarily violent, but always $S(X) \leqslant S(Y)$ if $X \prec Y$. Indeed, the early thermodynamicists understood the meaning of entropy as defined for equilibrium states. Of course, in the real world, one is often close to equilibrium without actually being there, and it makes sense to talk about entropy as a function of time, and even space, for situations close to equilibrium. We do a similar thing with respect to temperature, which has the same problem that temperature is only strictly defined for a homogeneous system in equilibrium. At some point the thought arose (and we confess our ignorance about how it arose and by whom) that it ought to be possible to define an entropy function rigorously for manifestly *non*equilibrium states in such a way that the numerical value of this function will increase with time as a system goes from one equilibrium state to another.

Despite the fact that most physicists believe in such a nonequilibrium entropy, it has so far proved impossible to define it in a clearly satisfactory way. (For example, Boltzmann's famous H-Theorem shows the steady increase of a certain function called H. This, however, is not the whole story, as Boltzmann himself knew; for one thing, $H \neq S$ in equilibrium (except for ideal gases), and, for another, no one has so far proved the increase without making severe assumptions, and then only for a short time interval (cf. Lanford 1975).) Even today, there is no universal agreement about what, precisely, one should try to prove (as an example of the ongoing discussion, see Lebowitz et al. (1999)).

It is not clear if entropy can be consistently extended to nonequilibrium situations in the desired way. After a century and a half of thought, including the rise of the science of statistical mechanics as a paradigm (which was not available to the early thermodynamicists and, therefore, outside their thoughts), we are far from success. It has to be added, however, that a great deal of progress in understanding the problem has been made recently (e.g. Gallavotti 1999).

If such a concept can be formulated precisely, will it have to involve the notion of atoms and statistical mechanical concepts, or can it be defined irrespective of models, as we have done for the entropy of equilibrium states? This is the question we pose.

There are several major problems to be overcome, and we list two of them.

(a) The problem of time reversibility. If the concept is going to depend upon mechanical models, we have to account for the fact that both classical and quantum mechanics are time reversible. This makes it difficult to construct a mechanical quantity that can only increase under classical or quantum mechanical time evolution. Indeed, this problem usually occupies center stage in most discussions of the subject, but it might, ultimately, not be the most difficult problem after all.

(b) In our view of the subject, a key role is played by the idea of a more or less arbitrary (peaceful or violent) interaction of the system under discussion with the rest of the Universe whose final result is a change of the system, the change in height of a weight, and nothing more. How can one model such an arbitrary interaction in a mechanical way? By means of an additional term in the Hamiltonian? That hardly seems like a reasonable way to model a sledgehammer that happens to fall on the system or a gorilla jumping up and down. Most discussions of entropy increase refer to the evolution of systems subject to a dynamical evolution that is usually Hamiltonian (possibly time dependent) or a mixture of Hamiltonian and stochastic evolution. This can hardly even cope with describing a steam engine, much less a random, violent external force.

As a matter of fact, most people would recognize (a) as the important problem, now and in the past. In (b) we interject a new note, which, to us, is possibly more difficult. There are several proposals for a resolution of the irreversibility problem, such as the large number ($10^{23} \approx \infty$) of atoms involved, or the 'sensitive dependence on initial conditions' (one can shoot a ball out of a cannon, but it is very difficult to shoot it back into the cannon's mouth). Problem (b), in contrast, has not received nearly as much attention.

2. An essential role in our story was played by Axioms A4 and A5, which require the possibility of having arbitrarily small samples of a given material and that these small samples behave in exactly the same way as a 1 kg sample. While this assumption is made in everyone's formulation of the second law, we have to recognize that absurdities will arise if we push the concept to its extreme. Eventually, the atomic nature of matter will reveal itself and entropy will cease to have a clear meaning. What protects us is the huge power of 10 (e.g. 10^{23}) that separates macroscopic physics from the realm of atoms.

Likewise, a huge power of 10 separates time scales that make physical sense in the ordinary macroscopic world and time scales (such as 10^{25} s = 10^7 times the age of the Universe) which are needed for atomic fluctuations to upset the time evolution one would obtain from macroscopic dynamics. One might say that one of the hidden assumptions in our (and everyone else's) analysis is that $\prec$ *is reproducible*, i.e. $X \prec Y$ either holds or it does not, and there are no hidden stochastic or probabilistic mechanisms that would make the list of pairs $X \prec Y$ 'fuzzy.'

One of the burgeoning areas of physics research is 'mesoscopics,' which deals with the interesting properties of tiny pieces of matter that might contain only a million atoms (= a cube of 100 atoms on a side) or less. At some point the second law has to get fuzzy and a significant open problem is to formulate a fuzzy version of what we have done in Lieb and Yngvason (1999). Of course, no amount of ingenuity with mesoscopic systems is allowed to violate the second law on the macroscopic level, and this will have to be taken into account. One possibility could be that an entropy function can still be defined for mesoscopic systems but that $\prec$ is fuzzy, with the consequence that entropy increases only on 'the average,' but in a totally unpredictable way—so that the occasional decrease of entropy cannot be utilized to violate the second law on the macroscopic level.

There are other problems as well. A simple system, such as a container of hydrogen gas, has states described by energy and volume. For a mesoscopic quantity of matter, this may not suffice to describe an equilibrium state. Another problem is the meaning of equilibrium and the implicit assumption we made that after the (violent) adiabatic process is over the system will eventually come to some equilibrium state in a time scale that is short compared to the age of the Universe. On the mesoscopic level, the achievement of equilibrium may be more delicate because a mesoscopic system might never settle down to a state with insignificant fluctuations that one would be pleased to call an equilibrium state.

To summarize, we have listed two (and there are surely more) areas in which more thought, both mathematical and physical, is needed: the extension of the second law and the entropy concept to (1) nonequilibrium situations and (2) mesoscopic and even atomic situations. One might object that the problems cannot be solved until the 'rules of the game' are made clear, but discovering the rules is part of the problem. That is sometimes inherent in mathematical physics, and that is one of the intellectual challenges of the field.

8.3 Some Remarks about Statistical Mechanics

We are frequently asked about the connection between our approach to the second law of thermodynamics and the statistical mechanical Boltzmann–Gibbs–Maxwell approach. Let us make it clear that we value statistical mechanics as much as any physicist. It is a powerful tool for understanding physical phenomena and for calculating many quantities, especially in systems at or near equilibrium. It is used to calculate entropy, specific and latent heats, phase transition properties, transport coefficients and so on, often with good accuracy. Important examples abound, such as Max Planck's 1901 realization (Planck 1901) that by staring into a furnace he could find Avogadro's number, or Linus Pauling's highly accurate back-of-the-envelope calculation of the residual entropy of ice (Pauling 1935) in 1935. But is statistical mechanics essential for the second law?

In any event, it is still beyond anyone's computational ability (except in idealized situations) to account for this very precise, essentially infinitely accurate law of physics from statistical mechanical principles. No exception has ever been found to the second law of thermodynamics—not even a tiny one. Like conservation of energy (the 'first' law) the existence of a law so precise and so independent of details of models must have a logical foundation that is independent of the fact that matter is composed of interacting particles. Our aim in Lieb and Yngvason (1999) was to explore that foundation. It was also our aim to try to formulate clear statements on the macroscopic level so that statistical mechanics can try to explain them in microscopic terms.

As Albert Einstein put it (Einstein 1970),

> A theory is the more impressive the greater the simplicity of its premises is, the more different kinds of things it relates, and the more extended is its area of applicability. Therefore the deep impression which classical thermodynamics made upon me. It is the only physical theory of universal content concerning which I am convinced that, within the framework of the applicability of its basic concepts, it will never be overthrown.

We maintain that the second law, as understood for equilibrium states of macroscopic systems, does not require statistical mechanics, or any other particular mechanics, for its existence. It does require certain properties of macroscopic systems, and statistical mechanics is one model that, hopefully, can give those properties, such as irreversibility. One should not confuse the existence, importance, and usefulness of the Boltzmann–Gibbs–Maxwell theory with its necessity on the macroscopic level as far as the second law is concerned. Another way to make the point is this. If the statistical mechanics of atoms is essential for the second law, then that law must imply something about atoms and their dynamics. Does the second law prove the existence of atoms in the way that light scattering, for example, tells us what Avogadro's number has to be? Does the law distinguish between classical and quantum mechanics? The answer to these and similar questions is 'no' and, if there were a direct connection, the late 19th-century wars about the existence of atoms would have been won much sooner. Alas, there is no such direct connection that we are aware of, despite the many examples in which atomic constants make an appearance at the macroscopic level such as Planck's radiation formula mentioned above, the Sackur–Tetrode equation, stability of matter with Coulomb forces, and so on. The second law, however, is not such an example.

References

Boyling, J. G. 1972 An axiomatic approach to classical thermodynamics. *Proc. R. Soc. Lond.* A **329**, 35–70.

Buchdahl, H. A. 1966 *The Concepts of Classical Thermodynamics*. Cambridge University Press.

Carathéodory, C. 1909 Untersuchung über die Grundlagen der Thermodynamik. *Math. Annalen* **67**, 355–386.

Cooper, J. L. B. 1967 The foundations of thermodynamics. *J. Math. Analysis Appl.* **17**, 172–193.

Duistermaat, J. J. 1968 Energy and entropy as real morphisms for addition and order. *Synthese* **18**, 327–393.

Einstein, A. 1970 Autobiographical notes. In *Albert Einstein: Philosopher-Scientist* (ed. P. A. Schilpp). Library of Living Philosophers, vol. VII, p. 33. Cambridge University Press.

Fermi, E. 1956 *Thermodynamics*, p. 101. Dover, New York.

Gallavotti, G. 1999 *Statistical Mechanics; a Short Treatise*. Springer Texts and Monographs in Physics.

Giles, R. 1964 *Mathematical Foundations of Thermodynamics*. Pergamon, Oxford.

Herstein, I. N. and Milnor, J. 1953 An axiomatic approach to measurable utility. *Econometrica* **21**, 291–297.

Lanford III, O. E. 1975 *Time evolution of large classical systems* (ed. J. Moser). Springer Lecture Notes in Physics, vol. 38, pp. 1–111.

Lebowitz, J. L., Prigogine, I. and Ruelle, D. 1999 Round table on irreversibility. In *Statistical Physics*, vol. XX, pp. 516–527, 528–539, 540–544. North-Holland.

Lieb, E. H. 1999 Some problems in statistical mechanics that I would like to see solved. 1998 IUPAP Boltzmann Prize Lecture. *Physica A* **263**, 491–499.

Lieb, E. H. and Yngvason, J. 1998 A guide to entropy and the second law of thermodynamics. *Notices Am. Math. Soc.* **45**, 571–581; mp_arc 98–339; arXiv math-ph/9805005. (This paper received the American Mathematical Society 2002 Levi Conant prize for 'the best expository paper published in either the Notices of the AMS or the Bulletin of the AMS in the preceding five years'.)

Lieb, E. H. and Yngvason, J. 1999 The physics and mathematics of the second law of thermodynamics. *Phys. Rep.* **310**, 1–96; Erratum **314**, 669; arXiv cond-mat/9708200; http://www.esi.ac.at/ESI-Preprints.html #469.

Lieb, E. H. and Yngvason, J. 2000a A fresh look at entropy and the second law of thermodynamics. *Phys. Today* **53**, 32–37; mp_arc 00-123; arXiv math-ph/0003028. (See also Letters to the Editor, *Phys. Today* **53** (2000), 11–14, 106.)

Lieb, E. H. and Yngvason, J. 2000b The mathematics of the second law of thermodynamics. In *Visions in Mathematics, Towards 2000* (ed. A. Alon, J. Bourgain, A. Connes, M. Gromov and V. Milman). GAFA, Geom. Funct. Anal. Special Volume—GAFA, pp. 334–358; mp_arc 00-332.

Lieb, E. H. and Yngvason, J. 2002 The mathematical structure of the second law of thermodynamics. In *Contemporary Developments in Mathematics 2001* (ed. A. J. de Jong et al.), pp. 89–129. International Press.

Pauling, L. 1935 The structure and entropy of ice and of other crystals with some randomness of atomic arrangement. *J. Am. Chem. Soc.* **57**, 2680–2684.

Planck, M. 1901 Über die Elementarquanta der Materie und der Elektrizität. *Annalen Phys.* **4**, 564–566.

Planck, M. 1926 Über die Begrundung des zweiten Hauptsatzes der Thermodynamik. *Sitzungsberichte der preußischen Akademie der Wissenschaften, Math. Phys. Klasse*, pp. 453–463.

Roberts, R. D. and Luce, F. S. 1968 Axiomatic thermodynamics and extensive measurement. *Synthese* **18**, 311–326.

Uffink, J. 2001 Bluff your way in the second law of thermodynamics. *Stud. Hist. Phil. Mod. Phys.* **32**, 305–394. (See also Uffink's contribution in this book, Chapter 7.)

PART 3
Entropy in Stochastic Processes

Chapter Nine

Large Deviations and Entropy

S. R. S. Varadhan
Courant Institute of Mathematical Sciences

We describe how entropy plays a central role in the calculation of probabilities of large deviations. This is illustrated by several examples.

9.1 Where Does Entropy Come From?

The entropy functional $h(p) = h(p_1, \ldots, p_m) = -\sum_j p_j \log p_j$, defined on the simplex of probability distributions on m points, i.e. on the simplex

$$\left\{ p = (p_1, \ldots, p_m) : p_i \geq 0, \ \sum_i p_i = 1 \right\},$$

is usually characterized abstractly as the only function that satisfies certain natural properties (Khinchin 1957). However, for a probabilist the natural way to meet entropy is in a combinatorial computation, while approximating factorials with the use of Stirling's formula (Feller 1957).

Let us a toss the mythical fair coin n times and observe the number of heads. The number of ways of getting k heads is $\binom{n}{k}$. If we use Stirling's approximation for factorials, then we obtain

$$\binom{n}{k} \simeq \frac{\sqrt{2\pi} e^{-n} n^{n+1/2}}{\sqrt{2\pi} e^{-k} k^{k+1/2} \sqrt{2\pi} e^{-(n-k)} (n-k)^{n-k+1/2}}$$

$$= \frac{\sqrt{n}}{\sqrt{2\pi k(n-k)}} \exp\left[n \left[-\frac{k}{n} \log \frac{k}{n} - \frac{n-k}{n} \log \frac{n-k}{n} \right] \right].$$

If we fix the proportion of heads and tails at approximately p and $q = 1 - p$, then

$$\binom{n}{k} \simeq \frac{1}{\sqrt{2\pi npq}} e^{nh(p,q)}.$$

The entropy is the factor in the exponential growth rate of the 'volume,' or in this case the number of distinct outcomes in the space of coin tosses that corresponds to a given number k of heads. The actual probability in the context of tossing a fair coin is

$$P_n(k) \simeq \frac{1}{\sqrt{2\pi npq}} e^{n[h(p,q) - \log 2]}.$$

Let us look at the situation when an 'unfair' coin with probabilities α for heads and $1 - \alpha$ for tails is tossed. Then

$$P_n(\alpha, k) \simeq \frac{1}{\sqrt{2\pi npq}} e^{n[h(p,q) + p\log\alpha + (1-p)\log(1-\alpha)]}.$$

The exponential constant $h(p, q) + p \log \alpha + (1 - p) \log(1 - \alpha)$ is a sum of two terms, both depending on $p = 1 - q = k/n$. The first term $h(p, q)$ is the 'volume' term that we have already seen and does not depend on α. The second term is the 'energy' term and is a function of p and α. The most likely state for k corresponds to the value of p that maximizes the combination $h(p, q) - E(p, q)$, which is $p = \alpha$. Deviations of p away from α are called 'large deviations.' Their probabilities decay exponentially as

$$\exp[-nI(\alpha; p)],$$

where

$$I(\alpha; p) = p \log \frac{p}{\alpha} + (1 - p) \log \frac{1 - p}{1 - \alpha} = E(p, q) - h(p, q). \quad (9.1)$$

The large-deviation rate function is the rate of decay given by (9.1). $I(\alpha; p) \geq 0$ and in fact $I(\alpha; p) > 0$ unless $p = \alpha$.

While it is difficult to extend the definition of the entropy functional $h(p)$ to arbitrary probability distributions, the 'relative entropy' $I_\alpha(p)$ has a definition that makes sense for any pair of probability distributions α and β on an arbitrary measurable space. If $\beta \ll \alpha$, with a Radon–Nikodym derivative

$$f(x) = \frac{d\beta}{d\alpha} \quad \text{and} \quad \int f(x) \log f(x) \, d\alpha(x) < \infty,$$

then

$$I(\alpha; \beta) = \int f(x) \log f(x) \, d\alpha(x) = \int \log f(x) \, d\beta(x). \quad (9.2)$$

Otherwise, $I_\alpha(\beta) = +\infty$.

There are large-deviation rate functions that at first glance do not seem to come directly from entropy. Cramér's Theorem (Cramér 1938) is an example. Consider a sequence $X_1, X_2, \ldots, X_n, \ldots$ of i.i.d. random variables. Assume that the moment-generating function

$$M(\theta) = E[\exp[\theta X]]$$

is finite for all $\theta \in \mathbb{R}$. Let $\phi(a)$ be the Legendre transform of $\psi(\theta) = \log M(\theta)$:

$$\phi(a) = \sup_\theta [a\theta - \psi(\theta)] = \sup_\theta [a\theta - \log M(\theta)]. \quad (9.3)$$

Then $\phi(a) \geq 0$ and $\phi(a) = 0$ if and only if $a = E[X]$. The 'law of large numbers' tells us that the probabilities

$$P\left[\frac{X_1 + \cdots + X_n}{n} \sim a\right]$$

LARGE DEVIATIONS AND ENTROPY

of large-deviations decay, away from $a = E[X]$. In fact, they do so exponentially, with an explicit rate constant

$$P\left[\frac{X_1 + \cdots + X_n}{n} \sim a\right] \sim \exp[-\phi(a)]$$

with $\phi(a)$ given by (9.3).

The precise formulation of the above relation is as follows. A sequence of probability distributions $\{P_n\}$ on a Polish space (complete separable metric space) $\mathcal{X}$ is said to satisfy a 'large-deviation principle' with rate n and rate function $I(x)$ if the following are valid.

1. The rate function $I(x)$ is lower semicontinuous, nonnegative and has level sets $K_\ell = \{x : I(x) \leq \ell\}$ that are compact in $\mathcal{X}$.

2. Moreover,

$$\limsup_{n\to\infty} \frac{1}{n} \log P_n(C) \leq -\inf_{x\in C} I(x) \quad \text{for closed } C \subset \mathcal{X}, \quad (9.4)$$

$$\liminf_{n\to\infty} \frac{1}{n} \log P_n(G) \geq -\inf_{x\in G} I(x) \quad \text{for open } G \subset \mathcal{X}. \quad (9.5)$$

It looks like that, in the context of Cramér's Theorem, Legendre transforms control large-deviation rate functions rather than entropy. But, as we shall see in the next section, this is deceptive.

One of the consequences of the validity of a large-deviation principle is a Laplace asymptotic formula for certain types of integrals. If P_n satisfies a large-deviation principle with a rate function $I(x)$ on a Polish space $\mathcal{X}$ and if $F : \mathcal{X} \to \mathbb{R}$ is a bounded continuous function, then

$$\lim_{n\to\infty} \frac{1}{n} \log E^{P_n}[\exp[nF(x)]] = \sup_{x\in\mathcal{X}}[F(x) - I(x)]. \quad (9.6)$$

9.2 Sanov's Theorem

In the multinomial case, $X_1, X_2, \ldots, X_n$ are n independent random variables that take values $1, 2, \ldots, m$ with probabilities $\pi_1, \pi_2, \ldots, \pi_m$ and the frequencies $f_1, f_2, \ldots, f_m$ are observed. The probabilities of large deviations for the empirical probabilities $\{f_j/n\}$ are

$$P_n\left[\left\{\frac{f_j}{n}\right\} \sim \{p_j\}\right] = \exp\left[-n\sum_j p_j \log \frac{p_j}{\pi_j} + o(n)\right]$$

$$= \exp[-nI(\{\pi_j\}; \{p_j\}) + o(n)].$$

This extends to the empirical distribution

$$\nu_n = \frac{1}{n}\sum_{r=1}^{n} \delta_{X_r} \quad (9.7)$$

of n identically distributed random variables on $\mathbb{R}$ with a common distribution μ viewed as a random measure on $\mathbb{R}$. The large-deviation rate function takes the form

$$P[\nu_n \simeq \nu] = \exp[-n\, I(\mu; \nu) + o(n)],$$

where $I(\mu; \nu)$ was defined in (9.2). The connection to Cramér's Theorem is provided by the contraction principle

$$\inf_{\nu:\int x\, d\nu = a} I(\mu; \nu) = \phi(a)$$

and the infimum is attained when

$$d\nu = \frac{e^{\theta x}}{M(\theta)}\, d\mu$$

and θ is chosen so that

$$\psi'(\theta) = \frac{M'(\theta)}{M(\theta)} = a.$$

9.3 What about Markov Chains?

Let $\pi_{i,j}$ be the transition probabilities of a Markov chain on a finite state space with k points. We want to compute the probabilities of large deviations for the empirical distribution ν_n of the proportion of times the chain spends at different sites. The rate function is easily computed as

$$\mathcal{I}_\pi(\nu) = \sup_V \left[\sum_j V(j)\nu(j) - \lambda(V) \right],$$

where the supremum is taken over all vectors $V = \{V(1), V(2), \ldots, V(k)\}$. Here

$$\lambda(V) = \log \rho(V)$$

and $\rho(V)$ is the eigenvalue with the largest modulus of the matrix

$$\pi_{i,j}^V = \pi_{i,j} e^{V(j)}$$

(which is positive because the matrix has nonnegative entries). It is not clear what the spectral radius of nonnegative matrices has to do with entropy. But again there is a connection. Instead of considering just the one-dimensional marginals $\nu_n(j)$, we could consider the frequencies of sites visited at consecutive times,

$$\nu_n^{(2)}(i, j) = \frac{1}{n} \sum_{r=0}^{n-1} \delta_{X_r}(i) \delta_{X_{r+1}}(j), \qquad (9.8)$$

LARGE DEVIATIONS AND ENTROPY

where we have adopted the convention that $X_n = X_0$. If F is the state space, then $v_n^{(2)}$ is a probability measure on $F \times F$ such that both marginals are equal to the empirical distribution v_n defined in (9.7). If $\mu(i)$ is the invariant measure for the chain, then $v_n^{(2)} \to \mu^{(2)}$ given by $\mu^{(2)}(i, j) = \mu(i)\pi_{i,j}$. We want to compute the rate function $I_2(q)$ defined for measures q on $F \times F$ with equal marginals that governs the probabilities of large deviations of the empirical frequencies $v_n^{(2)}(\cdot, \cdot)$ away from $\mu^{(2)}(\cdot, \cdot)$:

$$P[v_n^{(2)} \sim q] = \exp[-nI_2(q) + o(n)].$$

For each q define q_1 to be its marginal and $\hat{q}$ on $F \times F$ by $\hat{q}(i, j) = q_1(i)\pi_{i,j}$. Then

$$I_2(q) = I(\hat{q}; q) = \sum_{i,j} q(i, j) \log \frac{q(i, j)}{\hat{q}(i, j)} \qquad (9.9)$$

and we are back with our relative entropy. In particular,

$$\inf_{q:q_1=v} I_2(q) = I_\pi(v).$$

9.4 Gibbs Measures and Large Deviations

All of this can be understood in terms of large deviations of Gibbs measures in one dimension. Let F be a finite set and $\Omega = F^{\mathbb{Z}}$ be the set of maps $\omega : \mathbb{Z} \to F$ of the set of integers $\mathbb{Z}$ into F. $T : \Omega \to \Omega$ is the shift operator defined by $(T\omega)(n) = \omega(n+1)$. If $\varphi(\omega)$ is a local function of ω, i.e. a function depending only on a finite number of coordinates $\{\omega(j) : |j| \leq k\}$, the Gibbs measure P_φ is the (unique) limit as $N \to \infty$ of normalized probability measures $P_{N,\varphi}$ on $F^{[-N, N]}$, defined by

$$P_{N,\varphi}[\omega(-N)\ldots, \omega(N)] \simeq \exp\left[-\sum_{j=-N}^{N} \varphi(T^j\omega)\right].$$

In principle, for finite N, these measures depend on some boundary effects, arising from the edges in the summation $\sum_{j=-N}^{N} \varphi(T^j\omega)$. But, in one dimension as $N \to \infty$, these effects disappear and we get a unique, well defined, stationary probability measure P_φ on $\Omega = F^{\mathbb{Z}}$ corresponding to any local function φ. A special P, which we denote by P_0, corresponds to $\varphi \equiv 0$ and is the product measure on Ω of the uniform distribution on the finite set F. If Q is any stationary measure on Ω, we have its marginal distribution Q_N on $\Omega_N = F^{[-N, N]}$ with entropy $h_N(Q)$ given by

$$-\sum_{\omega_N \in \Omega_N} Q_N(\omega_N) \log Q(\omega_N)$$

and the Kolmogorov–Sinai entropy $H(Q)$ of Q, defined by

$$H(Q) = \lim_{N \to \infty} \frac{1}{2N+1} h_N(Q),$$

can be easily shown to exist (by considerations of subadditivity). The largest possible value for $H(Q)$ is $\log k$ and is achieved only when $Q = P_0$, the product of uniform distributions.

The 'free energy' $\psi_0(g)$ for a local function g (relative to P_0) is defined by

$$\psi_0(g) = \lim_{N \to \infty} \frac{1}{2N+1} \log E^{P_0}\left[\exp\left[\sum_{j=-N}^{N} g(T^j \omega)\right]\right]. \tag{9.10}$$

The empirical process $R_n(\omega)$ is defined by

$$R_n(\omega) = \frac{1}{n}[\delta_\omega + \delta_{T\omega} + \cdots + \delta_{T^{n-1}\omega}]$$

and is nearly stationary for large n. Moreover, the marginals of R_n are precisely the empirical distributions $v_n^{(1)}, v_n^{(2)}, \ldots$ with a negligible correction coming from the edges for large n. By the law of large numbers $R_n(\omega) \to P_0$ and the probabilities of large deviations from the limit P_0 are governed by the rate function

$$I_0(Q) = [\log k - H(Q)].$$

There is also a variational formula for the free energy, arising from the Laplace asymptotic formula (9.6)

$$\psi_0(\varphi) = \sup_{Q \in \mathcal{S}}[E^Q[\varphi(\omega)] - I_0(Q)],$$

where the supremum is taken over all translation-invariant probability measures Q on Ω. It is attained uniquely at some $Q = P_\varphi$ which is called the Gibbs measure with energy φ. The free energy relative to P_φ is defined as

$$\psi_\varphi(g) = \lim_{N \to \infty} \frac{1}{2N+1} \log E^{P_\varphi}\left[\exp\left[\sum_{j=-N}^{N} g(T^j \omega)\right]\right]$$

and is seen to equal

$$\psi_0(\varphi + g) - \psi_0(\varphi) = \sup_{Q \in \mathcal{S}}[E^{P_\varphi}[\varphi(\omega)] - I_\varphi(Q)]$$

with

$$I_\varphi(Q) = \psi_0(\varphi) - [E^Q[\varphi(\omega)] - I_0(Q)].$$

It is now clear that the large deviations of $R_n(\omega)$ under P_φ will be controlled by the rate function $I_\varphi(Q)$. If we are only interested in the marginals $v_n^{(1)}$ or $v_n^{(2)}$, the rate function will have to be contracted to yield

$$I_\varphi^{(1)}(\{p_j\}) = \inf_{Q \in \mathcal{S}_1(\{p_j\})} I_\varphi(Q), \tag{9.11}$$

where the infimum is taken over stationary (i.e. translation-invariant) probability measures Q on Ω with one-dimensional marginal $\{p_j\}$ or

$$I_\varphi^{(2)}(\{q_{i,j}\}) = \inf_{Q \in \mathcal{S}_2(\{q_{ij}\})} I_\varphi(Q), \qquad (9.12)$$

where the infimum is taken over stationary processes Q with two-dimensional marginal $\{q_{i,j}\}$.

If $\phi(\omega) = V(\omega(0))$, then P_φ is a product measure with $\pi_j \simeq e^{V(j)}$. The infimum in attained at the product measure $Q = \otimes^{\mathbb{Z}} \{p_j\}$ and $I_\varphi^{(1)}(\{p_j\})$ becomes the relative entropy $I(\{\pi_j\}; \{p_j\})$. We are back at Sanov's Theorem.

If $\phi(\omega) = V(\omega(0), \omega(1))$, then P_φ is Markov with transition a transition probability π that can be determined. Moreover, $\{q_{i,j}\}$ determines a unique Markov process Q_0 and $I_\varphi^{(2)}(\{q_{i,j}\})$, with the infimum attained at Q_0, is seen to equal $\mathcal{I}_2(q)$, which was defined in (9.9). See Lanford (1973) for more details.

9.5 Ventcel–Freidlin Theory

There are other large-deviation investigations, e.g. the Ventcel–Freidlin theory (Ventcel and Freidlin 1970). Here the context is very different. Consider a Markov process with generator

$$L_\varepsilon = \tfrac{1}{2}\varepsilon \Delta + b(x) \cdot \nabla$$

acting on smooth functions on $\mathbb{R}^d$. If the paths start from $x \in \mathbb{R}^d$, then they are given by the solution to the SDE

$$x_\varepsilon(t) = x + \int_0^t b(x_\varepsilon(s))\, ds + \sqrt{\varepsilon}\, \beta(t),$$

where $\beta(t)$ is the d-dimensional Brownian motion. As $\varepsilon \to 0$, $x_\varepsilon(t)$ converges to the solution of the ODE

$$x(t) = x + \int_0^t b(x(s))\, ds.$$

We are interested in the probabilities of large deviations, i.e.

$$P[x_\varepsilon(\cdot) \sim f(\cdot)],$$

where f is an arbitrary function $f(\cdot) : [0, T] \to \mathbb{R}^d$ with $f(0) = x$. These probabilities decay like $\exp[-(1/\varepsilon)\mathcal{I}(f)]$ and $\mathcal{I}(f)$ is calculated as

$$\mathcal{I}(f) = \frac{1}{2} \int_0^T \|f'(t) - b(f(t))\|^2\, dt.$$

This is also essentially an entropy calculation. One way to make sure that the solution converges to $f(t)$ is to replace the original generator by the new time-dependent generator

$$L_{\varepsilon, f} = \tfrac{1}{2}\varepsilon \Delta + f'(t) \cdot \nabla$$

corresponding to the SDE

$$y_\varepsilon(t) = x + \int_0^t f'(s)\,ds + \sqrt{\varepsilon}\,\beta(t).$$

The solution $y_\varepsilon(t)$ converges to $f(t)$. A formula, known to probabilists as Girsanov's formula, gives us the Radon–Nikodym derivative (on the σ-field up to time T) of the distribution μ_ε (on path space) of the solution of

$$dz(t) = b(t, z(t))\,dt + \sqrt{\varepsilon}\,d\beta(t)$$

relative to the distribution of λ_ε of $z(\cdot)$ given by $z(t) = \sqrt{\varepsilon}\beta(t)$, as

$$\frac{d\mu_\varepsilon}{d\lambda_\varepsilon}(z(\cdot)) = \exp\left[\frac{1}{\varepsilon}\int_0^T \langle b(t, z(t)), dz(t)\rangle - \frac{1}{2\varepsilon}\int_0^T \|b(t, z(t))\|^2\,dt\right],$$

where $\int_0^T \langle b(t, z(t)), dz(t)\rangle$ is the Itô stochastic integral. The relative entropy of the two measures P_ε of $x_\varepsilon(\cdot)$ and Q_ε of $y_\varepsilon(\cdot)$, which correspond to the choice of $b(t, x) = b(x)$ and $b(t, x) = f'(t)$ respectively, can now be easily computed as

$$I(P_\varepsilon; Q_\varepsilon) = E^{Q_\varepsilon}\left[\frac{1}{\varepsilon}\int_0^T \langle b(x(s)), dx(s)\rangle - \frac{1}{2\varepsilon}\int_0^T \|b(x(s))\|^2\,ds\right.$$
$$\left. - \frac{1}{\varepsilon}\int_0^T \langle f'(s), dx(s)\rangle + \frac{1}{2\varepsilon}\int_0^T \|f'(s)\|^2\,ds\right]$$
$$\simeq \frac{1}{2\varepsilon}\int_0^T |f'(s) - b(f(s))|^2\,ds.$$

9.6 Entropy and Large Deviations

In fact, it is not hard to see that any 'large deviation' has to be related to 'entropy.' If P is a measure and we want to estimate $P(A)$, then we can do it if we can estimate

$$\inf_{Q: Q(A)=1} I(P; Q) = \ell.$$

We have the inequality (Jensen) that says

$$E^Q[g] \leq I(P; Q) + \log E^P[\exp[g]]. \tag{9.13}$$

Taking $g = c\chi_A$ in (9.13), with $c > 0$, we get

$$Q(A) \leq \frac{1}{c}[I(P; Q) + \log[e^c P(A) + (1 - P(A))]$$

or

$$\ell \geq c - \log[1 - P(A) + e^c P(A)].$$

Letting $c \to \infty$ we get

$$\ell \geq -\log P(A)$$

or
$$P(A) \leq e^{-\ell}.$$

There is a partial converse. If $Q \ll P$ and $E^Q|\log dQ/dP - \ell|$ is small and $Q(A) = q$, then

$$P(A) \geq \int_A \frac{dP}{dQ} dQ = Q(A) \frac{1}{Q(A)} \int_A \exp\left[-\log \frac{dQ}{dP}\right] dQ$$

$$\geq q \exp\left[\frac{1}{q} \int_A \left[-\log \frac{dQ}{dP}\right] dQ\right]$$

$$\geq q \exp\left[-\ell - \frac{1}{q} E^Q \left|\log \frac{dQ}{dP} - \ell\right|\right].$$

Large deviations is then essentially estimating entropies, but the new measure Q has to be guessed. In a dynamical situation the dynamics has to be perturbed in order to get a Q with the right property. There are lots of choices and the entropy can be calculated by a Girsanov formula. Then the optimal choice becomes an optimal control problem with entropy as the cost functional (Dupuis and Ellis 1997; Fleming 1985).

Let us illustrate this by means of the examples in Sections 9.2–9.4.

If $X_1, \ldots, X_n, \ldots$ is a sequence of i.i.d. random variables with a common distribution α and we want the empirical distribution $(1/n) \sum_{r=1}^{n} \delta_{X_r}$ to be close to β, we can achieve this by changing the joint distribution of the $\{X_r\}$ from the product measure $P = \otimes^{\mathbb{Z}} \alpha$ to any stationary process Q with one-dimensional marginals equal to β. The relative entropy $I_n(P; Q)$ of Q with respect to P on the σ-field generated by the first n coordinates grows linearly in n and

$$\lim_{n \to \infty} \frac{1}{n} I_n(P; Q) = \mathcal{I}(P; Q)$$

exists and can be evaluated as

$$\mathcal{I}(P; Q) = E^Q[h(q_\omega, \alpha)],$$

where q_ω is the conditional distribution of X_1 given the past history $\{X_j : j \leq 0\}$ calculated under Q. The control problem reduces to the variational problem

$$\mathcal{I}(\beta) = \inf_{Q \in \mathcal{M}_\beta} [\mathcal{I}(P; Q)],$$

where $\mathcal{M}_\beta$ is the set of stationary processes Q, with one-dimensional marginals β. Self-consistency requires that for $Q \in \mathcal{M}_\beta$,

$$\int q_\omega dQ = \beta.$$

It is not hard to see at this point that, from the convexity of $\mathcal{I}(\cdot, \alpha)$, the infimum is attained when $Q = \otimes^{\mathbb{Z}} \beta$, and we arrive at Sanov's Theorem.

The advantage of this somewhat complex view point is that the principle is universal. If we replace the product measure P by a nicely mixing Markov chain with transition probabilities $\pi(x, dy)$, nothing changes except the formula for $\mathcal{I}(P; Q)$, which now reads

$$\mathcal{I}(P; Q) = E^Q[h(q_\omega, p_\omega)],$$

where, by our Markovian assumptions, $p_\omega = \pi(X_{-1}, dy)$. Let us carry out the optimization

$$\inf_{Q \in \mathcal{M}_\beta} \mathcal{I}(P; Q)$$

in two steps. We let Γ_β be the set of bivariate distributions γ with common marginals β and denote by $\mathcal{M}_\gamma^{(2)}$ the set of stationary processes Q such that its bivariate marginal at two consecutive times is γ. Then the infimum above can be carried out as

$$\inf_{\gamma \in \Gamma_\beta} \inf_{Q \in \mathcal{M}_\gamma} \mathcal{I}(P; Q).$$

Any $\gamma \in \Gamma_\beta$ can be disintegrated as $\beta(dx) \otimes \hat{\pi}(x, dy)$ through its marginal and conditional distribution. $\hat{\pi}(\cdot, \cdot)$ is a transition probability with β as an invariant measure, and therefore defines a canonical stationary Markov process Q_γ with (consecutive) bivariate marginals γ and one-dimensional marginals β. One can show that the inner infimum is attained at $Q = Q_\gamma$,

$$\mathcal{I}(Q_\gamma; P) = \mathcal{I}^{(2)}(\gamma) = \int I(\pi(x, \cdot), \hat{\pi}(x, \cdot)) \, \beta(dx),$$

and the rate function becomes

$$\mathcal{I}(\beta) = \inf_{\gamma \in \Gamma_\beta} \mathcal{I}^{(2)}(\gamma).$$

The duality relation

$$\sup_\beta \left[\int V(x) \, \beta(dx) - \mathcal{I}(\beta) \right] = \log \lambda(V),$$

where $\lambda(V)$ is the spectral radius of $\pi(x, dy) e^{V(x)}$, is just a piece of convex analysis and establishes the connection with Section 9.3. It goes more or less like this. We denote by

$$\Gamma = \bigcup_\beta \Gamma_\beta$$

the set of all bivariate distributions with equal marginals:

$$\sup_V \left[\int V(x)\,d\beta - \mathcal{I}(\beta) \right]$$
$$= \sup_\beta \sup_{\gamma \in \Gamma_\beta} \left[\int V(x)\,\gamma(dx, dy) - \mathcal{I}^{(2)}(\gamma) \right]$$
$$= \sup_{\gamma \in \Gamma} \left[\int V(x)\,\gamma(dx, dy) - \int \log \frac{\hat{\pi}(x, dy)}{\pi(x, dy)} \gamma(dx, dy) \right]$$
$$= \sup_\gamma \inf_U \left[\int [V(x) + U(x) - U(y)]\,\gamma(dx, dy) \right.$$
$$\left. - \int \log \frac{\hat{\pi}(x, dy)}{\pi(x, dy)} \gamma(dx, dy) \right]$$
$$= \inf_U \sup_\gamma \left[\int [V(x) + U(x) - U(y)]\,\gamma(dx, dy) \right.$$
$$\left. - \int \log \frac{\hat{\pi}(x, dy)}{\pi(x, dy)} \gamma(dx, dy) \right]$$
$$= \inf_U \left[\sup_x [V(x) + U(x) + \log \int e^{-U(y)} \pi(x, dy)] \right]$$
$$= \lambda(V).$$

Finally, the connection with the example in Section 9.5 is straightforward. We replace $b(x)$ with a new $c(t, x)$ and the SDE is

$$dx(t) = c(t, x(t))\,dt + \sqrt{\varepsilon}\,\beta(t).$$

We need c to be such that f is a solution of $\dot{x} = c(t, x(t))$ or $c(t, f(t)) = f'(t)$, which is really what we did.

9.7 Entropy and Analysis

Entropy plays a role in analysis as well. Let $p(t, x, y)$ be the transition probability density (relative to an invariant distribution q) of a Markov process. We have the Markov semigroup

$$(T_t f)(x) = \int f(y) p(t, x, y) q(dy)$$

acting on bounded measurable functions and the adjoint semigroup

$$(T_t^* g)(y) = \int g(x) p(t, x, y) q(dx)$$

acting on $L_1(q)$. Let L and L^* be the infinitesimal generators of the semigroups T_t and T_t^*, respectively. The function

$$u(t, x) = (T_t f)(x)$$

satisfies the evolution equation

$$\frac{\partial u}{\partial t} = Lu$$

and $g_t = T_t^* g_0$ satisfies the Fokker–Planck (Kolmogorov's forward) equation

$$\frac{\partial g}{\partial t} = L^* g$$

with L^* being the adjoint of L with respect to q, i.e.

$$\int (Lf) g \, dq = \int f(L^* g) \, dq.$$

The H-theorem asserts that the relative entropy $H(t) = \int g_t \log g_t \, dq$ is non-increasing in t. If we know that $H_0 = \int g_0 \log g_0 \, dq < \infty$, then

$$\int_0^\infty \left[-\frac{dH_t}{dt} \right] dt \leq H_0,$$

giving us a decay rate on $I(t) = -dH_t/dt$. An elementary calculation shows that

$$-\frac{d}{dt} H_t = -\frac{d}{dt} \int g_t \log g_t \, dq = -\int (L^* g_t) \log g_t \, dq = -\int g_t (L \log g_t) \, dq$$

$$= -2 \int g_t (L \log \sqrt{g_t}) \, dq \geq -2 \int g_t \frac{L \sqrt{g_t}}{\sqrt{g_t}} \, dq = 2\mathcal{D}(\sqrt{g_t}).$$

Here $\mathcal{D}$ is the Dirichlet form on $L_2(q)$, defined as

$$\mathcal{D}(g) = \langle g, -Lg \rangle_q = -\int g(x)(Lg)(x) q(dx) = -\int g(x)(L^* g)(x) q(dx),$$

and this provides us with a control

$$\int_0^\infty \mathcal{D}(\sqrt{g_t}) \, dt \leq \tfrac{1}{2} H_0.$$

Since usually $D(g) = 0$ only if g is a constant, a small value of the Dirichlet form implies that the state is close to equilibrium. This can be used to show that large systems with multiple equilibria approach local equilibria over large volumes with a length scale that is related to the time scale. If a log-Sobolev estimate is available, then this can be used to establish hydrodynamic scaling results. These typically derive a partial differential equation for the parameters that define the local equilibria as functions of rescaled space and time. We shall illustrate this in the next section. Even if there are no such inequalities available, one can still use the estimate on the Dirichlet form in very many ways (see, for instance, Guo et al. 1988).

9.8 Hydrodynamic Scaling: an Example

Hydrodynamic scaling refers to a wide class of phenomena where a large interacting system with conserved quantities and local interactions evolves over time. Because of the conserved quantities there will be multiple equilibria, which can usually be parametrized by the average values of the conserved quantities. The system will reach local equilibria first before evolving towards the global equilibrium rather slowly. If we speed up time, the local equilibria will be established over larger domains. At the appropriate rescaling of space and time, the evolution of the complex system can be fully described by describing the parametric label of the equilibrium as a function of space and time, which is determined for later times by an evolution equation from its initial values. This evolution equation, usually a parabolic or hyperbolic transport equation, is called the hydrodynamic limit of the complex interacting system.

Let us illustrate this by a familiar class of examples called simple exclusion processes. We will limit ourselves to the one-dimensional case. We start with the lattice $\mathbb{Z}$. At any given time we have particles at some of the sites and the associated variables $\eta(x)$, which are 1 if there is a particle at x and 0 otherwise. The term 'exclusion' signifies that we are not allowed to have more than one particle at the same time at any site. The particles wait for a random exponential waiting time (with mean 1) and then pick a random site to jump to. If the particle is at site x, then the probability of picking site y to jump to is $p(y-x)$. Of course, $\sum_{z \neq 0} p(z) = 1$. If the site y is selected, the particle jumps to the new site if it is empty. If the chosen site has a particle and is therefore unavailable, the jump is disallowed and the particle waits at the original site for a new exponential time. All of this can be specified by writing down the generator L acting on functions $f(\eta)$ that depend on the configuration η:

$$(Lf)(\eta) = \sum_{x,y} p(y-x)\eta(x)(1-\eta(y))[f(\eta^{x,y}) - f(\eta)].$$

Here $\eta^{x,y}$ is the new configuration obtained by exchanging the situations at x and y, i.e.

$$\eta^{x,y}(z) = \begin{cases} \eta(x) & \text{if } z = y, \\ \eta(y) & \text{if } z = x, \\ \eta(z) & \text{otherwise.} \end{cases} \qquad (9.14)$$

For simplicity, let us look at the periodic case where the lattice $\mathbb{Z}$ is replaced by $\mathbb{Z}_N$ of N sites with periodic boundary conditions. Let us also suppose that $p(z)$ is symmetric, i.e. $p(z) = p(-z)$. Let the initial configuration, which depends on N and could be random, be such that

$$\lim_{N \to \infty} \frac{1}{N} \sum_x J\left(\frac{x}{N}\right) \eta_0(x) = \int_T J(z) \rho_0(z)\, dz \qquad (9.15)$$

for some deterministic $0 \leqslant \rho_0(z) \leqslant 1$. In the random case the above limit is interpreted as limit in probability. Here the continuum limit of the rescaled lattice is the circle T, which is viewed as the unit interval with end points identified. Then the evolution has the property that, at time $N^2 t$,

$$\lim_{N \to \infty} \frac{1}{N} \sum_x J\left(\frac{x}{N}\right) \eta_{N^2 t}(x) = \int_T J(z) \rho(t, z) \, dz$$

in probability, where $\rho(\cdot, \cdot)$ is the solution of the heat equation

$$\rho_t = \tfrac{1}{2} \sigma^2 \rho_{xx} \qquad (9.16)$$

with initial condition $\rho(0, z) = \rho_0(z)$. Here $\sigma^2 = \sum_z z^2 p(z)$. In this context the heat equation (9.16) is called the hydrodynamic limit of the symmetric simple exclusion process.

It is fairly easy to establish this fact by a direct computation of two moments remembering that the speeded-up process has generator $N^2 L$. One expects that in the speeded-up time scale, for any local function $f(\eta)$, denoting by $f_x(\eta)$ its spatial translate by the lattice variable x,

$$\lim_{N \to \infty} \frac{1}{N} \int_0^T \left[\sum_x f_x(\eta_{N^2 s}) \right] ds = \int_0^T \int_T \hat{f}(\rho(t, z)) \, dz \, dt \qquad (9.17)$$

with $\hat{f}(\rho) = E^\rho[f(\eta)]$. Here E^ρ refers to the expectation with respect to the Bernoulli product measure on $\{0, 1\}^{\mathbb{Z}}$ with $E^\rho[\eta(x)] = \rho$. This result will attest to the fact that in the time scale N^2 the system is locally in a Bernoulli state over spatial scales that involve N lattice sites. This can be deduced from the estimate on the Dirichlet form in the following way: if we use the uniform distribution on 2^N sites as our q, then the relative entropy of any initial state with respect to q is at most $N \log 2$. Therefore, in the speeded-up time scale, since the Dirichlet form inherits a factor of N^2,

$$\int_0^\infty \mathcal{D}(\sqrt{g_t}) \, dt \leqslant \frac{\log 2}{2N}.$$

By convexity, if

$$\bar{g} = \frac{1}{T} \int_0^T g_t \, dt,$$

then

$$\mathcal{D}(\sqrt{\bar{g}}) \leqslant \frac{\log 2}{2NT}.$$

We want to use this estimate to show that on a large block of εN sites the conditional distribution under $\bar{q}$ of the configurations given the mean number m of particles in that block is close to the Bernoulli measure μ_m on that block. This can be controlled if we can control the relative entropy of the restriction of $\bar{q}$ to a block of size εN to a mixture of μ_m on that block. We have to use a mixture, because the average density m can be random under $\bar{q}$. Since the Dirichlet form is additive in volume, the restriction of $\bar{q}$ to a typical block of length εN will have a Dirichlet form of about $(\varepsilon \log 2)/N$. The log-Sobolev inequality is known to hold with a constant proportional to the square of the block size. Therefore, the relative entropy of the restriction of $\bar{q}$ to a typical block of size εN will have a relative entropy of order $\varepsilon^3 N$ relative to a suitable mixture of Bernoulli distribution in that block. The specific entropy is roughly ε^3 per site. This can be controlled uniformly in N provided ε is small. This is enough to prove (9.17).

We now turn to the totally asymmetric case where $p(1) = 1$ and $p(z) = 0$ for $z \neq 1$. In this case time is rescaled by a factor of N and we assume that the initial condition (9.15) holds as before. Then

$$\lim_{N \to \infty} \frac{1}{N} \sum_x J\left(\frac{x}{N}\right) \eta_{Nt}(x) = \int_T J(z) \rho(t, z) \, dz, \qquad (9.18)$$

where $\rho(t, z)$ is now a weak solution of Burgers equation

$$\frac{\partial \rho}{\partial t} + \frac{\partial [\rho(1 - \rho)]}{\partial z} = 0 \qquad (9.19)$$

on $[0, T] \times T$ with initial condition

$$\rho(0, z) = \rho_0(z).$$

There is no uniqueness for solutions of Burgers equation (9.19) in the class of weak solutions. See Chapter 6 by Dafermos (2003) in this book. While there is uniqueness within the class of regular solutions, in general, regular solutions do not exist or exist only up to a certain time when shocks develop. The hydrodynamic limit (9.18) is singled out as the unique weak solution satisfying the entropy condition (Rezakhanlou 1991), i.e. for every convex 'entropy' h and the corresponding 'flux' g, defined by

$$h'(\rho)(1 - 2\rho) = g'(\rho),$$

one has

$$\frac{\partial h(\rho)}{\partial t} + \frac{\partial g(\rho)}{\partial z} \leqslant 0 \qquad (9.20)$$

as a distribution. A special case is the entropy

$$h(\rho) = \rho \log \rho + (1 - \rho) \log(1 - \rho)$$

with corresponding flux

$$g(\rho) = \rho - \rho(1 - \rho) \log \frac{\rho}{1 - \rho}.$$

It turns out (Jensen 2000) that this particular entropy controls the probabilities of large deviations. If we want the empirical density to be close to an arbitrary

$\rho(\cdot, \cdot)$ in $[0, T]$, which may not be the entropy solution, this will be a large deviation. The rate function $\mathcal{I}(\rho)$ for such a large deviation is finite only if $\rho = \rho(t, z)$ is a weak solution of equation (9.19), and the distribution

$$\mu = \frac{\partial h(\rho)}{\partial t} + \frac{\partial g(\rho)}{\partial z}$$

is a signed measure of bounded variation. While the condition for the weak solution $\rho(\cdot, \cdot)$ to be an entropy solution is $\mu \leqslant 0$, the large-deviation rate function is given by

$$\mathcal{I}(\rho(\cdot, \cdot)) = \mu^+[[0, T] \times \boldsymbol{T}].$$

References

Cramér, H. 1938 Sur un noveau théorème—limites de la théorie des probabilités. *Actualités Scientifiques et Industrielles* **736**, 5–23. *Colloque consacré à la théorie des probabilités*, Vol. 3. Hermann, Paris

Dafermos, C. 2003 Entropy for hyperbolic conservation laws. In *Entropy* (ed. A. Greven, G. Keller and G. Warnecke), pp. 107–120. Princeton University Press.

Dupuis, P. and Ellis, R. S. 1997 *A Weak Convergence Approach to the Theory of Large Deviations*. Wiley Series in Probability and Statistics (Probability and Statistics Section). Wiley-Interscience.

Feller, W. 1957 *An Introduction to Probability Theory and Its Applications*, Vol. I, 2nd edn. Wiley, Chapman and Hall.

Fleming, W. H. 1985 A stochastic control approach to some large deviations problems. *Recent Mathematical Methods in Dynamic Programming* (Rome, 1984), pp. 52–66. Lecture Notes in Mathematics, vol. 1119. Springer.

Guo, M. Z., Papanicolaou, G. C. and Varadhan, S. R. S. 1988 Nonlinear diffusion limit for a system with nearest neighbor interactions. *Commun. Math. Phys.* **118**(1), 31–59.

Jensen, L. 2000 Large deviations of the asymmetric simple exclusion process. PhD thesis, New York University.

Khinchin, A. I. 1957 *Mathematical Foundations of Information Theory* (translated by R. A. Silverman and M. D. Friedman). Dover, New York.

Lanford III, O. E. 1973 Entropy and equilibrium states in classical statistical mechanics. In *Statistical Mechanics and Mathematical Problems* (ed. A. Lenard), pp. 1–113. Lecture Notes in Physics, 20. Springer.

Rezakhanlou, F. 1991 Hydrodynamic limit for attractive particle systems on Z^d. *Commun. Math. Phys.* **140**, 417–448.

Ventcel, A. D. and Freidlin, M. I. 1970 Small random perturbations of dynamical systems. *Russ. Math. Surv.* (translation of *Usp. Mat. Nauk*) **25**, 3–55.

Chapter Ten

Relative Entropy for Random Motion in a Random Medium

F. den Hollander
EURANDOM

In this chapter we explain with a concrete example how the notion of *relative entropy* is used to study random motion in a random medium. We consider an infinite system of particles on the integer lattice that (1) migrate to the right with a random delay, and (2) branch along the way according to a random law depending on their position. The initial configuration has one particle at each site. With the help of large-deviation theory, we compute the exponential growth rate of the average number of particles in a large box (the global growth rate) and the average number of particles at the origin (the local growth rate). Both these growth rates are expressed in terms of a variational problem. An analysis and comparison of these variational problems reveals various interesting phase transitions as a function of the underlying parameters. In the calculations *relative entropy* is used to compute the weight of the various growth strategies of the system and to identify the optimal growth strategy that determines the growth rate.

10.1 Introduction

10.1.1 Motivation

The contributions in this book deal with the different faces that entropy has in probability theory, ergodic theory, dynamical systems, information theory, statistical physics, and thermodynamics. In this chapter we focus on an application of *relative entropy* in probability theory and statistical physics, namely, to population growth in random media.

Over the past 10 years or so, I have been working with Andreas Greven on a program in which we study random processes that interact either with themselves or with a random medium and we try to describe their evolution using large-deviation theory, entropy techniques, and variational calculus. We have so far looked at random polymers, random walks in random potentials, systems of interacting diffusions, and population growth in random media (see den Hollander 2000, Part B, and references therein). In order to explain what the program is about, I have chosen the simplest example studied so far, namely, particles doing a one-sided migration on the integer lattice and branching according to a local offspring distribution that is chosen randomly and independently for different sites. This example is typical for the type of ideas and techniques that also arise in more complicated examples. This chapter is a review of the work in Baillon et al. (1993, 1994) and Greven and den Hollander (1991a,b, 1992, 1994).

We begin by formulating *Sanov's Theorem in large-deviation theory* (see den Hollander 2000, Theorem II.36), which is the main conceptual tool in what follows. Let $(Z_i)_{i \in \mathbb{N}}$ be i.i.d. random variables on a countable state space $\mathbb{S}$ with marginal probability law μ. For $n \in \mathbb{N}$, let

$$L_n = \frac{1}{n} \sum_{i=1}^{n} \delta_{Z_i} \tag{10.1}$$

be the empirical measure associated with $Z_1, \ldots, Z_n$. This is a random element of $\mathcal{P}(\mathbb{S})$, the set of probability measures on $\mathbb{S}$, which is endowed with the total variation distance. By the ergodic theorem, L_n tends to μ in $\mathcal{P}(\mathbb{S})$ as $n \to \infty$. Sanov's Theorem says that the family $(L_n)_{n \in \mathbb{N}}$ satisfies the large-deviation principle on $\mathcal{P}(\mathbb{S})$, i.e.

$$\left. \begin{array}{l} \limsup_{n \to \infty} \dfrac{1}{n} \log \mu^{\mathbb{N}}(L_n \in C) \leqslant - \inf_{\nu \in C} H(\nu \mid \mu) \quad \forall C \subset \mathcal{P}(\mathbb{S}) \text{ closed,} \\[6pt] \liminf_{n \to \infty} \dfrac{1}{n} \log \mu^{\mathbb{N}}(L_n \in O) \geqslant - \inf_{\nu \in O} H(\nu \mid \mu) \quad \forall O \subset \mathcal{P}(\mathbb{S}) \text{ open,} \end{array} \right\} \tag{10.2}$$

where the rate function $H(\cdot \mid \mu) \colon \mathcal{P}(\mathbb{S}) \to [0, \infty]$ is given by

$$H(\nu \mid \mu) = \sum_{s \in \mathbb{S}} \nu(s) \log \frac{\nu(s)}{\mu(s)}$$

$$= \text{the } \textit{relative entropy} \text{ of } \nu \text{ w.r.t. } \mu. \tag{10.3}$$

What (10.2) says is that for all $\nu \in \mathcal{P}(\mathbb{S})$ where H is continuous,

$$\mu^{\mathbb{N}}(L_n \approx \nu) = \exp[-H(\nu \mid \mu) n + o(n)], \quad n \to \infty. \tag{10.4}$$

Thus, Sanov's Theorem gives us a precise description of the large-deviation probabilities of L_n away from its typical value μ in terms of the rate function H. See Sections 3.3 and 9.2 for other appearances of Sanov's Theorem in this book.

It is important to note that $H(\nu \mid \mu) \geqslant 0$ with equality if and only if $\nu = \mu$. Moreover, $\nu \mapsto H(\nu \mid \mu)$ is strictly convex, lower semi-continuous, and has compact level sets (see den Hollander 2000, Lemma II.39). We may think of $H(\nu \mid \mu)$ as a kind of distance between ν and μ: the larger this distance the more costly the large deviation.

10.1.2 A Branching Random Walk in a Random Environment

In Sections 10.1.2–10.1.6 we describe the work in Baillon et al. (1993) and Greven and den Hollander (1991a).

With each $x \in \mathbb{Z}$ is associated a random probability measure F_x on $\mathbb{N}_0 = \mathbb{N} \cup \{0\}$, called the offspring distribution at site x. The sequence

$$F = \{F_x\}_{x \in \mathbb{Z}} \tag{10.5}$$

is i.i.d. with common distribution α. For fixed F, define a discrete-time Markov process $(\eta_n)_{n \in \mathbb{N}_0}$ on $\mathbb{N}_0^{\mathbb{Z}}$, with the interpretation

$$\left.\begin{aligned}\eta_n &= \{\eta_n(x)\}_{x \in \mathbb{Z}}, \\ \eta_n(x) &= \text{number of particles at site } x \text{ at time } n,\end{aligned}\right\} \tag{10.6}$$

by specifying its one-step transition mechanism as follows. Start from initial state $\eta_0 \equiv 1$. Given the state η_n at time n, the state η_{n+1} at time $n+1$ arises in two steps.

  (i) Each particle is independently replaced by a new generation. The size of a new generation descending from a particle at site x has distribution F_x. All particles branch independently.

  (ii) Immediately after creation, each new particle decides to jump one lattice spacing to the right with probability h or stand still with probability $1-h$, with $h \in (0, 1)$.

The sequence F plays the role of a *random environment* and stays fixed during the evolution. The parameter h is the *drift of the migration*. We write $\mathbb{P}$, $\mathbb{E}$ and P, E to denote the probability and expectation for the environment and the migration processes, respectively.

Let b_x denote the average offspring at site x and let β denote the distribution of b_x induced by α. Write $\mathbb{B}$ to denote the support of β. We assume that

$$\left.\begin{aligned}&\mathbb{B} \text{ is countable and not a singleton,} \\ &\mathbb{B} \text{ is bounded away from 0 and } \infty,\end{aligned}\right\} \tag{10.7}$$

and write

$$M = \text{the supremum of } \mathbb{B}. \tag{10.8}$$

10.1.3 Particle Densities and Growth Rates

We will consider the following two particle densities:

$$\left.\begin{aligned}d_n &= (\mathbb{E} \times E)(\eta_n(0)) = \textit{global} \text{ particle density at time } n, \\ \hat{d}_n(F) &= E(\eta_n(0) \mid F) = \textit{local} \text{ particle density at time } n.\end{aligned}\right\} \tag{10.9}$$

Here, note that for each n the sequence η_n is stationary and ergodic under $\mathbb{P} \times P$, so that by the ergodic theorem we have

$$d_n = \lim_{N \to \infty} \frac{1}{2N+1} \sum_{x=-N}^{N} \eta_n(x) \quad (\mathbb{P} \times P)\text{-a.s.}, \tag{10.10}$$

showing that d_n is indeed the particle density in a large box. We will be interested in computing the exponential growth rates

$$\lambda = \lim_{n \to \infty} \frac{1}{n} \log d_n, \quad \hat{\lambda}(F) = \lim_{n \to \infty} \frac{1}{n} \log \hat{d}_n(F). \tag{10.11}$$

These are identified in Theorems 10.1 and 10.2 below.

For $\theta \in [0, 1]$, let B_θ and π_θ denote, respectively, the Bernoulli and the geometric distributions with parameter θ given by

$$\left.\begin{aligned} B_\theta(0) &= 1 - \theta, & B_\theta(1) &= \theta, \\ \pi_\theta(i) &= \theta(1-\theta)^{i-1}, & i &\in \mathbb{N}. \end{aligned}\right\} \tag{10.12}$$

Theorem 10.1 (Baillon et al. 1993). *For $h \in (0, 1)$ and β as in (10.7):*

$$\lambda = \lambda(\beta, h) \tag{10.13}$$

with

$$\lambda(\beta, h) = \sup_{\theta \in (0,1]} \left[\theta \left\{ \sup_{\nu \in \mathcal{M}_\theta} [f(\nu) - H(\nu \mid \pi_\theta)] \right\} - H(B_\theta \mid B_h) \right], \tag{10.14}$$

where

$$\left.\begin{aligned} \mathcal{M}_\theta &= \left\{ \nu \in \mathcal{P}(\mathbb{N}) : \sum_{i \in \mathbb{N}} i\nu(i) = \frac{1}{\theta} \right\}, \\ f(\nu) &= \sum_{i \in \mathbb{N}} \nu(i) \log \sum_{j \in \mathbb{B}} \beta(j) j^i. \end{aligned}\right\} \tag{10.15}$$

Theorem 10.2 (Greven and den Hollander 1991a). *For $h \in (0, 1)$ and β as in (10.7):*

$$\hat{\lambda}(F) = \hat{\lambda}(\beta, h) \quad P\text{-a.s.} \tag{10.16}$$

with

$$\hat{\lambda}(\beta, h) = \sup_{\theta \in (0,1]} \left[\theta \left\{ \sup_{\nu \in \hat{\mathcal{M}}_{\theta,\beta}} [\hat{f}(\nu) - H(\nu \mid \pi_\theta \times \beta)] \right\} - H(B_\theta \mid B_h) \right], \tag{10.17}$$

where

$$\left.\begin{aligned} \hat{\mathcal{M}}_{\theta,\beta} &= \Big\{ \nu \in \mathcal{P}(\mathbb{N} \times \mathbb{B}) : \\ &\sum_{i \in \mathbb{N}, j \in \mathbb{B}} i\nu(i,j) = \frac{1}{\theta}, \sum_{i \in \mathbb{N}} \nu(i,j) = \beta(j) \,\forall j \in \mathbb{B} \Big\}, \\ \hat{f}(\nu) &= \sum_{i \in \mathbb{N}, j \in \mathbb{B}} \nu(i,j) \log j^i. \end{aligned}\right\} \tag{10.18}$$

Note that the variational problems in (10.14) and (10.17) are similar in structure, but different in detail.

The link between the global and the local growth rates is expressed by the following theorem.

Theorem 10.3 (Baillon et al. 1993; Greven and den Hollander 1991a). *For $h \in (0, 1)$ and β as in (10.7):*

$$\lambda(\beta, h) = \sup_{\gamma \in \mathcal{P}(\mathbb{B})} [\hat{\lambda}(\gamma, h) - H(\gamma \mid \beta)]. \tag{10.19}$$

10.1.4 Interpretation of the Main Theorems

What drives Theorems 10.1 and 10.2 is a close interplay between the migration process and the environment. The idea is that, for large n, *the population predominantly consists of those particles whose history happens to be best adapted to the environment*. Here, 'best adapted' means that the particles have a *path of descent* that spends a lot of time on sites x where b_x is large, and little time on sites x where b_x is small, and does so in a way that is not too unlikely. Indeed, such particles are part of a family that produces the most offspring and therefore dominates the population ('survival of the fittest'). Note that when we say population, we must distinguish between two types of population:

$$\left. \begin{array}{l} \text{the } global \text{ population (the population in a large box),} \\ \text{the } local \text{ population (the population at the origin).} \end{array} \right\} \tag{10.20}$$

Two particles drawn randomly from these two populations will have different paths of descent.

We now explain where the two variational problems in Theorems 10.1 and 10.2 come from. We give only the main idea. The details will be explained in Section 10.4.

Let us first look at the global variational problem. Note that this variational problem has a two-layer structure: it consists of an outer and an inner variational problem. The reason for this is as follows. What happens is that the path of descent of a particle may assume any *empirical drift* θ and any *empirical local time law* ν, i.e. the empirical measure for, respectively, the size of the steps in the path and the time spent at the sites that are visited successively. These are the two variables appearing in (10.14). The cost of adopting θ is $H(B_\theta \mid B_h)$, the cost of adopting ν given θ is $H(\nu \mid \pi_\theta)$. Conditional on adopting ν, the path of descent produces offspring at rate $f(\nu)$ in (10.15), which is the growth factor on a site whose local time law is ν when averaged over the local environment at that site. Thus we see where (10.14) comes from: the growth rate is determined by the optimal choice for θ and ν in a competition between cost and gain. Note that the constraint in $\mathcal{M}_\theta$ in (10.15) says that the average local time at a site must be compatible with the drift, so we have a constrained variational problem.

For the local variational problem the structure is the same, except that the environment is fixed. In this case ν plays the role of the *empirical joint local time law and local environment law*, keeping track of which local time occurs on top of which local environment. The cost of adopting ν given θ is $H(\nu \mid \pi_\theta \times \beta)$, and conditioned on ν the path of descent produces offspring at rate $\hat{f}(\nu)$ in (10.18), which is the growth factor on a site whose joint local time law and local environment law are ν. This explains (10.17). Note that the set $\hat{M}_{\beta,\theta}$ in (10.18) has two constraints: the first again says that the average local time at a site must be compatible with the drift, the second says that the projection onto the environment coordinate must equal β.

The explanation of the link established in Theorem 10.3 is as follows. The global particle density is the average over the random environment of the local particle density (recall (10.9)). Effectively, this means that for the global growth rate the random environment participates in the competition between cost and gain, while for the local growth rate it does not. The cost for the environment of adopting an *empirical local environment law* γ is $H(\gamma \mid \beta)$. Conditioned on adopting γ, the growth rate is $\hat{\lambda}(\gamma, h)$. Thus, (10.19) says that the global growth rate is determined by the optimal choice for γ: most of the population in a large box comes from those stretches in the box where the environment is optimal.

It turns out that (10.14) and (10.17) have unique maximizers. In fact, it can be proved that the path of descent of a typical particle drawn from the global population has an empirical drift and an empirical local time law that converges to the unique maximizers of the variational problems, and that the path of descent of a typical particle drawn from the local population has an empirical joint local time law and local environment law that converges to the unique maximizers of the variational problems. This is related to Csiszár's conditional limit theorem 3.20 cited in Section 3.5. Also, (10.19) has a unique maximizer, and along the path of descent the empirical local environment law converges to this maximizer.

10.1.5 Solution of the Variational Problems

The variational problems in Theorems 10.1 and 10.2 can be solved explicitly. Indeed, they involve maximization of functionals containing relative entropy under linear constraints, which can be achieved with the help of exponential families of probability measures via the Gibbs–Jaynes Principle cited in Section 3.4.

Define, for $r \geq 0$,

$$\left. \begin{aligned} G(r) &= \sum_{j \in \mathbb{B}} \beta(j) \left(\frac{e^{-r}[j/M]}{1 - e^{-r}[j/M]} \right), \\ \hat{G}(r) &= \exp\left[\sum_{j \in \mathbb{B}} \beta(j) \log\left(\frac{e^{-r}[j/M]}{1 - e^{-r}[j/M]} \right) \right]. \end{aligned} \right\} \quad (10.21)$$

Define
$$h_c = \lim_{r \downarrow 0} \frac{1}{1 + G(r)}, \qquad \theta_c = \lim_{r \downarrow 0} \frac{1}{-[\log G]'(r)}, \qquad (10.22)$$

and similarly for $\hat{G}$. Note that

$$-[\log G]'(r) > 1 + G(r) = -[\log \hat{G}]'(r) > 1 + \hat{G}(r) \quad \forall r \geqslant 0 \quad (10.23)$$

when all quantities are finite at $r = 0$, implying that

$$\theta_c < \hat{\theta}_c = h_c < \hat{h}_c. \qquad (10.24)$$

To express our solution in a compact form we need two more quantities, $r^* = r^*(\beta, h)$ and $\theta^* = \theta^*(\beta, h)$, defined as follows:

$$\left.\begin{aligned} h \leqslant h_c: &\quad r^* = 0, \\ h > h_c: &\quad r^* \text{ is the unique solution of } h = \frac{1}{1 + G(r)}, \\ h \leqslant h_c: &\quad \theta^* = 0, \\ h > h_c: &\quad \theta^* = \frac{1}{-[\log G]'(r^*)}, \end{aligned}\right\} \qquad (10.25)$$

and similarly for $\hat{G}$.

Theorem 10.4 (Baillon et al. (1993) and Greven and den Hollander (1991a); see Figures 10.1 and 10.2). *Fix β subject to (10.7).*

(i) *For $h \in (0, 1)$:*

$$\left.\begin{aligned} \lambda(\beta, h) &= \log[M(1-h)] + r^*(\beta, h), \\ \hat{\lambda}(\beta, h) &= \log[M(1-h)] + \hat{r}^*(\beta, h). \end{aligned}\right\} \qquad (10.26)$$

(ii) *$h \mapsto \lambda(\beta, h)$ is continuous and strictly decreasing on $(0, 1)$, analytic on $(0, h_c)$ and $(h_c, 1)$, while at the boundary points:*

$$\lambda(\beta, 0) = \log M, \qquad \lambda(\beta, 1) = \log \sum_{j \in \mathbb{B}} \beta(j) \, j. \qquad (10.27)$$

(iii) *$h \mapsto \hat{\lambda}(\beta, h)$ is continuous and strictly decreasing on $(0, 1)$, analytic on $(0, \hat{h}_c)$ and $(\hat{h}_c, 1)$, while at the boundary points:*

$$\hat{\lambda}(\beta, 0) = \log M, \qquad \hat{\lambda}(\beta, 1) = \sum_{j \in \mathbb{B}} \beta(j) \log j. \qquad (10.28)$$

(iv) *$\lambda(\beta, h) \geqslant \hat{\lambda}(\beta, h)$ with strict inequality if and only if $h > h_c$.*

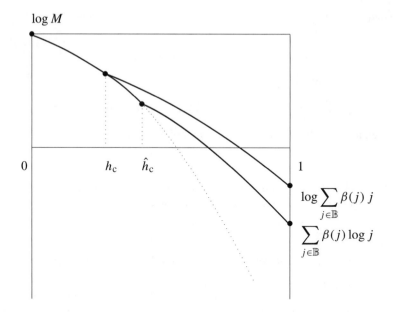

Figure 10.1. Qualitative picture of $h \mapsto \lambda(\beta, h)$ (upper solid curve), $h \mapsto \hat{\lambda}(\beta, h)$ (lower solid curve) and $h \mapsto \log[M(1-h)]$ (dotted curve). The solid curves split off the dotted curve at h_c and $\hat{h}_c$, respectively.

(v) If $h_c > 0$, then

$$\frac{\partial}{\partial h}\lambda(\beta, h_c+) - \frac{\partial}{\partial h}\lambda(\beta, h_c-) = \frac{\theta_c}{h_c(1-h_c)}. \tag{10.29}$$

Similarly for $\hat{\lambda}(\beta, h)$, $\hat{h}_c$, $\hat{\theta}_c$.

(vi) For $h \in (0, 1)$, $\theta^*(\beta, h)$ and $\hat{\theta}^*(\beta, h)$ are the maximizers of the outer variational problem in (10.14) and (10.17), respectively.

(vii) On $(h_c, 1)$, $h \mapsto \theta^*(\beta, h)$ is strictly increasing, analytic and satisfies $\theta_c < \theta^*(\beta, h) < h$. Similarly for $\hat{\theta}^*(\beta, h)$, $\hat{h}_c$, $\hat{\theta}_c$.

10.1.6 Phase Transitions

Theorem 10.4 shows that our system exhibits various interesting phase transitions. We now interpret these phase transitions for the case where β, in addition to (10.7), has the following properties (already assumed in Figures 10.1 and 10.2).

(I) $\log M > 0 > \log \sum_{j \in \mathbb{B}} \beta(j) j$: both solid curves in Figure 10.1 cross zero.

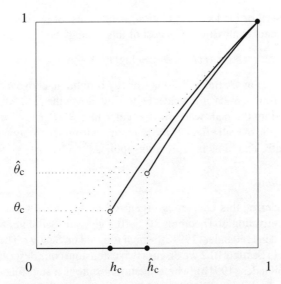

Figure 10.2. Qualitative picture of $h \mapsto \theta^*(\beta, h)$ (upper solid curve) and $h \mapsto \hat{\theta}^*(\beta, h)$ (lower solid curve). The solid curves jump from 0 to θ_c and $\hat{\theta}_c$ at h_c and $\hat{h}_c$, respectively.

(II) $\log[M(1 - \hat{h}_c)] > 0$: both solid curves in Figure 10.1 split off the dashed curve before crossing zero.

(III) $\sum_{j \in \mathbb{B}} \beta(j)(1 - [j/m])^{-2} < \infty$: both solid curves in Figure 10.2 jump.

Under these assumptions we have the following.

(I) *Survival versus extinction.* Figure 10.1 shows that there is a critical value for h at which $\lambda(\beta, h)$ changes sign. Below this value the population grows (global survival), above this value the population decays (global extinction). The same interpretation applies for $\hat{\lambda}(\beta, h)$.

(II) *Clumping.* Figure 10.1 shows that there is an intermediate range of h-values where $\lambda(\beta, h) > 0 > \hat{\lambda}(\beta, h)$, i.e. global survival but local extinction. This means that the particles are strongly clustering together (clumping): the density of populated sites decays to zero, but the population on these sites grows so fast that the overall particle density still grows.

(III) *Localization versus delocalization.* Figure 10.2 shows that for $h \leq h_c$ we have $\theta^*(\beta, h) = 0$, meaning that the typical path of descent moves at sublinear speed (global localization), while for $h > h_c$ we have $\theta^*(\beta, h) > 0$, meaning that the typical path of descent moves at linear speed (global delocalization). At $h = h_c$ the speed makes a jump of size $\theta_c > 0$. The dashed curve in Figure 10.1 corresponds to the strategy where the typical path of descent of a particle drawn from the global or the local population spends most of its time on sites where the environment is maximal. The cost of this strategy is

$$H(B_0 \mid B_h) = \log(1/(1-h))$$

and the gain is $\log M$. Below h_c this strategy is optimal, above h_c it is not. The random environment has the tendency to slow down the path where it is large and to speed up the path where it is small. Since $\theta^*(\beta, h) < h$, as shown in Figure 10.2, the overall effect of the random environment is apparently to slow down the path. The same interpretation applies for $\hat{\theta}^*(\beta, h), \hat{h}_c, \hat{\theta}_c$.

10.1.7 Outline

The remainder of this chapter is organized as follows. In Section 10.4 we sketch the derivation of Theorems 10.1–10.3 given in Baillon et al. (1993) and Greven and den Hollander (1991a), which is based on Sanov's Theorem in Section 10.1.2. In Section 10.2 we discuss two extensions: one described in Greven and den Hollander (1991b), where the initial state is a step function, and one described in Baillon et al. (1994) and Greven and den Hollander (1992, 1994), where the migration runs both forwards and backwards. In Section 10.3 we close by formulating some open problems and making some general remarks. The proof of Theorem 10.4 relies on a straightforward calculus of variations applied to the two variational problems in (10.14) and (10.17), and will be omitted. We refer the reader to Baillon et al. (1993) and Greven and den Hollander (1991a) for details.

10.2 Two Extensions

1. In Greven and den Hollander (1991b) the same model as in Sections 10.1 and 10.4 is considered, but with a different initial state, namely,

$$\eta_0(x) = \begin{cases} 1 & \text{for } x \leqslant 0, \\ 0 & \text{for } x > 0. \end{cases} \tag{10.30}$$

In this situation we get *wave front propagation*. The exponential growth rate at a site moving at speed $\tau \geqslant 0$ is computed, again in terms of a variational problem arising via Sanov's Theorem. The result reveals two characteristic wave front speeds: τ_1, the speed of the front of *zero growth* (the speed of the right-most particle), and τ_2, the speed of the front of *maximal growth* (the speed up to which the growth rate equals $\hat{\lambda}(\beta, h)$). The latter speed exhibits a phase transition, changing from zero to positive as the drift in the migration crosses a certain critical threshold.

2. In Baillon et al. (1994) and Greven and den Hollander (1992, 1994), the same model as in Sections 10.1 and 10.4 is considered, but with the migration running

both forwards and backwards: particles jump one lattice spacing to the right with probability $\frac{1}{2}(1+h)$ or one lattice spacing to the left with probability $\frac{1}{2}(1-h)$, where $h \in (0, 1)$ is again the drift parameter. The analogues of Theorems 10.1–10.4 are proved, but the analysis is considerably more complex than in the one-sided case. The reason is that the underlying local time process no longer has an i.i.d. structure. The global and the local growth rate come out as variational problems, but this time a more refined version of Sanov's Theorem is needed, where the empirical measure in (10.1) is replaced by the so-called empirical process, i.e. the empirical measure built from a sequence of random variables rather than from single random variables (see Section 3.6). It turns out that the variational problems can be solved in terms of certain random continued fractions. Though the latter may seem a bit prohibitive, they can actually be used to derive a phase diagram similar to Figures 10.1 and 10.2.

10.3 Conclusion

The model described in this chapter is an example of an interacting particle system in a random medium. It is closely related to the class of problems studied in Sznitman's book (Sznitman 1998), which revolves around large-deviation theory for Brownian motion among Poissonian obstacles. The distinction between local and global corresponds to what in the physics literature is often called quenched and annealed, namely, in a fixed and an averaged medium, respectively.

An interesting open problem is to prove the law of large numbers:

$$\frac{\eta_n(0)}{E(\eta_n(0) \mid F)} \to 1 \qquad (\mathbb{P} \times P)\text{-a.s. as } n \to \infty. \tag{10.31}$$

Indeed, only then does $\hat{\lambda}(F)$ in (10.11) truly deserve the name of local growth rate, in the same way that λ in (10.11) deserves the name of global growth rate because of (10.10). No doubt (10.31) holds true because the particles at the origin at time n come from many different ancestors at time 0, but the proof has not been worked out.

One of the major challenges is to extend the results described above to higher dimensions: particles migrate and branch on $\mathbb{Z}^d$ with $d \geqslant 2$. Qualitatively, we expect a phase diagram similar to Figures 10.1 and 10.2, but it is not clear *a priori* that all the phase transitions survive in higher dimensions. In any case, the variational study has not been carried through in any detail.

Large-deviation theory is the crossroads of probability theory, ergodic theory, and statistical physics. The sketch in Section 10.4 shows that both the global and the local growth rates are the result of an optimal selection from possible growth strategies (parametrized by θ and ν) for the path of descent of a typical particle (i.e. a particle drawn randomly from the global and the local populations, respectively). *Relative entropy is the key notion, since it determines the weight of the different growth strategies and leads to the variational problems embodying the selection.*

As is evident from Chapters 3 and 9, relative entropy shows up in various different contexts, but each time its role is one of selection. The model described in this chapter is but one illustration of this role. The rich behavior seen in Figures 10.1 and 10.2 shows that relative entropy is capable of describing subtle phenomena like phase transitions, which makes it a versatile tool indeed.

10.4 Appendix: Sketch of the Derivation of the Main Theorems

In this appendix we work out the mathematical details of the interpretation offered in Section 10.1.4. The calculation is somewhat technical and can be skipped by those readers who are only interested in the main line of argument of the chapter.

10.4.1 Local Times of Random Walk

Let
$$\left. \begin{array}{l} S_0 = 0, \quad S_n = X_1 + \cdots + X_n, \quad n \in \mathbb{N}, \\ (X_i) \text{ i.i.d. with } P_h(X_i = 1) = 1 - P_h(X_i = 0) = h \end{array} \right\} \quad (10.32)$$

denote the random walk with drift h that serves as the underlying migration process of the particles. For $x \in \mathbb{N}_0$ and $n \in \mathbb{N}_0$, let

$$\ell_n(x) = |\{0 < i \leqslant n : S_i = x\}| \quad (10.33)$$

denote its local time at site x up to time n. The following representation of the local particle density in (10.9) is the starting point for our explanation. We write P_h and E_h to denote probability and expectation w.r.t. the random walk.

Lemma 10.5. $\hat{d}_n(F) = E_h\left(\prod_{y \in \mathbb{N}_0} [b_{-y}]^{\ell_n(y)} \right).$

Proof. Fix F. From the evolution mechanism of our process (recall steps (i) and (ii) in Section 10.1.2) we obtain the recursion relation

$$E(\eta_n(x) \mid F) = (1-h) b_x E(\eta_{n-1}(x) \mid F) + h b_{x-1} E(\eta_{n-1}(x-1) \mid F). \quad (10.34)$$

In the notation of (10.32) this may be rewritten as

$$\hat{d}_n(x, F) = E_h(b_{x-X_1} \hat{d}_{n-1}(x - X_1, F)), \quad (10.35)$$

where $\hat{d}_n(x, F)$ is the local particle density at time n at site x. Iteration of (10.35) gives

$$\hat{d}_n(x, F) = E_h\left(\prod_{i=1}^{n} b_{x-S_i} \hat{d}_0(x - S_n, F) \right). \quad (10.36)$$

Finally, put $x = 0$, substitute $\hat{d}_0(\cdot, F) \equiv 1$ and write

$$\prod_{i=1}^{n} b_{-S_i} = \prod_{y \in \mathbb{N}_0} [b_{-y}]^{\ell_n(y)}.$$

□

Next, we rewrite Lemma 10.5 in a form that is more appropriate for the large-deviation analysis to be carried out in Section 10.4.2. For $x \in \mathbb{N}_0$, let

$$\ell(x) = \lim_{n \to \infty} \ell_n(x) = |\{0 < i < \infty : S_i = x\}| \qquad (10.37)$$

denote the total local time at site x.

Lemma 10.6. *For $h \in (0, 1)$:*

$$\hat{d}_n(F) = \sum_{x \in \mathbb{N}_0} P_h(S_n = x) E_h(x, n, \tilde{F}), \qquad (10.38)$$

where $\tilde{F}$ is the reversed environment defined by $\tilde{F}_y = F_{-y}$, $y \in \mathbb{Z}$, and

$$E_h(x, n, F) = E_h\left(\prod_{y=0}^{x} [b_y]^{\ell(y)} \,\bigg|\, \sum_{y=0}^{x} \ell(y) = n \right). \qquad (10.39)$$

Proof. Rewrite the representation in Lemma 10.5 as in (10.38) with

$$E_h(x, n, F) = E_h\left(\prod_{y=0}^{x} [b_y]^{\ell(y)} \,\bigg|\, S_n = x \right), \qquad (10.40)$$

using that $\ell_n(y) = 0$ for all $y > x$ on the event $\{S_n = x\}$. Add $S_{n+1} = x + 1$ to the condition in (10.40), note that

$$\{S_n = x, S_{n+1} = x + 1\} = \left\{ \sum_{y=0}^{x} \ell(y) = n \right\},$$

and use the fact that on this event $\ell_n(y) = \ell(y)$ for all $0 \leqslant y \leqslant x$. □

Note that in (10.39) we have reversed the medium for convenience. Further note that the conditional expectation in (10.40) is actually independent of h, because the condition $\{S_n = x\}$ fixes the number of steps to the right up to time n to be equal to x: every walk subject to this constraint has the same weight $h^x(1-h)^{n-x}$, which reduces the conditional probability to the uniform probability on all paths subject to this constraint.

10.4.2 Large Deviations and Growth Rates

Lemma 10.6 is the starting point for our large-deviation analysis.

Local growth rate

We begin by writing (10.38) as an integral:

$$\hat{d}_n(F) = n \int_{\theta \in (0,1]} d\theta \, P_h(S_n = \lceil \theta n \rceil) E_h(\lceil \theta n \rceil, n, \tilde{F}) + (1-h)^n [b_0]^n. \tag{10.41}$$

(Here, $\lceil \cdot \rceil$ denotes the upper integer part.) The last term is harmless because $\lambda \geq \log[M(1-h)]$ (recall (10.8), (10.11), (10.16), and (10.26)). Next, under the integral we may replace E_h by E_θ, i.e.

$$E_h(\lceil \theta n \rceil, n, \tilde{F}) = E_\theta(\lceil \theta n \rceil, n, \tilde{F}) \tag{10.42}$$

by the remark following the proof of Lemma 10.6. Next, Sanov's Theorem in Section 10.1.2 tells us that

$$\lim_{n \to \infty} \frac{1}{n} \log P_h(S_n = \lceil \theta n \rceil) = -H(B_\theta \mid B_h). \tag{10.43}$$

Indeed, pick $\mathbb{S} = \{0, 1\}$, $\mu = B_h$, $L_n = (1/n) \sum_{i=1}^{n} \delta_{X_i}$ and note that $\{S_n = \lceil \theta n \rceil\} = \{L_n = B_{\lceil \theta n \rceil / n}\}$. Next, we will show that

$$\lim_{n \to \infty} \frac{1}{n} \log E_\theta(\lceil \theta n \rceil, n, \tilde{F}) = -\hat{J}_\beta(\theta) \quad \mathbb{P}\text{-a.s.}, \tag{10.44}$$

where $\hat{J}_\beta(\theta)$ is the inner variational problem in (10.17). Theorem 10.2 follows from (10.41), (10.43), and (10.44) with the help of Varadhan's Lemma (see den Hollander 2000, Theorem III.13).

To prove (10.44), we again use Sanov's Theorem. For $N \in \mathbb{N}$, let

$$L_N = \frac{1}{N} \sum_{x=0}^{N-1} \delta_{(\ell(x), b_x)} \tag{10.45}$$

denote the empirical distribution over the N-interval of the total local time process and the random environment jointly. In terms of this quantity, we may write

$$\prod_{y=0}^{\lceil \theta n \rceil} [b_y]^{\ell(y)} = \exp\left[\sum_{y=0}^{\lceil \theta n \rceil} \ell(y) \log b_y\right]$$

$$= \exp\left[K_n \sum_{i \in \mathbb{N}, j \in \mathbb{B}} L_{K_n}(i, j) i \log j\right]$$

$$= \exp[K_n \hat{f}(L_{K_n})] \tag{10.46}$$

with $\hat{f}$ defined in (10.18) and $K_n = \lceil \theta n \rceil + 1$. Let

$$\hat{\mathcal{A}}_n = \left\{ \nu \in \mathcal{P}(\mathbb{N} \times \mathbb{B}) : \sum_{i \in \mathbb{N}, j \in \mathbb{B}} i \nu(i, j) = \frac{n}{K_n} \right\}. \tag{10.47}$$

Since $\{\sum_{y=0}^{\lceil\theta n\rceil} \ell(y)\} = \{L_{K_n} \in \hat{\mathcal{A}}_n\}$, we get from (10.39), (10.42), and (10.46) that

$$E_\theta(\lceil\theta n\rceil, n, \tilde{F}) = \int e^{K_n \hat{f}(\nu)} P_\theta(L_{K_n} \in d\nu \mid L_{K_n} \in \hat{\mathcal{A}}_n). \qquad (10.48)$$

Now, under P_θ the sequence $(\ell(x), b_x)_{x \in \mathbb{N}_0}$ is i.i.d. with state space $\mathbb{S} = \mathbb{N} \times \mathbb{B}$ and marginal law $\mu = \pi_\theta \times \beta$. Therefore, $(L_N)_{N \in \mathbb{N}}$ satisfies the large-deviation principle on $\mathcal{P}(\mathbb{N} \times \mathbb{B})$ with rate function $H(\nu \mid \pi_\theta \times \beta)$. Hence (10.44) follows from (10.48) via Varadhan's Lemma.

There are two technical points here that need consideration. First, we are in fact using Sanov's Theorem under two conditions, namely,

$$\left.\begin{array}{c} \displaystyle\sum_{i \in \mathbb{N}, j \in \mathbb{B}} i L_{K_n}(i, j) = \frac{n}{K_n}, \\[12pt] \displaystyle\sum_{i \in \mathbb{N}} L_{K_n}(i, j) = \frac{1}{K_n} \sum_{x=0}^{K_n - 1} \delta_{b_x}(j) \quad \forall j \in \mathbb{B}, \end{array}\right\} \qquad (10.49)$$

with the last empirical measure fixed because F is fixed. However, as $n \to \infty$ the right-hand sides of (10.49) tend to $1/\theta$ and $\beta(j)$, respectively (by the ergodic theorem applied to F). The fact that this translates into the two restrictions in $\hat{\mathcal{M}}_{\theta,\beta}$ defined in (10.18) requires some continuity arguments. Second, the application of Varadhan's Lemma requires certain regularity properties, which can be obtained via approximation. We refer the reader to Baillon et al. (1993) and Greven and den Hollander (1991a) for details.

Global growth rate

Since, by (10.9),

$$d_n = \mathbb{E}(\hat{d}_n(F)), \qquad (10.50)$$

Lemma 10.5 also gives us a representation of the global particle density. We can therefore follow the same line of argument as above. Because of (10.50), we put

$$E_h(\lceil\theta n\rceil, n) = \mathbb{E}(E_h(\lceil\theta n\rceil, n, \tilde{F})). \qquad (10.51)$$

It suffices to show that the analogue of (10.44) holds:

$$\lim_{n \to \infty} \frac{1}{n} \log E_\theta(\lceil\theta n\rceil, n) = -J_\beta(\theta) \qquad (10.52)$$

with $J_\beta(\theta)$ the inner variational problem in (10.14). For $N \in \mathbb{N}$, let

$$L_N = \frac{1}{N} \sum_{x=0}^{N-1} \delta_{\ell(x)}. \qquad (10.53)$$

We have

$$\mathbb{E}(E_h(\lceil \theta n \rceil, n, \tilde{F})) = \mathbb{E}(E_\theta(\lceil \theta n \rceil, n, \tilde{F}))$$

$$= E_\theta \left(\mathbb{E}\left(\prod_{y=0}^{\lceil \theta n \rceil} [b_{-y}]^{\ell(y)} \right) \, \Bigg| \, \sum_{y=0}^{\lceil \theta n \rceil} \ell(y) = n \right)$$

$$= E_\theta \left(\prod_{y=0}^{\lceil \theta n \rceil} \sum_{j \in \mathbb{B}} \beta(j) j^{\ell(y)} \, \Bigg| \, \sum_{y=0}^{\lceil \theta n \rceil} \ell(y) = n \right) \quad (10.54)$$

$$= \int e^{K_n f(L_{K_n})} P_\theta(L_{K_n} \in d\nu \mid L_n \in \mathcal{A}_n) \quad (10.55)$$

with f defined in (10.15), $K_n = \lceil \theta n \rceil + 1$ and

$$\mathcal{A}_n = \left\{ \nu \in \mathcal{P}(\mathbb{N}) : \sum_{i \in \mathbb{N}} i \nu(i) = \frac{n}{K_n} \right\}. \quad (10.56)$$

Under P_θ the sequence $(\ell(x))_{x \in \mathbb{N}_0}$ is i.i.d. with state space $\mathbb{S} = \mathbb{N}$ and marginal law $\mu = \pi_\theta$. Therefore, $(L_N)_{N \in \mathbb{N}}$ satisfies the large-deviation principle on $\mathcal{P}(\mathbb{N})$ with rate function $H(\nu \mid \pi_\theta)$. Hence (10.52) follows from (10.51) and (10.55) via Varadhan's Lemma.

This completes the sketch of the proof of Theorems 10.1 and 10.2.

10.4.3 Relation between the Global and the Local Growth Rate

The proof of Theorem 10.3 runs as follows. For any $\nu \in \hat{\mathcal{M}}_{\theta, \gamma}$, we have the identity

$$H(\nu \mid \pi_\theta \times \gamma) + H(\gamma \mid \beta) = H(\nu \mid \pi_\theta \times \beta) \quad (10.57)$$

because $\sum_{i \in \mathbb{N}} \nu(i, j) = \gamma(j) \, \forall j \in \mathbb{B}$. Hence the right-hand side of (10.19) equals

$$\sup_{\gamma \in \mathcal{P}(\mathbb{B})} \sup_{\nu \in \hat{\mathcal{M}}_{\theta, \gamma}} [\hat{f}(\nu) - H(\nu \mid \pi_\theta \times \beta)]. \quad (10.58)$$

Let $\bar{\nu}(i) = \sum_{j \in \mathbb{B}} \nu(i, j)$. A little computation gives

$$[\hat{f}(\nu) - H(\nu \mid \pi_\theta \times \beta)] = [f(\bar{\nu}) - H(\bar{\nu} \mid \pi_\theta)] - D \quad (10.59)$$

with

$$D = \sum_{i \in \mathbb{N}} \bar{\nu}(i) H(\mu_i \mid \mu_i^*), \quad (10.60)$$

where $\mu_i(j) = \nu(i, j) / \bar{\nu}(i)$ and $\mu_i^*(j) = \beta(j) j^i / \sum_{j \in \mathbb{B}} \beta(j) j^i$. Since

$$\left\{ \bar{\nu} \in \mathcal{P}(\mathbb{N}) : \nu \in \bigcup_{\gamma \in \mathcal{P}(\mathbb{B})} \hat{\mathcal{M}}_{\theta, \gamma} \right\} = \mathcal{M}_\theta, \quad (10.61)$$

there is no constraint on (μ_i), only on $\bar{\nu}$. Hence the supremum in (10.58) after we substitute (10.59) is taken at $(\mu_i) = (\mu_i^*)$ where $D = 0$. Therefore the right-hand side of (10.19) equals

$$\sup_{\bar{\nu} \in \mathcal{M}_\theta} [f(\bar{\nu}) - H(\bar{\nu} \mid \pi_\theta)], \qquad (10.62)$$

which is the same as the left-hand side of (10.19).

References

Baillon, J.-B., Clément, Ph., Greven, A. and den Hollander, F. 1993 A variational approach to branching random walk in random environment. *Ann. Prob.* **21**, 290–317.

Baillon, J.-B., Clément, Ph., Greven, A. and den Hollander, F. 1994 On a variational problem for an infinite particle system in a random medium. I. The global growth rate. *J. Reine Angew. Math.* **454**, 181–217.

den Hollander, F. 2000 *Large Deviations*. Fields Institute Monographs 14. American Mathematical Society, Providence, RI.

Georgii, H.-O. 2003 Probabilistic aspects of entropy. In *Entropy* (ed. A. Greven, G. Keller and G. Warnecke), pp. 37–54. Princeton University Press.

Greven, A. and den Hollander, F. 1991a Population growth in random media. I. Variational formula and phase diagram. *J. Statist. Phys.* **65**, 1123–1146.

Greven, A. and den Hollander, F. 1991b Population growth in random media. II. Wave front propagation. *J. Statist. Phys.* **65**, 1147–1154.

Greven, A. and den Hollander, F. 1992 Branching random walk in random environment: phase transitions for local and global growth rates. *Prob. Theory Relat. Fields* **91**, 195–249.

Greven, A. and den Hollander, F. 1994 On a variational problem for an infinite particle system in a random medium. II. The local growth rate. *Probab. Theory Relat. Fields* **100**, 301–328.

Sznitman, A.-S. 1998 *Brownian Motion, Obstacles and Random Media*. Springer.

Varadhan, S. R. S. 2003 Large deviations and entropy. In *Entropy* (ed. A. Greven, G. Keller and G. Warnecke), pp. 199–214. Princeton University Press.

Chapter Eleven

Metastability and Entropy

E. Olivieri
II Università di Roma, Tor Vergata

We review the connections between metastability and entropy. First, we give a general introduction and describe various attempts to describe metastability; in particular, we relate metastability to the maximum entropy principle under suitable constraints. Then we analyze a class of stochastic Ising models at low temperature and, more generally, Metropolis Markov chains in the Freidlin–Wentzell regime. There, metastable behavior is associated with some large-deviation phenomena. The intrinsic stochasticity of the escape from metastability is discussed and the time entropy associated with the transition is analyzed.

11.1 Introduction

Metastability is related to entropy. The notion of entropy enters into every attempt to explain metastability. Indeed, one can argue that metastable states are determined by *the maximum entropy* (or minimum free energy) principle under suitable constraints. The decay from the metastable to the stable state is a typical thermodynamic *irreversible process* towards the absolute maximum of entropy (or absolute minimum of free energy). The transition towards stability is a large-deviation phenomenon that can be described in terms of suitable *rate functions*. The exit from metastability to stability is, in general, intrinsically random. For physically relevant systems the transition mechanism involves what can be called *time entropy*: the transition takes place after long random waiting times inside suitable *permanence sets*. A large set of individually not very likely paths is preferred over a single more probable path. So the key concepts of these connections are thermodynamic equilibrium under constraints, local minima of the free energy, large deviations, rate functions, and time entropy.

Let us briefly list the main features of metastable behavior. Metastability is a dynamical phenomenon often associated with first-order phase transitions. Examples are supercooled vapors and liquids, supersaturated vapors, supersaturated solutions, and ferromagnets in that part of the hysteresis loop where the magnetization is opposite to the external magnetic field. Let us take the example of a supersaturated vapor. Consider a vapor below its critical temperature, near its condensation point. Compress, isothermically, a certain amount of vapor, free of impurities, up to the saturated vapor pressure (at the corresponding temperature). Then continue to slowly increase the pressure, trying to avoid significant density gradients inside the sample. With such careful experimentation we can prepare what is called a supersaturated vapor: indeed, we observe that the system is still in a pure gaseous phase. It persists in this situation of apparent equilibrium for a very long time; this is called a 'metastable state,' as opposed to a stable state, which, for the given values of the thermodynamic parameters, would correspond to coexistence of liquid and vapor. The stationary situation with a pure phase that we have described above persists until an external perturbation or a spontaneous fluctuation induces the nucleation of the liquid phase, starting an irreversible process that leads to a final stable state where liquid and vapor are segregated, and coexist at the saturated vapor pressure. The lifetime of the metastable state decreases as the degree of supersaturation increases, up to a threshold value for the pressure (the spinodal point) where the gas becomes unstable.

The above-described behavior is typical of an evolution that is *conservative* in the sense that it preserves the number of molecules. Another example of a metastability phenomenon concerns ferromagnetic systems below the Curie temperature T_C. As we shall see, this case is naturally described in a *nonconservative* context. In the phase diagram of an ideal ferromagnet we have that the magnetization, which we denote by m, always has the same sign as the external magnetic field h, when h is nonvanishing. Moreover, for a temperature $T < T_C$, the phenomenon of spontaneous magnetization takes place:

$$\lim_{h\to 0^+} m(h) > 0, \qquad \lim_{h\to 0^-} m(h) < 0. \tag{11.1}$$

On the other hand, for $T > T_C$ we have a *paramagnetic* behavior:

$$\lim_{h\to 0} m(h) = 0. \tag{11.2}$$

The existence of a spontaneous magnetization can be interpreted as coexistence, for the same value of the magnetic field $h = 0$, of two different phases with opposite magnetization (see, for example, Ellis 1985, Theorem V.6.1). What is observed for a real ferromagnet is the *hysteresis loop*.

For $T < T_C$, change the value of the magnetic field h from a small negative to a small positive value. Our magnet persists, apparently in equilibrium, in a state with magnetization opposite to the external magnetic field h. This situation corresponds to a metastable state since the stable state would have magnetization parallel to the field. We can continue to increase h up to a limiting value h^*, called the *coercive field*, representing the instability threshold at which we observe a decay to the stable state with positive magnetization. In a similar way we can cover the rest of the hysteresis loop. The properties of the system along the metastable arcs are similar to the ones that we have already described in the case of fluids: the stationary situation persisting for a long time up to an irreversible decay towards the stable state. The main difference here is that we do not have the conservation law we had before, namely, fixed total number of molecules: the corresponding quantity, i.e. the total number of 'up' spins, is now not fixed at all. From the point of view of equilibrium statistical mechanics, the previously described situation for a fluid corresponds to the *canonical ensemble*, whereas the present situation for a magnet corresponds to the *grand-canonical ensemble*.

The chapter is organized as follows. In Sections 11.2 and 11.3 we discuss mean-field theory and in particular the van der Waals and Curie–Weiss models. In Section 11.4 we make a comparison between mean-field and short-range systems, showing the necessity of a dynamical approach to metastability. In Section 11.5 we introduce the approach to metastability based on restricted ensembles. In Section 11.6 the pathwise approach is presented. Section 11.7 deals with metastability and nucleation for stochastic Ising model. In Section 11.8 we discuss the first-exit problem for general reversible Freidlin–Wentzell Markov chains. Section 11.9 contains a description of the tube of typical exit trajectories; the role of time entropy for the best exit mechanism is discussed.

11.2 van der Waals Theory

Let us start by briefly discussing metastability in the framework of a static, mean-field, point of view: van der Waals theory for fluids and Curie–Weiss theory for ferromagnets.

The semiphenomenological van der Waals equation of state for 'real fluids' is given by

$$(P + a/v^2)(v - b) = kT, \qquad (11.3)$$

where P is the pressure, T is the absolute temperature, v is the specific volume, k is Boltzmann's constant, and a, b are positive constants depending on the particular fluid under consideration (see, for example, Huang 1963, Chapter 2, for more details). The van der Waals equation of state can be deduced in the set-up of equilibrium statistical mechanics by introducing some drastic assumptions, in particular, by supposing that each molecule interacts in the same way with all other molecules, in such a way that the pair potential does not decay with the mutual distance. This corresponds to a range of the interaction that equals the linear dimension of the container.

Looking at the expression (11.3), it is easy to see that there exists a critical temperature T_C such that, below this value, the pressure P is not monotone as a function of v: it exhibits the so-called van der Waals loops. The unphysical feature given by the presence of intervals with negative compressibility corresponds to the nonconvex behavior of the Helmholtz free energy $f = f(v, T) = u - Ts$, where u is the internal energy and s is the thermodynamic entropy. f is related to P by the thermodynamic relationship coming from the second principle, as

$$P(v, T) = -\frac{\partial f(v, T)}{\partial v}.$$

It is well known that, under general hypotheses on the decay at infinity of the interactions, the Helmholtz free energy, which can be deduced in the framework of equilibrium statistical mechanics, is a convex function of v (see Ellis 1985, Chapter 4). Therefore, it is clear that nonconvexity is a consequence of the (unphysical) assumption that the intermolecular potential does not decay with the distance. Maxwell was able to 'correct' the unpleasant nonconvex behavior of f with the following ad hoc argument.

Fix $T < T_C$ and neglect the dependence on T in the formulae. The 'true' expression $f^{\text{true}}(v)$ of the free energy as a function of v is obtained by the van der Waals expression $f^{\text{vdW}}(v)$ via the so-called 'common tangent construction.' This corresponds to putting

$$f^{\text{true}}(v) = (CEf^{\text{vdW}})(v), \qquad (11.4)$$

where by CEf we denote the 'convex envelope' of the function f, namely, the maximal convex function that is everywhere less than or equal to f. If f has an interval of nonconvexity, then its convex envelope will contain an interval of linearity, say, $[v_l, v_g]$ with $v_g > v_l$. Maxwell's justification for the introduction of the convex envelope, based on the notion of a first-order phase transition, is that the hypothesis of spatial homogeneity, implicit in the derivation of the van der Waals equation, has to be abandoned and substituted with the assumption of phase segregation between liquid and vapor, with specific volumes v_l and v_g, respectively. Thus, between v_l and v_g, $f^{\text{true}}(v)$ is given by a convex combination of the values $f^{\text{vdW}}(v_l)$, $f^{\text{vdW}}(v_g)$ corresponding to the pure liquid and the pure vapor phases, respectively. These convex combinations correspond to different fractions of the total volume occupied by the liquid. It is easily seen that taking the convex envelope for f^{vdW} corresponds to Maxwell's equal area rule applied to the pressure P, which gives rise to a horizontal plateau in the graph of the isotherm $P(v)$. This plateau is the classical way of seeing the appearance of a first-order phase transition.

Let v^- and v^+ ($v_l < v^- < v^+ < v_g$) be the values of the specific volume corresponding to the inflection points of the graph of f^{vdW}. Between v_l and v^- and between v^+ and v_g the function $f^{\text{vdW}}(v)$ is convex. It is natural to interpret the arcs of the graph of the function $f^{\text{vdW}}(v)$ corresponding to the intervals $[v_l, v^-]$ and $[v^+, v_g]$ as describing the metastable states. So, for instance, the 'metastable branch' $[v^+, v_g]$ arises as the analytic continuation of the (stable) vapor isotherm beyond the condensation point.

11.3 Curie–Weiss Theory

Let us now recall the essential aspects of the Curie–Weiss model. Consider a system of N spins $(\sigma_i)_{i=1,\dots,N}$, where $\sigma_i \in \{-1,+1\}$. The energy associated with a configuration σ in $\Omega_N \equiv \{-1,+1\}^N$ is

$$H_N(\sigma) = -\frac{J}{2N}\sum_{i,j=1}^{N}\sigma_i\sigma_j - h\sum_{i=1}^{N}\sigma_i. \qquad (11.5)$$

Let

$$m_N(\sigma) = \frac{1}{N}\sum_{i=1}^{N}\sigma_i$$

be the magnetization. Let us define the (grand-canonical) free energy

$$F_N(\beta,h) = -\frac{1}{\beta N}\log Z_N(\beta,h),$$

where $\beta = 1/kT$ and $Z_N(\beta,h) = \sum_{\sigma \in \Omega_N} e^{-\beta H_N(\sigma)}$ is the (grand-canonical) partition function. One gets

$$\lim_{N\to\infty} F_N(\beta,h) = \min_m f(\beta,m,h), \qquad (11.6)$$

where $f(\beta,m,h)$ is the canonical free energy given by

$$f(\beta,m,h) = -\tfrac{1}{2}\beta J m^2 - \beta h m + \mathcal{E}(m) \qquad (11.7)$$

and $\mathcal{E}(m)$ is the Bernoulli entropy given by

$$\mathcal{E}(m) = \frac{1+m}{2}\log\left(\frac{1+m}{2}\right) + \frac{1-m}{2}\log\left(\frac{1-m}{2}\right). \qquad (11.8)$$

It is easy to see that for $\beta > \beta_c \equiv 1/J$ there exists $h_0 = h_0(\beta)$ such that for $|h| < h_0(\beta)$ the canonical free energy $f(\beta,m,h)$ as a function of m has a double-well structure with a global and local minimum; indeed, the absolute minimum is unique when the external magnetic field h is different from zero and has the same sign as h. Again, the nonconvexity of f is a consequence of the fact that the interaction has a range of the order of the volume. The absolute minimum is associated with the stable equilibrium, whereas the secondary (local) minimum, with magnetization opposite to the field h, is naturally interpreted as the metastable state.

11.4 Comparison between Mean-Field and Short-Range Models

Let us now summarize the main ideas about metastability arising in the framework of the mean-field approach to phase transitions:

- metastability can be described via the equilibrium Gibbs formalism;

- metastability is associated with the existence of local minima of the free energy;

- the metastable branch can be obtained via an analytic continuation beyond the condensation point.

Unfortunately, for realistic systems with short-range interactions, the above features, at least in a naive form, fail, and in that case we need new ideas to describe metastability.

In both the van der Waals and the Curie–Weiss models, metastability is associated with nonconvex (unphysical) behavior of the canonical free energy. Moreover, there is a general result, due to Lanford and Ruelle, saying that when the interaction decays sufficiently fast at infinity, a translationally invariant Gibbs measure necessarily satisfies the variational principle of thermodynamics (see Georgii 1988; Lanford and Ruelle 1969, Chapter 5). Thus a state that is metastable in the sense that (1) it is an equilibrium state, namely, a translationally invariant Gibbs measure, and (2) it does not correspond to an absolute minimum of the free energy functional cannot exist.

Let us discuss in more detail the variational principle of thermodynamics in the framework of statistical mechanics of short-range systems. To this end let us now introduce the standard, short-range Ising model, which will also be considered for other purposes in the present chapter. We start from the finite volume set-up. Let Λ be a finite cube contained in the lattice $\mathbb{Z}^d$. The space of configurations is $\mathcal{X}_\Lambda := \{-1, +1\}^\Lambda$. Thus, a configuration $\sigma \in \mathcal{X}_\Lambda$ is a function on Λ with values in $\{-1, +1\}$. For $i \in \Lambda$ the value $\sigma(i)$ is called the spin at the site i. We set $\partial^+ \Lambda = \{i \notin \Lambda : \exists j \in \Lambda \text{ with } |i-j| = 1\}$ to denote the external boundary of Λ. Given the ferromagnetic coupling constant $\frac{1}{2}J > 0$, the external magnetic field $\frac{1}{2}h \in \mathbb{R}$, the inverse temperature $\beta > 0$ and a boundary configuration $\tau \in \mathcal{X}_{\partial^+\Lambda}$, the energy associated with the configuration σ is

$$H_\Lambda^\tau(\sigma) = -\tfrac{1}{2}J \sum_{\substack{i,j \in \Lambda \\ |i-j|=1}} \sigma(i)\sigma(j) - \tfrac{1}{2}h \sum_{i \in \Lambda} \sigma(i) - \tfrac{1}{2}J \sum_{\substack{i \in \Lambda, j \in \partial^+\Lambda: \\ |i-j|=1}} J\sigma(i)\tau(j).$$

(11.9)

The (grand-canonical) Gibbs measure is

$$\mu_\Lambda^\tau(\sigma) = \frac{\exp(-\beta H_\Lambda^\tau(\sigma))}{Z_\Lambda^\tau} \quad \text{with } Z_\Lambda^\tau = \sum_{\sigma \in \mathcal{X}_\Lambda} \exp(-\beta H_\Lambda^\tau(\sigma)). \quad (11.10)$$

Denote by $\mathcal{M}_0(\Omega)$ the set of translationally invariant Borel probability measures on $\Omega = \{-1, +1\}^{\mathbb{Z}^d}$ endowed with the product topology of discrete topologies on the factors $\{-1, +1\}$. The subset of $\mathcal{M}_0(\Omega)$, obtained as the set of limits of μ_Λ^τ for $\Lambda \to \mathbb{Z}^d$ corresponding to different sequences of boundary conditions τ, are called infinite volume Gibbs measures.

The *mean entropy* is the functional s on $\mathcal{M}_0(\Omega)$ defined by

$$s(\nu) := \lim_{\Lambda \to \mathbb{Z}^d} -\frac{1}{|\Lambda|} \sum_{\sigma_\Lambda \in \Omega_\Lambda} \nu(\sigma_\Lambda) \log[\nu(\sigma_\Lambda)]. \tag{11.11}$$

It is possible to show (see Ellis 1985, Chapter 4) that the above definition is well posed: the limit exists and it defines an upper semicontinuous functional on $\mathcal{M}_0(\Omega)$. The mean energy E is a functional on $\mathcal{M}_0(\Omega)$ defined by

$$E(\nu) := \int \nu(d\sigma) \left(-\tfrac{1}{2} h \sigma_O - \tfrac{1}{2} J \sum_{\substack{i \neq O \\ |i|=1}} \sigma_i \sigma_O \right). \tag{11.12}$$

We denote by F be the free energy functional on $\mathcal{M}_0(\Omega)$ defined by

$$F(\nu) := E(\nu) - \frac{1}{\beta} s(\nu). \tag{11.13}$$

Then the above statement about the impossibility of defining metastable states in the framework of the Gibbs equilibrium formalism can be formulated in the following way. The functional F has a minimum in $\mathcal{M}_0(\Omega)$ and the minima are precisely the infinite volume translationally invariant Gibbs measures.

Next, we quote the fundamental result due to Isakov (1984) concerning the problem of analytically continuing the free energy beyond the condensation point of a first-order phase transition. It says that, for a ferromagnetic short-range Ising system in $d \geqslant 2$ dimensions at low enough temperature, the free energy (and hence the magnetization $m = m(\beta, h)$) has an essential singularity at $h = 0$ forbidding any real analytic continuation. Even for short-range systems, a sort of extrapolation beyond the condensation point can be associated with metastability in a dynamical context (see Schonmann and Shlosman 1998), but this is done in a more complicated and deeper way.

We can conclude the above critical analysis by saying that metastability has to be seen as a genuine dynamical phenomenon. In any case both static and dynamic properties have to be explained. In particular, one has to give a definition of the lifetime of the metastable state and to give a description of the mechanism of transition towards the stable state. This mainly means that one has to characterize the 'nucleation pattern' and, in particular, to find the shape and the size of the 'critical nucleus.'

We refer to Penrose and Lebowitz (1987) and Olivieri and Scoppola (1996b) for a review of known rigorous results on metastability.

11.5 The 'Restricted Ensemble'

The first attempt to formulate a rigorous approach to metastability was made by Penrose and Lebowitz (1971). These authors characterize metastable states on the basis of the following three conditions.

(i) Only one thermodynamic phase is present.

(ii) The lifetime of a metastable state is very large in the sense that a system that starts in this state is likely to take a long time to get out.

(iii) Decay from a metastable to a stable state is an irreversible process in the sense that the return time to a metastable state is much longer than the decay time.

The main idea introduced by Penrose and Lebowitz (1971) is that of a 'restricted ensemble.' An initial state is considered, given by a probability measure μ on the phase space Γ, and a law governing time evolution is assigned. For example, for a continuous system of classical particles it is natural to assume a deterministic time evolution given by the Liouville equation. The initial measure μ, describing the metastable state, has to be chosen according to the above three criteria. The idea is to choose μ with the canonical Gibbs prescription, but with a substantial modification: zero probability is given to the class of configurations producing phase segregation. In other words, μ is Gibbsian but restricted to a suitable subset R of Γ. For instance, in the case of a supersaturated vapor, in R the density is constrained to an interval of values forbidding the appearance of the liquid phase.

Penrose and Lebowitz (1971) consider the case of Kac potentials, namely, long-range small-intensity interactions, which represent the acceptable version of mean-field theory. Indeed, using Kac potentials it is possible to rigorously justify the van der Waals theory with Maxwell's equal-area rule in the framework of statistical mechanics. The idea is to perform the so-called van der Waals limit, which consists of first taking the thermodynamic limit and then decreasing to zero the value of a scale parameter γ that fixes, at the same time, the intensity and the inverse of the range of the interaction. On the 'mesoscopic scale' γ^{-1}, the system has a mean-field behavior, but on the macroscopic scale, much larger than γ^{-1}, it has the correct thermodynamic behavior, fulfilling, in particular, the convexity properties of the thermodynamic functions. In Penrose and Lebowitz (1971), for a continuous system of classical particles interacting with Kac potentials and evolving according to the Liouville equation, the authors, after suitably choosing the set R are able to reproduce, in the van der Waals limit, the metastable branch of the van der Waals theory, predicting, in that limit, an infinite lifetime.

It is also possible to extend the restricted Gibbs formalism to the infinite volume situation (see Capocaccia et al. 1974; Sewell 1980). The restricted Gibbs measure corresponding to a translation-invariant subset R of the space of infinite volume configurations can be identified as the one minimizing the free energy functional (11.13) when ν varies in the set $\mathcal{M}_0(R)$ of the translation-invariant probability measures on R (see Sewell 1980).

11.6 The Pathwise Approach

The point of view developed in Penrose and Lebowitz (1971) can be called the 'evolution of ensembles.' In Cassandro et al. (1984) a different point of view has been proposed, which can be called the 'pathwise approach to metastability.' The situation analyzed in the context of the pathwise approach is that of a stochastic process with a unique stationary measure that, for suitable initial conditions, behaves for a 'very long time' as if it were described by another stationary measure, and unpredictably undergoes the proper transition.

A typical example is an Itô stochastic differential equation in $\mathbb{R}^d$ of gradient type

$$X_t^{\varepsilon,x} = x - \int_0^t \nabla V(X_s^{\varepsilon,x})\,\mathrm{d}s + \varepsilon W_t \qquad (11.14)$$

when the potential V has a nonsymmetric double-well structure with exactly three critical points: a local minimum m, an absolute minimum M, and a saddle z. The noise intensity ε is small and the initial point x is chosen in the secondary well around m. W_t is the standard Wiener process in $\mathbb{R}^d$. The system will spend a large time randomly oscillating in the secondary well and, eventually, it will tunnel towards the deeper well around M. The basic idea of the pathwise approach is to single out typical trajectories and to look at their statistics. The most natural statistics along a path are time-averages and the proposal consists of characterizing metastability through the following behavior: stability–sharp transition–stability. The 'unpredictability' of the transition manifests itself through an approximate exponential random time.

We want to stress that, in the framework of the evolution of ensembles, it is quite difficult to distinguish between the case when single trajectories behave as above (metastability) and the case when they evolve smoothly, and very slowly towards equilibrium, which has nothing to do with metastability. The pathwise approach is one proposal for this clarification. The main ingredient is the existence of two time scales. On the first time scale the apparent stabilization is achieved, which may be called a 'thermalization' process. On the longer scale, one finally observes the 'tunnelling,' that is, the escape from the metastable situation.

11.7 Stochastic Ising Model. Metastability and Nucleation

Let us now introduce an example of a class of stochastic Markovian dynamics: the so-called stochastic Ising models.

This example is the (discrete-time) Glauber–Metropolis dynamics, i.e. a Markov chain on the state space $\mathcal{X}_\Lambda$ whose transition probabilities are given, for $\sigma \neq \sigma'$, by

$$P_\Lambda^\tau(\sigma, \sigma') = \begin{cases} 0 & \text{if } \sigma' \neq \sigma^{(x)} \text{ for all } x \in \Lambda, \\ \dfrac{1}{|\Lambda|} \exp\{-\beta[\Delta_x H_\Lambda^\tau(\sigma) \vee 0]\} & \text{if } \sigma' = \sigma^{(x)} \text{ for some } x, \end{cases} \qquad (11.15)$$

where, recalling the definition (11.9), $\Delta_x H_\Lambda^\tau(\sigma) = H_\Lambda^\tau(\sigma^{(x)}) - H_\Lambda^\tau(\sigma)$ and

$$\sigma^{(x)}(y) = \begin{cases} \sigma(y) & \text{if } x \neq y, \\ -\sigma(x) & \text{if } x = y. \end{cases} \qquad (11.16)$$

For $\sigma' = \sigma$ we set

$$P_\Lambda^\tau(\sigma, \sigma) = 1 - \sum_{\sigma' \in \mathcal{X}_\Lambda, \sigma' \neq \sigma} P_\Lambda^\tau(\sigma, \sigma').$$

The above Markov chain is *reversible* with respect to the (finite volume) Gibbs measure $\mu_\Lambda^\tau(\sigma)$ in the sense that

$$\mu_\Lambda^\tau(\sigma) P_\Lambda^\tau(\sigma, \sigma') = \mu_\Lambda^\tau(\sigma') P_\Lambda^\tau(\sigma', \sigma), \quad \forall \sigma, \sigma', \qquad (11.17)$$

so that μ_Λ^τ is the unique invariant measure for the chain. Continuous time versions of (11.15) may also be considered.

Let us now discuss metastability in the framework of the stochastic Ising model from the point of view of the pathwise approach. We will consider a system described by (11.9), enclosed in a two-dimensional torus Λ (square with periodic boundary conditions) with side length L under Metropolis dynamics given by (11.15). We consider the situation for which $0 < h < 2J < h(L-2)$ are fixed and we take the limit of large inverse temperature β. Physically, this corresponds to analyzing local aspects of nucleation at low temperatures. Mathematically, we have a Markov chain in the so-called Freidlin–Wentzell regime (see Freidlin and Wentzell 1984, Chapter 6).

It is easily seen that, with our choice of the parameters, the unique ground state is given by the configuration $+\underline{1}$, where all the spins are up. The natural candidate to describe metastable situation is the configuration $-\underline{1}$, where all spins are down. We denote by σ_t the stochastic trajectory.

Given a set of configurations $Q \subset \mathcal{X}_\Lambda$, we denote by τ_Q the first hitting time of Q, namely, $\tau_Q = \min\{t > 0 : \sigma_t \in Q\}$. We are interested in the asymptotic behavior, for large β, of the first hitting time $\tau_{+\underline{1}}$ of the configuration $+\underline{1}$ starting from the configuration $-\underline{1}$. In the present case, the main feature at the basis of metastable behavior is the presence of a critical length l^*: square clusters of $+1$ with sides smaller than l^* have a tendency to shrink (their energy decreases when shrinking), whereas square clusters with sides larger than l^* tend to grow (the energy decreases when growing) (see Neves and Schonmann (1991) and Olivieri and Scoppola (1996b) for more details). It is reasonable to expect that the critical droplet is the 'gate' to the stable situation. More specifically, a particularly interesting set of configurations which we call 'critical configurations' and denote by $\mathcal{P}$ is the set of configurations in which the $+1$ spins are precisely the ones contained in a polygonal given by a rectangle with side lengths l^*, $l^* - 1$ plus a unit square protuberance attached to one of the longest sides. Here $l^* := \lceil 2J/h \rceil$ represents the 'critical length' ($\lceil \cdot \rceil$ denotes the upper integer part).

We denote by G the formation energy of a critical 'droplet':

$$G = H(\mathcal{P}) - H(\underline{-1}) = 4Jl^* - h(l^*)^2 + h(l^* - 1).$$

Let ϑ_{-1} be the last instant at which $\sigma_t = \underline{-1}$ prior to $\tau_{+\underline{1}}$, and let $\vartheta_{-\underline{1},\mathcal{P}}$ be the first instant after $\vartheta_{-\underline{1}}$ at which σ_t visits $\mathcal{P}$. The following results (among others) have been proved by Neves and Schonmann (1991).

Theorem. *Let $h < 2J$, $L \geqslant l^* + 3$. For every $\delta > 0$:*

$$\lim_{\beta \to \infty} P_{-\underline{1}}(\exp[\beta(G - \delta)] < \tau_{+\underline{1}} < \exp[\beta(G + \delta)]) = 1, \quad (11.18)$$

$$\lim_{\beta \to \infty} P_{-\underline{1}}(\vartheta_{-\underline{1},\mathcal{P}} < \tau_{+\underline{1}}) = 1. \quad (11.19)$$

We refer to Neves and Schonmann (1991) and Olivieri and Scoppola (1995) for a proof. Here we only note that the estimate

$$P_{-\underline{1}}(\exp[\beta(G - \delta)] < \tau_{+\underline{1}}) \sim 1$$

is an easy consequence of reversibility and of the fact that $\mathcal{P}$ is the 'global saddle' between $-\underline{1}$ and $+\underline{1}$. On the other hand, to get

$$P_{-\underline{1}}(\exp[\beta(G + \delta)] > \tau_{+\underline{1}}) \sim 1$$

we need a lower bound on the probability to bypass the energy barrier between $-\underline{1}$ and $+\underline{1}$. To this end we have to understand in detail what is the typical mechanism for the tunnelling; technically, one has to construct an efficient tunnelling event. It turns out that the efficient event has to take place during a sufficiently long interval of time. Random waiting times in suitable 'permanence sets' will be needed, giving rise to a dynamical phenomenon that can be described in terms of the concept of 'time entropy.'

By combining the reversibility with the construction of an efficient tunnelling event, it is not difficult to get the above theorem. One can also show that $\tau_{+\underline{1}}/\mathbb{E}\tau_{+\underline{1}}$ converges, as $\beta \to \infty$, to a mean one exponential random time. The tunnelling is satisfactorily described as the exit from a 'basin of metastability' given, roughly speaking, by the set of subcritical configurations, namely, the ones which, taken as initial condition, give rise to $\tau_{-\underline{1}} < \tau_{+\underline{1}}$ with high probability.

It is also possible to give more precise and explicit results on the first excursion between $-\underline{1}$ and $+\underline{1}$ by specifying the typical tube of trajectories between $\vartheta_{-\underline{1}}$

and $\tau_{+\underline{1}}$. It turns out that the typical first excursion between $-\underline{1}$ and $+\underline{1}$ is made through the growth of the $+1$ phase along a sequence of square or quasisquare clusters. More precisely, let $\mathcal{R}(l_1, l_2)$ be the set of configurations whose plus spins are precisely the ones sitting on the sites internal to a rectangle $R(l_1, l_2)$ with edges parallel to the lattice axes (and vertices on the dual lattice $\mathbb{Z}^2 + (\frac{1}{2}, \frac{1}{2})$) of lengths l_1, l_2, respectively. $\mathcal{R}(l_1, l_2)$ corresponds to a local minimum for the energy. Its *basin* $\mathcal{B}(l_1, l_2)$ corresponds to the set of all configurations attracted by $\mathcal{R}(l_1, l_2)$ w.r.t. the $\beta = \infty$ dynamics, where only spin flips decreasing the energy are allowed. It is possible to show that, with a probability approaching 1 as $\beta \to \infty$, our process starting from $-\underline{1}$ visits, between $\vartheta_{-\underline{1}}$ and $\vartheta_{-\underline{1},\mathcal{P}}$, a growing sequence of rectangular configurations $\mathcal{R}(l_1, l_2)$ with $\underline{l} \equiv (l_1, l_2)$ belonging to $\mathcal{L}_1 = \{\underline{l} \equiv (l_1, l_2) : 1 \leqslant l_1 \wedge l_2 \leqslant l^* - 1, \ |l_1 - l_2| \leqslant 1\}$. Then, between $\vartheta_{-\underline{1},\mathcal{P}}$ and $\vartheta_{+\underline{1}}$, it continues to monotonically grow through rectangles $R(l_1, l_2)$ with (l_1, l_2) belonging to $\mathcal{L}_2 = \{(l_1, l_2) \in (\mathbb{Z}_+)^2 : l_1 \wedge l_2 \geqslant l^*\}$. The crucial point is that, during the above-described 'nucleation pattern,' when the process visits a rectangle it stays inside a suitable subset of its basin for an exponentially (in β) long *random* time.

11.8 First-Exit Problem for General Markov Chains

In order to explain the above results and to clarify the general mechanism behind them, let us now introduce a general set-up and give a few useful definitions. The following sections will be technically more complicated than the previous ones. We refer to Olivieri and Scoppola (1995) for more details. Consider a Markov chain $\{X_t\}_{t \geqslant 1}$ on a finite state space S. Suppose a given energy function $H : S \to \mathbb{R}$. The transition probabilities $P(x, y)$ of the Markov chain are given by $P(x, y) = q(x, y) \exp(-\beta[\{(H(y) - H(x)\} \vee 0])$, where q is an irreducible Markov kernel satisfying $q(x, y) = q(y, x)$. It is clear that our Metropolis stochastic Ising model is a particular case of the above setting. We denote by P_x the law of the process starting at $x \in A$. A *path* ω is a sequence $\omega := (\omega_1, \ldots, \omega_N)$, with $P(\omega_j, \omega_{j+1}) > 0$ when $j = 1, \ldots, N - 1$. We write $\omega : x \to y$ to denote a path joining x to y. Given $Q \subset S$, its *boundary* is $\partial Q = \{y \notin Q : \exists x \in Q \text{ with } q(x, y) > 0\}$ and its *ground* is

$$F(Q) = \left\{ x \in Q : \min_{y \in Q} H(y) = H(x) \right\}. \tag{11.20}$$

A set $Q \subset S$ is *connected* if for all $x, x' \in Q$ there exists a path $\omega : x \to x'$ contained in Q.

A set A that is either a singleton $\{x\}$ or a connected set satisfying

$$\max_{x \in A} H(x) < \min_{z \in \partial A} H(z)$$

is called a *cycle*. The *depth* $\Delta(A)$ of a nontrivial cycle A represents the difference between the minimal energy on the boundary ∂A and its absolute minimum in

METASTABILITY AND ENTROPY

A: $\Delta(A) = H(F(\partial A)) - H(F(A))$. It is possible to show (see Olivieri and Scoppola 1995) that cycles satisfy the following properties.

(i) Cycles are ordered by inclusion: for any two cycles A_1, A_2 with $A_1 \cap A_2 \neq \emptyset$ either $A_1 \subseteq A_2$ or $A_2 \subseteq A_1$.

(ii) For all $x, y \in A$, $P_x(\tau_y < \tau_{\partial A})$ tends to 1 as $\beta \to \infty$ (we recall that τ_Q is the first hitting time of the set Q).

In other words it is very likely that the process visits all the states in A before exiting. The general results about the first exit from a cycle A are

$$\forall x \in A, \delta > 0: \quad P_x(\tau_{\partial A} \in [e^{\beta(\Delta(A)-\delta)}, e^{\beta(\Delta(A)+\delta)}]) \to 1 \quad \text{as } \beta \to \infty$$

and

$$\forall x \in A : P_x(X_{\tau_{\partial A}} \in F(\partial A)) \to 1 \quad \text{as } \beta \to \infty;$$

namely, with high probability the exit from a cycle A takes place on the set of minima of the energy on ∂A (we refer to Olivieri and Scoppola (1995) for the proofs). Moreover, it is possible to characterize the typical 'tube of exit.' One can show, using reversibility, that the typical tube of exit is the time-reversed of the typical tube of first descent to the ground of the cycle (see Olivieri and Scoppola 1995; Schonmann 1992).

Let us first analyze a very simple situation: the one where A is completely attracted by a unique equilibrium point w.r.t. the $\beta = \infty$ dynamics. We begin describing the analogue for the case of the Itô equation with small noise which was analyzed by Freidlin and Wentzell (1984, p. 108). Consider the process defined by (11.14). Suppose that $V \in C^\infty(\mathbb{R}^d, \mathbb{R})$ and let D be a domain with regular boundary ∂D, such that there exists a unique critical point $x_0 \in D$ being the absolute minimum for V in $\bar{D}$, so that, denoting by $X_t^{0,x}$ the unique solution of the deterministic equation obtained from (11.14) after setting $\varepsilon = 0$, we have $\forall x \in \bar{D}$: $\lim_{t \to \infty} X_t^{0,x} = x_0$. Moreover, we suppose that there exists a unique point $y_0 \in \partial D : \min_{y \in \partial D} V(y) = V(y_0)$. Let $B_r(y) = \{x \in \mathbb{R}^d : |x - y| < r\}$ and $\Delta V = V(y_0) - V(x_0)$. It is possible to show that

$$\forall x \in D, \delta > 0: \quad P_x(\tau_{\partial D} \in [e^{(\Delta V - \delta)/\varepsilon^2}, e^{(\Delta V + \delta)/\varepsilon^2}]) \to 1 \quad \text{as } \varepsilon \to 0$$

and

$$\forall x \in D, \delta > 0: \quad P_x(X_{\tau_{\partial D}}^{\varepsilon,x} \in B_\delta(y_0)) \to 1 \quad \text{as } \varepsilon \to 0.$$

Given a sufficiently small γ, let $\psi(t) := X_{-t}^{0,y_0}$ be the time-reversal of the deterministic trajectory with $\varepsilon = 0$, starting at y_0 and let

$$t_0 = \inf\{t : X_t^{0,y_0} \in B_\gamma(x_0)\} \quad \text{and} \quad \vartheta_\varepsilon = \sup\{t : X_t^{\varepsilon,x} \in B_\gamma(x_0)\}.$$

The following result holds (see Freidlin and Wentzell 1984, p. 116):

$$\lim_{\varepsilon \to 0} P_x \left(\sup_{\vartheta_\varepsilon < t < \tau_{\partial D}} |X_t^{\varepsilon,x} - \psi(t - \vartheta_\varepsilon - t_0)| \right) = 0. \qquad (11.21)$$

In other words, during the first excursion to ∂D, after the last visit to a small sphere $B_\gamma(x_0)$, the process follows, with high probability, a trajectory close to the suitably translated time-reversed of the deterministic motion with $\varepsilon = 0$, starting from y_0. Thus we have to wait for a very long time ($\sim e^{\Delta V/\varepsilon^2}$) to see the exit, but when it takes place it is relatively fast: it requires a finite time independent of ε. The fact that the typical length of $\tau_{\partial D}$ is exponentially long in $1/\varepsilon^2$ is related to the smallness of the probability of an excursion outside D in a finite time.

A similar behavior can be shown to occur in the context of Metropolis Markov chains. Consider a cycle A that is 'one-well,' in the sense that $F(A) = \{x_0\}$, namely, there exists a unique minimum for H inside A, attracting the whole of A w.r.t. the dynamics with $\beta = \infty$. Moreover, suppose that $F(\partial A) = \{y_0\}$. It can be proved that the first excursion between x_0 and ∂A takes place along an 'uphill path without hesitations' given by the time-reversed of a downhill descent from y_0 to x_0.

In general, when a cycle A contains many attractors and internal saddles, the first excursion from $F(A)$ to $F(\partial A)$ involves the permanence for long (exponentially in β), random times inside suitable subcycles of A. A mechanism of exit without hesitations turns out to have a very small probability. Instead of following a single optimal path, it is much more convenient for the dynamics to use a mechanism involving a large set of paths. An *energy–entropy balance* argument enters into the game; a large set of unlikely individual trajectories (corresponding to exponentially long hesitation) is more convenient than a fast exit realized by more likely but not numerous enough individual trajectories.

11.9 The First Descent Tube of Trajectories

As we will now explain, the natural way of posing the problem of the description of the first exit from a domain containing many attractors involves the so-called decomposition into maximal cycles (see Catoni and Cerf 1995/97; Olivieri and Scoppola 1996a; Trouvé 1996). Given a cycle A, consider the partition into maximal (by inclusion) cycles of $A \setminus F(A)$. We write

$$A \setminus F(A) = \{A_1, A_2, \ldots, A_N\}, \qquad (11.22)$$

where A_i can be either a singleton $\{x\}$ or a nontrivial cycle. Using a 'hydrodynamic' image, suppose that we introduce water into the 'energy landscape' from $F(\partial A)$, after having produced a hole in $F(A)$; the singletons among the A_i are 'wet' by the flowing of water, whereas the nontrivial cycles A_i are filled lakes. The completely attracted case (one-well) gives rise to a trivial partition (11.22), where every A_i is a singleton and there are no lakes.

Let us now go back to the description of the set of typical trajectories during the first descent from a point $y \in A$ to the ground $F(A)$. The case of the first descent from $F(\partial A)$ to $F(A)$ is similar. Using time reversibility this gives rise to the set of typical trajectories during the first excursion between $F(A)$ and ∂A (see Olivieri and Scoppola (1995) and Schonmann (1992) for more details).

It turns out that the natural objects that enter into the description of the first descent to $F(A)$ are the 'generalized paths' $\Gamma = \{\gamma_1, \gamma_2, \ldots, \gamma_i, \ldots\}$ with $\gamma_i \in \{A_1, \ldots, A_N\}$. We write $A_i = x$ or $A_i = C$ to denote whether A_i is a singleton or a nontrivial cycle (a lake), respectively. Olivieri and Scoppola (1995) show that, with a probability tending to 1 as $\beta \to \infty$, the set of trajectories followed by the process during the first descent between $y \in A \setminus F(A)$ and $F(A)$ are contained in a so-called 'standard cascade.'

To simplify the exposition, let us assume the hypothesis of the 'unique successor': we suppose that the energy function H is invertible and $\forall x \in A$ there exists at most one point $y \in A$ with $H(y) < H(x)$ and $q(x, y) > 0$. With these hypotheses $F(A)$, $F(\partial A)$ are unique points x_0, y_0, respectively. Moreover, the $\beta = \infty$ dynamics is deterministic.

A standard cascade is characterized by a generalized path of the form:

$$\Gamma = y, y_1, y_2, \ldots, y_{k_1}, C_1, y_{k_1+1}, \ldots, C_2, \ldots$$

with $H(y_1) > H(y_2) > \cdots > H(y_{k_1})$, $y_{k_1} \in \partial C_1$, $y_{k_1+1} \in F(\partial C_1)$ and so on, up to the ground state x_0. In other words, we have a downhill path $\omega_1 = y, y_1, y_2, \ldots, y_{k_1}$, entering a lake C_1, then exiting from C_1 through a minimal saddle, then following another downhill path ω_2, doing the same w.r.t. another cycle C_2, and so on. The C_i are 'permanence sets.' A standard cascade is a set of trajectories that first follow, without hesitations, the downhill path ω_1, then stay inside C_1 for a random time of order $e^{\beta \Delta(C_1)}$, then exit from C_1 through $F(\partial C_1)$ and follow a downhill path ω_2 and so on. In a sense, a standard cascade generalizes in an obvious manner the typical way of descending from $y \in A$ to $F(A)$ in the one-well case, which in that case is simply downhill. The distinguishing feature, in the general case with many attractors, is given by the exponentially long waiting times inside the permanence sets C_i. In other words, the mechanism is intrinsically random. The random hesitations correspond to a time entropy governing the mechanism of transition. The natural tendency of the system to use a large number of possible different paths causes the process to spend long times in the permanence sets versus exit without hesitations.

It is impossible to give more specifications about the history of the process inside the C_i without losing in probability, namely still remaining in a set of trajectories of full measure. When the process enters a C_i, it visits all the points in C_i many times before exiting. A sort of equilibrium is established before the exit. The following result holds true and can be deduced by the methods of Olivieri and Scoppola (1995).

Given a cycle A, let $\Theta(A)$ be the maximal depth of the cycles A' contained in A that do not entirely contain the ground $F(A)$. Then

$$\left| \frac{P_x(X_t \in B \mid \tau_{\partial A} > t)}{\mu_A(B)} - 1 \right| < o(\beta),$$

$$\forall B \subset A, \ \forall x \in A, \ t = e^{k\beta}, \ \Theta(A) < k < \Delta(A), \quad (11.23)$$

where $o(\beta)$ is an infinitesimal quantity for large β and μ_A is the conditional Gibbs measure defined by

$$\mu_A(x) = \frac{\mu(x)\mathbf{1}_A(x)}{\mu(A)} \quad \text{with } \mu(x) = \frac{\exp(-\beta H(x))}{\sum_{z \in S} \exp(-\beta H(z))}. \quad (11.24)$$

We say that $e^{\Theta(A)\beta}$ is the *thermalization time* inside A, which is much smaller than the typical exit time $e^{\Delta(A)\beta}$.

The above statements constitute a dynamical justification for the restricted ensemble. Indeed, consider the stochastic Ising model in the above-described regime: Λ, h fixed, $\beta \to \infty$. Recall that $\mathcal{P}$ is the set of saddle configurations corresponding to a critical droplet and representing the gate to stability. We have that before the first hitting time of $\mathcal{P}$ the process visits many times all the states contained in the cycle A^* given as the maximal connected set of configurations containing $-\underline{1}$ with energy less than $H(\mathcal{P})$. By applying the result given in (11.23) we get that before $\tau_{\mathcal{P}}$ the behavior of our process is well described by the measure μ_{A^*}, which coincides with the restricted Gibbs ensemble corresponding to A^*. The restricted ensemble describes, more generally, the apparent equilibrium, which is established inside a permanence set C_i when, during the first descent to the ground of a cycle, our process enters C_i. In other words, the system tends to minimize the free energy *conditioned* to stay in a permanence set, until it exits because of a large fluctuation and goes to a lower-energy situation. Moreover, the properly renormalized exit time tends to be exponentially distributed. When we apply this statement to a stochastic Ising model we get, using time reversal, the information that when, during the first excursion from $-\underline{1}$ to $+\underline{1}$, our process enters some basin $\mathcal{B}(l_1, l_2)$ (see Section 11.7) it performs random oscillations around $\mathcal{R}(l_1, l_2)$ for a very long time before continuing to grow. A nucleation without these random oscillations around the growing rectangles would have a very small probability.

11.10 Concluding Remarks

We have showed that a purely static approach to metastability does not work for short-range systems, even to explain the static properties of metastable states. However, some ideas, such as the identification of metastable states as the ones minimizing the free energy under suitable constraints, have a dynamical counterpart. Let us discuss the case of a finite volume stochastic Ising model at low temperature. As we have already said, when our system enters a nontrivial cycle A, it then stays there for an exponentially long time and its behavior in A is well described by the Gibbs equilibrium restricted to A. The system, before exiting from A, tries to minimize the free energy under the constraint to belong to A. The exit is almost unpredictable. This situation is reminiscent of quite usual examples of thermodynamic irreversible transformations with increase of entropy (decrease of free energy). Consider, for example, a box $\Lambda = \Lambda_1 \cup \Lambda_2$, where Λ_1 and Λ_2 are two equal cubes separated by a diaphragm and suppose to start with a situation where Λ_1 and Λ_2 contain certain quantities of the same gas at equilibrium, corresponding to different densities ρ_1 and ρ_2. If we remove the diaphragm, the system starts an irreversible transformation towards an equilibrium state with equal suitable density in Λ_1 and Λ_2.

In the case of transition between metastability and stability for a stochastic Ising model, we first have a Gibbs equilibrium situation under the constraint that the state belongs to the cycle A^*; then, when this constraint is spontaneously removed because of a large fluctuation (nucleation), the system performs a transition into another much larger and deeper cycle containing $+\underline{1}$. In our dynamical model we can give a sense to the statement that this transition to stability is an irreversible transformation: the return time is much longer than the decay time. We can also consider the possibility to induce nucleation via an external disturbance so that the situation becomes even more similar to the removal of the diaphragm.

References

Capocaccia, D., Cassandro, M. and Olivieri, E. 1974 A study of metastability in the Ising model. *Commun. Math. Phys.* **39**, 185–205.

Cassandro, M., Galves, A., Olivieri, E. and Vares, M. E. 1984 Metastable behavior of stochastic dynamics: a pathwise approach. *J. Statist. Phys.* **35**, 603–634.

Catoni, O. and Cerf, R. 1995/97 The exit path of a Markov chain with rare transitions. *ESAIM Prob. Statist.* **1**, 95–144 (electronic).

Ellis, R. S. 1985 *Entropy, Large Deviations and Statistical Mechanics*. Springer.

Freidlin, M. I. and Wentzell, A. P. 1984 *Random Perturbations of Dynamical Systems*. Springer.

Georgii, H.-O. 1988 *Gibbs Measures and Phase Transitions*. Walter de Gruyter, Berlin and New York.

Huang, K. 1963 *Statistical Mechanics*. Wiley.

Isakov, S. N. 1984 Nonanalytic features of the first order phase transition in the Ising model. *Commun. Math. Phys.* **95**, 427–443.

Lanford III, O. E. and Ruelle, D. 1969 Observables at infinity and states with short range correlations in statistical mechanics. *Commun. Math. Phys.* **13**, 194–215.

Neves, E. J. and Schonmann, R. H. 1991 Behavior of droplets for a class of Glauber dynamics at very low temperatures. *Commun. Math. Phys.* **137**, 209–230.

Olivieri, E. and Scoppola, E. 1995 Markov chains with exponentially small transition probabilities: first exit problem from a general domain. I. The reversible case. *J. Statist. Phys.* **79**, 613–647.

Olivieri, E. and Scoppola, E. 1996a Markov chains with exponentially small transition probabilities: first exit problem from a general domain. II. The general case. *J. Statist. Phys.* **84**, 987–1041.

Olivieri, E. and Scoppola, E. 1996b Metastability and typical exit paths in stochastic dynamics. In *Proc. 2nd European Congress of Mathematics, Budapest, 22–26 July*. Birkhäuser.

Penrose, O. and Lebowitz, J. L. 1971 Rigorous treatment of metastable states in van der Waals theory. *J. Statist. Phys.* **3**, 211–236.

Penrose, O. and Lebowitz, J. L. 1987 Towards a rigorous molecular theory of metastability. In *Fluctuation Phenomena* (ed. E. W. Montroll and J. L. Lebowitz), 2nd edn. North-Holland.

Schonmann, R. H. 1992 The pattern of escape from metastability of a stochastic Ising model. *Commun. Math. Phys.* **147**, 231–240.

Schonmann, R. H. and Shlosman, S. B. 1998 Wulff droplets and the metastable relaxation of kinetic Ising models. *Commun. Math. Phys.* **194**, 389–462.

Sewell, G. L. 1980 Stability, equilibrium and metastability in statistical mechanics. *Phys. Rep.* **57**, 307–342

Trouvé, A. 1996 Cycle decompositions and simulated annealing. *SIAM J. Control Optim.* **34**, 966–986.

Chapter Twelve

Entropy Production in Driven Spatially Extended Systems

Christian Maes
Instituut voor Theoretische Fysica, K. U. Leuven

This is a short review of the statistical mechanical definition of entropy production for systems composed of a large number of interacting components. Emphasis is on open systems driven away from equilibrium where the entropy production can be identified with a logarithmic ratio of microstate multiplicities of the original macrostate with respect to the time-reversed state. A special role is taken by Gibbs measures for the stationary spatio-temporal distribution of trajectories. The mean entropy production is always nonnegative and it is zero only when the system is in equilibrium. The fluctuations of the entropy production satisfy a symmetry first observed in Evans et al. (1993) and then derived in Gallavotti and Cohen (1995a,b) for the phase-space contraction rate in a class of strongly chaotic dynamical systems. Aspects of the general framework are illustrated via a bulk driven diffusive lattice gas.

12.1 Introduction

The production of entropy in spatially extended systems was already discussed, both at length and in depth, by the founding fathers of thermodynamics and statistical mechanics. It is instructive to divide the discussion into two (very unequal) parts.

Firstly, there is the situation of a perfectly closed (or truly isolated) classical mechanical system undergoing a Hamiltonian dynamics and evolving towards equilibrium from nonequilibrium initial conditions. That is known as the approach or the convergence or the relaxation towards equilibrium. The issue here is to understand the usual second law of thermodynamics that associates an arrow of time to the macroscopic behavior leading to equilibrium and to derive from the reversible microscopic laws the irreversible kinetic and hydrodynamic equations describing this evolution on the appropriate space-time scale.

Secondly, there are the many phenomena of dissipation in open driven systems. Here one considers a stationary situation of a system or medium in contact with several reservoirs that interact with the system. Think of heat baths or particle reservoirs at different temperatures or concentrations through which a nonequilibrium state is maintained in the system or also of particle systems subject to an external driving field. As a result, a stationary current is installed and entropy is produced.

The plan is that I start by recalling some of the main ingredients in the qualitative understanding of macroscopic irreversibility for the first scenario, relaxation to equilibrium. This has attracted the most attention in the past and it is better understood. On the other hand, for the nonequilibrium steady-state scenario the literature is mostly limited to the situation close to equilibrium and Section 12.3 will remind us of the phenomenology of entropy production. Then, I get a chance to say something new when in Section 12.4 I give a Boltzmann-like definition of entropy production in a simplified context. Formulae (12.3)–(12.7) are the most important. The general framework is presented in Sections 12.5 and 12.6. The application in the final Section 12.7 is to the asymmetric exclusion process.

This contribution uses and studies aspects of thermodynamics, statistical mechanics, and probability theory with a variety of language and tools. One application and source of inspiration is, however, not included—that of the theory of smooth dynamical systems. Yet, some of the work that I will present here was started in Maes (1999) (see also Kurchan 1998; Lebowitz and Spohn 1999) from trying to understand the Gallavotti–Cohen results in Gallavotti and Cohen (1995a,b) on the fluctuations of the phase-space contraction in dynamical systems with very strong hyperbolicity assumptions and the 'new theoretical ideas in nonequilibrium statistical mechanics' à la Ruelle (see Ruelle 1999). While the underlying ideas here (see Sections 12.4 and 12.6) are somewhat different, many of the formal relations and results in fact coincide with what is obtained before in the context of chaotic dynamics.

The present chapter is, except for Sections 12.4 and 12.7.2, a short review and the many details and proofs are omitted, due to lack of space. More can be found in Maes (1999), Maes and Redig (2000), and Maes et al. (2000a,b, 2001a,b), and I am especially grateful to Frank Redig for many discussions.

12.2 Approach to Equilibrium

Here I only remind the reader about the problem of irreversibility in closed Hamiltonian systems and about its resolution. There is nothing new here (see, for example, Bricmont 1996; Goldstein 2003; Jaynes 1992; Lebowitz 1999).

What is the problem? The microscopic dynamics is invariant under time reversal. Why then do we always see real individual macroscopic systems taking a particular course in time, evolving towards equilibrium? What gives the direction of time in macroscopic behavior?

The resolution of this paradox has many sides and it can easily distract you for quite some time. Yet the beginning is easy and it says that there is absolutely no problem, and that it is a matter of counting and you should not expect otherwise: equilibrium is what you should expect. Here the basic ingredient is that there is a huge difference in scales between the microworld and the macroworld and that irreversibility belongs to the macroworld.

12.2.1 Boltzmann Entropy

Take a system of N (large!) components $x(i), i = 1, \ldots, N$, each having n possible values. A microstate $x = (x(i), i = 1, \ldots, N)$ is one of the n^N possible configurations. The set of all these configurations (or, when additional constraints are given, some subset of this) is the microscopic phase space Ω. From a microscopic perspective each element of Ω is equivalent and we remain indifferent in the estimates of plausibility: each x has the same probability of being realized. From a macroscopic point of view, some distinction can be made. For this we make a choice of macroscopic variables which are approximately additive (and, in the presence of a 'natural' dynamics, are locally conserved). The macrovariable partitions Ω in a number of subsets. Say, for $\Omega = \{+1, -1\}^N$ and macrovariable $X_N(x) \equiv \sum_i x(i)/N$, each of these subsets will contain between 1 and, for N very large, almost 2^N elements. So, we all expect that randomly picking an element from Ω gives us a configuration that 'typically' has zero X_N (with standard error $1/\sqrt{N}$). In other words, while all microstates are equivalent, some microstates are more typical from a macroscopic point of view.

To connect the microscopic world with macroscopic behavior Boltzmann introduced the entropy

$$S(X) \equiv \log W(X),$$

where $W(X)$ is the number of microstates x for which $X_N(x) = X$. The logarithm is very convenient and makes $S(X)$ extensive. In the same way (and by some abuse of notation), we speak about the entropy S of a microstate x:

$$S(x) \equiv S(X_N(x)).$$

Of course, the phase space for realistic systems is a bit more complicated. We should be speaking about a (classical) Hamiltonian dynamics for N point particles enclosed in a finite box with x corresponding to the positions and momenta of all the particles. The phase space Ω is now the collection of all these microstates constrained to given values of certain macrovariables Y_N. As we have a closed system with energy conservation, we can keep in mind that the total energy $Y_N(x) = E$ is thus fixed. We are interested in the evolution of a second class of macrovariables X_N. Let us think here about the macroscopic variable that corresponds to the spatial density profile, say the number of particles to the left of our box. Yet, for this more complicated phase space, more or less the same counting procedure can be followed. In this same language, microstates have a larger entropy when they correspond to a macroscopic value that can be microscopically realized in more ways. Now, it is important to get a feeling of the huge differences that can arise when the system is large. For air, under normal circumstances, there are about 10^{26} molecules in a box of 1 m^3. If we divide this box into 1 cm^3 pieces, we get $10^{60\,000\,000\,000\,000\,000\,000\,000\,000}$ possible arrangements that would correspond to a homogeneous density profile (about the same number of particles in each piece) while only 10^6 possibilities of having all of them in just one such piece of the box. When considering these huge numbers and reasoning about plausibilities just as we do in everyday life, it appears that only a great conspiracy of initial conditions or dynamics can lead to an effective decrease in entropy: that is, we expect that $S(x_t)$ grows as x_t follows the microscopic trajectory and equilibrium corresponds to maximal entropy. The law of increase of entropy is a statistical law, but one with a 'moral certainty.'

It must be added that larger entropy does not necessarily mean less ordered or more spatially homogeneous. It depends on the type of interaction and/or the value of the energy that is fixed. It suffices to think of gravity (where clustering of mass is typical) or of molecular interactions at low energy leading to ordered structures.

12.2.2 Initial Conditions

Note that the counting argument above is symmetric in time, valid as such for prediction as well as for retrodiction. For the same reason that we predict approach to equilibrium for the future evolution, we must retrodict equilibrium as well. That is a consequence of the dynamical reversibility of the microscopic dynamics, which we accept. So the real surprise is to observe nonequilibrium initial conditions. When trying to see the origin of these special low-entropy initial conditions we soon get in a chain of arguments leading to cosmological questions. Observations of cosmological processes today and in the past tell us how improbable indeed (in the sense of low entropy) the initial conditions must have been. Here input from the standard model of physical cosmology is necessary. The upshot is summarized by Richard Feynman, 'it is necessary to add to the physical laws the hypothesis that in the past the Universe was more ordered, in the technical sense, than it is today...to make an understanding of the irreversibility', or, much earlier, by Ludwig Boltzmann, 'That in nature the transition from a probable to an improbable state does not take place as often as the converse, can be explained by assuming a very improbable initial state of the entire universe surrounding us. This is a reasonable assumption to make, since it enables us to explain the facts of experience, and one should not expect to be able to deduce it from anything more fundamental.' I refer to the pleasant Chapter 7 in Penrose (1989) for some specific analysis.

12.3 Phenomenology of Steady-State Entropy Production

The definition of entropy production for nonequilibrium steady states cannot be completely arbitrary. Standard treatments are, however, largely restricted to close-to-equilibrium phenomenology.

Entropy production appears in the close-to-equilibrium thermodynamics of steady-state irreversible processes as the product of thermodynamic fluxes and thermodynamic forces. The forces are produced by gradients in intensive variables, the so-called affinities that are maintained throughout. One thinks of gradients in chemical or electrostatic potential or in temperature. The fluxes are currents in the conjugate extensive variables. In linear response, they are linear combinations of the affinities. In this way, linear transport coefficients are defined that are functions of the intensive variables that locally characterize the state of the system.

As an example take a cylindrical solid material through which a stationary heat current J_Q is maintained by coupling the material at its right and left ends to reservoirs at temperature T_r and T_l, respectively. Assuming a linear relation between the affinity $\nabla 1/T$ and the current J_Q with a linear response coefficient $L_Q = L_Q(T)$, one arrives at Fourier's law

$$J_Q = L_Q \nabla \left(\frac{1}{T} \right) = -h \nabla T,$$

where $h \equiv L_Q(T)/T^2$ is the heat conductivity. Here is expressed the empirical finding (close to equilibrium) of the proportionality of the heat current to the temperature gradient. The mean entropy production (MEP) in the material is then

$$\text{MEP} = J_Q \nabla \frac{1}{T} = hT^2 \left[\nabla \left(\frac{1}{T} \right) \right]^2.$$

Of course, the (close-to-equilibrium) entropy of the material remains constant (its macrostate is unchanged) and all produced entropy is carried away by the entropy current to be poured into the reservoirs. (The relation between entropy current and entropy production is established in the so-called entropy balance equation.) If we integrate the entropy production over the material, we see that the entropy of the reservoirs is changed by

$$\frac{dS}{dt} = a J_E \left(\frac{1}{T_r} - \frac{1}{T_l} \right),$$

where a is the cross-sectional area of the cylinder. To maintain the steady state we will have to carry away this extra entropy outside the coupled system. I refer to Balian (1982) for the standard treatment and to Bonetto et al. (2000) and Eckmann et al. (1999) for more recent studies.

12.4 Multiplicity under Constraints

An elementary mathematical clarification of the connection between Boltzmann entropy and thermodynamic equilibrium entropy goes by considering the multiplicity of microstates for a particular macroscopic observation (see also Georgii 2003).

Take N particles each of which can be in a certain phase space cell i (for example, having energy E_i), $i = 1, \ldots, n$. In total, there are n^N microstates.

The number of such microstates with m_1 particles in cell 1, ..., and m_n particles in cell n ($m_1 + \cdots + m_n = N$), is given by the multinomial coefficient

$$W_N(m_1, m_2, \ldots, m_n) = \frac{N!}{m_1! \cdots m_n!}.$$

With proportions $p_i \equiv m_i/N$ and via Stirling's formula,

$$\lim_N \frac{1}{N} \ln W_N(p_1 N, \ldots, p_n N) = -\sum_{i=1}^n p_i \ln p_i, \tag{12.1}$$

where the right-hand side is the Shannon entropy of the probability measure (p_i). Imagine now as a further constraint that

$$\sum_{i=1}^n p_i E_i = E.$$

Then, the multiplicities for which the entropy (12.1) is maximal are

$$p_i = \frac{e^{-\beta E_i}}{Z_\beta}$$

with $\beta = \beta(E)$ found from the constraint. This maximal entropy is the equilibrium or Gibbs entropy

$$S(E) = \ln Z_\beta + \beta E.$$

We can now move E to a new value $E + dE$ while supposing that the energy levels E_i do not change. Obviously, the maximal entropy does change and it is doing so according to

$$dS = \beta \, dE,$$

which is Clausius' formula $dS = \delta Q/T$ for the change in equilibrium entropy from a heat transfer δQ at reservoir temperature $T = 1/\beta$ for fixed volume.

I suggest a very similar scheme for entropy production. I sketch it here without explicit reference to dynamics just to emphasize, in this highly simplified setting, the structure of the definition that later, in all its glory, will follow. Divide the spatially extended system into N cells. To each cell is associated one of two possible current values -1 or $+1$ (left or right moving). Write $\sigma = (\sigma(i) = \pm 1, i = 1, \ldots, N)$ for the configuration. The cells are observed at p instants of time $t_1 < \cdots < t_p$, while the total (time-summed) current j is held fixed:

$$\sum_{r=1}^p \sum_{i=1}^N \sigma_{t_r}(i) = jNp \neq 0. \tag{12.2}$$

More generally, it will be through the form of the microscopic current that the dynamics will explicitly enter. The system is supposed to be stationary.

To observe a space-time window, select $M \leq N$ of these cells at $T \leq p$ instants of time and let $\eta \equiv (\eta_{t_r}(i), i = 1, \ldots, M, r = 1, \ldots, T)$ be a fixed history for them. Let $W_{M,T}^{(N,p)}(\eta, j)$ be the number of trajectories $(\sigma_{t_r}(i))$ so that $\sigma_{t_r}(i) = \eta_{t_r}(i)$ for $i = 1, \ldots, M; r = 1, \ldots, p$ and so that (12.2) is maintained. For any second history η' in the space-time window, consider the logarithmic ratio

$$\Phi_{M,T}^{(N,p)}(\eta, \eta', j) \equiv \ln \frac{W_{M,T}^{(N,p)}(\eta, j)}{W_{M,T}^{(N,p)}(\eta', j)}. \tag{12.3}$$

Evaluating this for large N and p, while fixing all the rest, gives as limit after some elementary calculations involving binomial coefficients,

$$\Phi_{M,T}(\eta, \eta', j) = \frac{\lambda(j)}{2} \sum_{i=1}^{M} \sum_{r=1}^{T} [\eta_{t_r}(i) - \eta'_{t_r}(i)] \tag{12.4}$$

with thermodynamic force $\lambda(j) \equiv \ln(1+j)/(1-j)$. The probability of a fixed history η is here therefore proportional to the exponential of the dissipation function

$$\Phi_{M,T}(\eta, j) \equiv \frac{\lambda(j)}{2} \sum_{i=1}^{M} \sum_{r=1}^{T} \eta_{t_r}(i). \tag{12.5}$$

Denoting the (Shannon) entropy (12.1) of this space-time law by $S(j)$, we have $dS(j) = -\lambda(j) \, dj/2$ upon changing the value of the current j and that is why I have called $\lambda(j)$ the thermodynamic force.

For large MT, the local current fluctuations

$$P_{M,T}^{(j)}[w] \equiv \text{Prob}\left[\sum_{i=1}^{M} \sum_{r=1}^{T} \eta_{t_r}(i) = wMT\right]$$

$$= \exp(\Phi(w, j) + S(w))/Z_{\lambda(j)}$$

$$= \exp\left[-MT\left[\frac{1}{2} w \ln \frac{(1-j)(1+w)}{(1+j)(1-w)} + \frac{1}{2} \ln \frac{1-w^2}{1-j^2}\right]\right] \tag{12.6}$$

with $\Phi(w, j) \equiv MT\lambda(j)w/2$ and $S(w) \geq 0$ the entropy, satisfy the symmetry

$$P_{M,T}^{(j)}[w] = P_{M,T}^{(j)}[-w]e^{MT\lambda(j)w}$$

without being Gaussian because only the dissipation term $\Phi(w, j)$ is not time-reversal invariant (changing w into $-w$).

The entropy production $\bar{S}(\eta, j)$ for a fixed history η is defined as in (12.3) but with the second history $\eta' = \bar{\eta}$ the time reversal of η: take $\eta'_{t_r} = -\eta_{t_{T+1-r}}$ and

$$\bar{S}(\eta, j) \equiv \Phi_{M,T}(\eta, \bar{\eta}, j). \tag{12.7}$$

Then, twice the dissipation $2\Phi_{M,T}(\eta, j) = \bar{S}(\eta, j)$ equals the entropy production in the history η. For large MT, overwhelmingly, $\sum_{i=1}^{M} \sum_{r=1}^{T} \eta_{t_r}(i) =$

jMT is most probable and hence $\bar{S}(\eta, j)/MT$ converges to the mean entropy production (density)

$$\text{MEP}(j) = \lambda(j)j = \frac{1+j}{2} \ln \frac{1+j}{1-j} + \frac{1-j}{2} \ln \frac{1-j}{1+j}.$$

That is, the mean entropy production is positive, it is time-reversal invariant (changing j to $-j$) and it equals the relative entropy between the original distribution and its time reversal. MEP(j) is minimal for $j = 0$ but for fixed j the most probable history η has maximal entropy production $\bar{S}(\eta, j)$ (maximum entropy production principle). Close to equilibrium, that is for $|j|$ small, the velocity fluctuations are Gaussian and MEP(j) $\approx 2j^2$ (linear response regime).

12.5 Gibbs Measures with an Involution

The following must be considered as a program, the context that we should keep in mind for more concrete realizations. The two theorems that we give below are for Gibbs distributions for lattice spin systems; they are easy to prove here but they will have to be replaced (and proved again) for more complicated nonequilibrium steady states. This will be illustrated in Section 12.7.

I write K for a finite set and take $\Omega \equiv K^{\mathbb{Z}^{d+1}}$. This $(d+1)$-dimensional configuration space Ω plays the role of pathspace; its elements $\sigma \equiv (\sigma_t(i), (t, i) \in \mathbb{Z} \times \mathbb{Z}^d)$ are space-time trajectories with values $\sigma_t(i) \in K$ at the space-time point (t, i). Consider regular space-time cubes $V_{T,L} \equiv \{(t, i) \in \mathbb{Z}^{d+1} : |t| \leqslant T, |i| \leqslant L\}$ centered around the origin, where T stipulates the temporal and L the spatial extension. Let $\Theta_{T,L}$ be the involution on Ω that reverses the time:

$$(\Theta_{T,L}\sigma)_t(i) \equiv \begin{cases} \sigma_{-t}(i) & \text{if } (t, i) \in V_{T,L}, \\ \sigma_t(i) & \text{otherwise.} \end{cases} \tag{12.8}$$

(I could also have included a kinematical time reversal π as involution on K and then write $(\Theta^{\pi}_{T,L}\sigma)_t(i) \equiv \pi(\sigma_{-t}(i))$ for $(t, i) \in V_{T,L}$ but, for simplicity, I will stick here to the choice $\pi = $ identity.) Given a local function f on Ω it is possible to find large enough T_o, L_o so that for all $T \geqslant T_o, L \geqslant L_o$, $f(\sigma) = f(\sigma_t(i), (t, i) \in V_{T,L})$. There is therefore no ambiguity in writing $f\Theta$ for the new (time-reversed) function and similarly, for a probability measure μ on Ω, to write $\mu\Theta$ for the new probability measure with expectations $\mu\Theta(f) \equiv \mu(f\Theta)$.

Next, consider a translation-invariant space-time interaction potential $U = (U_A)$ parametrized by the finite subsets A of $\mathbb{Z}^{d+1}$. Each U_A is a function of the variables $\sigma_t(i), (t, i) \in A$, and let me assume, just for convenience, that $U_A \equiv 0$ whenever the diameter of the set A is larger than a finite radius. I define the associated variable entropy production in a finite set Λ as

$$\bar{S}_\Lambda \equiv \bar{S}^{(U)}_\Lambda \equiv \sum_{A \subset \Lambda} [U_A \Theta - U_A]. \tag{12.9}$$

The idea is again that nonequilibrium steady states are characterized by a breaking of time-reversal invariance and that the entropy production precisely picks up that term in the space-time interaction that is the source of that breaking. Clearly, $\bar{S}_\Lambda$ is asymmetric under time reversal Θ and it is measurable from the values of the variables in the set Λ. Definition (12.9) is the analogue of (12.3)–(12.7). In all realistic realizations of this definition, $\bar{S}_\Lambda$ can be interpreted as a sum over products of fields and currents. The reader is probably waiting to see some dynamics and it may seem strange to speak about such an entropy production but I only deal with the general framework here and one illustration is contained in Section 12.7.

Let μ be a translation-invariant probability measure on Ω. I define the mean entropy production (MEP) in μ as the expectation

$$\text{MEP}(U, \mu) \equiv \lim_\Lambda \frac{1}{|\Lambda|} \mu(\bar{S}_\Lambda). \tag{12.10}$$

Limits are understood in the sense of increasing cubes $\Lambda = V_{T,L}$. It is important to realize that you do not in fact need to take the μ-average in the above but for a fixed (large enough) T, $\bar{S}/|\Lambda|$ will become μ-almost surely equal to the MEP if μ is ergodic with respect to spatial translations.

Let μ be a translation-invariant Gibbs measure for the potential (U_A). It has an entropy density $s(\mu)$ that equals the entropy density $s(\mu\Theta)$ of the time-reversed Gibbs distribution. Yet, μ and $\mu\Theta$ need not be equal; they can be discriminated via their relative entropy density $s(\mu \mid \mu\Theta) = s(\mu\Theta \mid \mu)$ (see Georgii 2003, Chapter 3 in this book).

Theorem 12.1 (MEP). *Let μ be a translation-invariant Gibbs measure for the potential (U_A). Then*

$$\text{MEP}(U, \mu) = s(\mu \mid \mu\Theta) \geqslant 0 \tag{12.11}$$

with equality if and only if the potentials U and $U\Theta$ are physically equivalent.

Remarks.

- The identification of the MEP with the relative entropy density was announced at the very end of Section 12.4.
- There can be no spontaneous breaking of time-reversal symmetry: if the MEP is zero, then μ and $\mu\Theta$ must be Gibbs measures for the same potential, which means that (U_A) and $(U_A\Theta)$ must be physically equivalent. This property of 'no current without heat' is established for infinite interacting-particle systems in Maes and Redig (2000) and Maes et al. (2000b, 2001a).

The entropy production fluctuates locally and I now present the local fluctuation theorem (LFT) for Gibbs measures with an involution. Take T and L

large and fix the volume $V \equiv V_{T,L}$ and a subset $\Lambda \equiv V_{T,L'}$ with $L' \ll L$. For a function f of the variables in V, define the expectation

$$\mathbb{E}_V(f) \equiv \frac{1}{Z_V} \sum_{\sigma \in K^V} f(\sigma) \exp\left(-\sum_{A \subset V} U_A(\sigma)\right)$$

with Z_V the normalizing partition function. There is no need for translation invariance here and the potential (U_A) is allowed to contain time-reversal-invariant hard-core interactions. I write

$$Z_{V \setminus \Lambda}(\sigma_\Lambda) \equiv \sum_{\sigma_{\Lambda^c} \in K^{V \setminus \Lambda}} \exp\left(-\sum_{A \subset V, A \cap \Lambda^c \neq \emptyset} U_A(\sigma_{\Lambda^c} \sigma_\Lambda)\right)$$

for the partition function in $\Lambda^c \equiv V \setminus \Lambda$ with boundary condition σ_Λ in Λ. Put

$$F_\Lambda^V(\sigma_\Lambda) \equiv \ln \frac{Z_{V \setminus \Lambda}(\Theta \sigma_\Lambda)}{Z_{V \setminus \Lambda}(\sigma_\Lambda)},$$

which, for local interactions, depends on the variables in (the interior boundary $\partial \Lambda$ of) Λ.

Put $R_\Lambda^V \equiv \bar{S}_\Lambda - F_\Lambda^V$.

Theorem 12.2 (LFT).

(i) *For every function G*

$$\mathbb{E}_V[G(-R_\Lambda^V)] = \mathbb{E}_V[G(R_\Lambda^V) e^{-R_\Lambda^V}], \qquad (12.12)$$

$$\mathbb{E}_V[G(-\bar{S}_\Lambda)] = \mathbb{E}_V[G(\bar{S}_\Lambda) e^{-\bar{S}_\Lambda + F_\Lambda^V}]. \qquad (12.13)$$

(ii) *For every family (λ_A) of complex numbers*

$$\mathbb{E}_V\left[\exp\left(-\sum_{A \subset \Lambda} \lambda_A [U_A \Theta - U_A]\right)\right]$$
$$= \mathbb{E}_V\left[\exp\left(-\sum_{A \subset \Lambda} (1 - \lambda_A)[U_A \Theta - U_A]\right) e^{F_\Lambda^V}\right]. \quad (12.14)$$

Remarks.

1. Theorem 12.2 describes a symmetry in the *local* fluctuations. The volume V can be much bigger (or even infinite) than the volume Λ in which we observe the fluctuations of the entropy production. This is physically rather important as we never observe global fluctuations. Notice, however, that the price to be paid is the correction term F_Λ^V in (12.13).

2. This fluctuation symmetry (12.12) and (12.13) was already mentioned below (12.6).

3. I take it to be understood that, for sufficiently local interactions, the potential F_Λ^V is of the order of the boundary of Λ. Thus, the correction to an exact symmetry in the distribution of the local entropy production (12.9) is of lower order. Therefore, asymptotically in a logarithmic sense, (12.13) leads directly to the symmetry (12.12) for the distribution of the entropy production $\bar{S}_\Lambda$ itself, a symmetry first uncovered in Evans et al. (1993), Gallavotti (1999), and Gallavotti and Cohen (1995a,b) for the large-deviation rate function of the phase-space contraction in the Sinai–Ruelle–Bowen (SRB) state of reversible dissipative mixing Anosov diffeomorphisms. One should not, however, take for granted that F_Λ^V is uniformly bounded by $|\partial\Lambda|$. In many of the models to which one wishes to apply the above scheme, one really needs to prove that it concerns here just a 'boundary term.' One source of problems can be that time is continuous, and an unbounded number of changes can happen in any finite interval. This is less of a problem for jump processes where one uses the fact that the Poisson process has all exponential moments but when working on noncompact phase spaces, the 'boundary term' F_Λ^V can possibly have nonexisting exponential moments and thus really can change the large-deviation rate function.

4. The main point about Theorem 12.2 is that the statements are very general and follow almost directly from identifying the variable entropy production with the part in the interaction that breaks the time-reversal invariance. It is to be hoped that this is useful for the general construction of nonequilibrium statistical mechanics. The identity (12.14) generates, by suitable differentiation, equalities between correlation functions. A discussion with an application to the Onsager reciprocity relations is contained in Gallavotti (1996) and Maes (1999). I like to compare (12.14) with the Ward identities as we know them from quantum field theory; the mathematical origin is very similar and nonperturbative, liberating us from close-to-equilibrium assumptions. Applications of Ward identities in statistical mechanics can be found in Simon (1993).

12.6 The Gibbs Hypothesis

The ambition in the above framework for the study of nonequilibrium driven systems is to use the standard Gibbs formalism and in particular its fluctuation theory in the 'current'-ensemble. The mathematics will therefore not deviate significantly from what is, for example, found in the contributions to this book of den Hollander (Chapter 10), Varadhan (Chapter 9), and Georgii (Chapter 3).

A frequently asked question is where these Gibbs measures come from or how they should be connected with a dynamics. Here are two short answers.

12.6.1 Pathspace Measure Construction

The first answer is illustrated below in Section 12.7 and many more examples can be found in Maes (1999), Maes and Redig (2000), Maes et al. (2000a,b, 2001a), Lebowitz and Spohn (1999), Kurchan (1998), and Eckmann et al. (1999). One constructs the Gibbs measure explicitly as the pathspace measure of a stationary process. Easiest is the case of stochastic dynamics. Just remember how the standard Ising model in one dimension can be constructed via the transfer matrix as the pathspace measure of a Markov chain on $\{-1, +1\}$. In general, it is not really necessary to realize the pathspace measure as a Gibbs distribution as we know it say for lattice spin systems; what is needed is to understand how the pathspace measure is governed by a space-time action which is approximately local and additive in space-time. Technically, this is a matter of setting up the appropriate Girsanov formula and I refer to the standard treatments in Lipster and Shiryayev (1978) and Brémaud (1981). Interestingly enough, this method of constructing the process via a Gibbsian space-time distribution has also led to new existence and uniqueness results for classes of diffusion processes (see, for example, Minlos et al. 2001).

Deterministic dynamics seem more of a problem here but, for example, we have learnt from Gallavotti (1996) and Gallavotti and Cohen (1995a,b) how the Gibbsian structure can be exploited on the level of the symbolic dynamics for sufficiently strongly chaotic dynamics. To treat deterministic dynamics via physical coarse-graining is explained in Maes and Netocny (2003).

12.6.2 Space-Time Equilibrium

The first answer has the advantage of being explicit and directly connected to model dynamics that one may have in mind. I believe, however, that the second answer is more to the point. The reason to use Gibbs measures here is exactly the same as for using Gibbs measures in the usual classical or quantum equilibrium statistical mechanics. All that changes is the type of ensemble, because we must work with the currents as macro-observables. Of course, the microscopic currents are functions of the trajectories that depend on the dynamics. To do this on space-time is the only price that must be paid, but it is a necessary one. Gibbs measures appear then as solutions of a variational principle but that need not be restricted to equilibrium conditions. In this way, the logic behind projects as the minimum entropy production principle is inverted: instead of trying to get a variational characterization of the stationary state starting from an approximate expression for the entropy production and using ad hoc assumptions, the statistical mechanical definition and properties of the entropy production are found from the Gibbsian character of the steady state. The essential observation on nonequilibrium steady states that confirms the Gibbsian picture on the appropriate scale of description is that for typical space-time boundary conditions on any chosen space-time window, we see again the same steady state inside the window!

12.7 Asymmetric Exclusion Processes

I illustrate the previous generalities using one concrete model, that of a bulk driven diffusive lattice gas. That consists of particles hopping randomly to nearby vacancies on a lattice in a preferred direction. Firstly, about the mean entropy production, to check that it coincides with what one should expect. Secondly, about the local entropy production fluctuations.

12.7.1 MEP for ASEP

I start with the asymmetric exclusion process (ASEP) on a one-dimensional ring $\{1, 2, \ldots, \ell\}$ (periodic boundary conditions). Each site i of the ring is either occupied by a particle (denoted by $\eta(i) = 1$) or is empty ($\eta(i) = 0$). The particle configuration η is subject to a Markovian particle conserving asymmetric hopping dynamics with rates

$$c(i, i+1, \eta) = \tfrac{1}{2}\eta(i)(1-\eta(i+1))e^{E/2} + \tfrac{1}{2}\eta(i+1)(1-\eta(i))e^{-E/2} \quad (12.15)$$

for changing the configuration from η to $\eta^{i,i+1}$ obtained by exchanging the occupations at sites i and $i + 1$. In words, particles hop to nearest-neighbor sites when there is a vacancy at a rate that depends on the direction. E is now an external driving field. We can easily obtain the space-time interaction from the standard Girsanov formula for Markov chains. The entropy production is then the relative action under time reversal and its expectation in a steady state is the MEP.

The product (or Bernoulli) measure ρ_u with uniform density $u \in [0, 1]$ is a stationary (nonreversible) measure for this dynamics. If we now consider a trajectory $(\eta_t, t \in [-T, T])$ of the stationary process in which at a certain time, when the configuration is η, a particle hops from site i to $i + 1$, then the time-reversed trajectory shows a particle jumping from $i + 1$ to i. The contribution of this event to the entropy production is therefore

$$\ln c(i, i+1, \eta) - \ln c(i, i+1, \eta^{i,i+1}) = E[\eta(i)(1-\eta(i+1)) - \eta(i+1)(1-\eta(i))]. \quad (12.16)$$

This equation must be understood from (12.9): it represents the local breaking of the time-reversal invariance in the space-time interaction due to the external field. This jump in the trajectory itself happens with a rate $c(i, i + 1, \eta)$ and therefore the mean entropy production equals

$$\begin{aligned} \mathrm{MEP}(u, E) &= \int c(0, 1, \eta) \ln \frac{c(0, 1, \eta)}{c(0, 1, \eta^{01})} \rho_u(d\eta) \\ &= Eu(1-u)\sinh(\tfrac{1}{2}E). \end{aligned} \quad (12.17)$$

This is correct: the MEP is the product of the field E with the current $j(u, E)$, where the current $j(u, E)$ is the expected net number of particles passing through a given bond:

$$j(u, E) \equiv \int \rho_u(d\eta) c(i, i+1, \eta)[\eta(i)(1 - \eta(i+1)) - \eta(i+1)(1 - \eta(i))]$$
$$= u(1 - u) \sinh(\tfrac{1}{2} E).$$

In quadratic approximation (that is close to equilibrium), $j(u, E) \approx u(1 - u) E/2$ and

$$\text{MEP}(u, E) \approx \frac{j(u, E)^2}{h_c}, \qquad (12.18)$$

which resembles the dissipated heat through a conductor in an electric field E with Ohmic conductivity

$$h_c \equiv \tfrac{1}{2} u(1 - u) = \tfrac{1}{4} \rho_u([\xi(0)(1 - \xi(1)) - \xi(1)(1 - \xi(0))]^2)$$

given in terms of the variance of the microscopic current (at $E = 0$).

12.7.2 LFT for ASEP

I take a $(2 + 1)$-dimensional set-up. There are (spatial) squares

$$V_0 \equiv [-L, L]^2 \cap \mathbb{Z}^2 \quad \text{and} \quad \Lambda_0 \equiv [-L', L']^2 \cap \mathbb{Z}^2$$

with $L' < L$ large, and a continuous time interval $[-T, T]$ in which we observe the ASEP with the external field E in the horizontal direction. This is the two-dimensional analogue of Section 12.7.1 but now on V_0 with periodic boundary conditions and hopping rates

$$c(i, j, \eta) \equiv \tfrac{1}{2} e^{E/2} \eta(i)(1 - \eta(j)) + \tfrac{1}{2} e^{-E/2} \eta(j)(1 - \eta(i))$$

for a horizontal bond $\langle ij \rangle$, $j = i + e_1$, with e_1 the unit vector in the positive horizontal direction, and

$$c(i, j, \eta) \equiv \tfrac{1}{2}[\eta(i)(1 - \eta(j)) + \eta(j)(1 - \eta(i))]$$

for a vertical bond $\langle ij \rangle$ with $j = i \pm e_2$. As stationary measure I take again ρ_u (Bernoulli with density u) and $(\eta_s(i), s \in [-T, T], i \in V_0)$ denotes the stationary process.

I show what becomes of relations (12.12) and (12.13).

Let $\mathbb{E}_V^E[\cdot]$ denote the expectation w.r.t. the process in $V \equiv [-T, T] \times V_0$ with (pathspace) law P_V^E. For a function f measurable from $\Lambda \equiv [-T, T] \times \Lambda_0$

to $\mathbb{R}$,

$$\mathbb{E}_V^E[f\Theta_{T,L'}] = \mathbb{E}_V^E[f\Theta_{T,L}]$$
$$= \mathbb{E}_V^E\left[f\frac{\mathrm{d}(P_V^E\Theta_{T,L})}{\mathrm{d}P_V^E}\right]$$
$$= \mathbb{E}_V^E\left[f\frac{\mathrm{d}P_V^{-E}|_\Lambda}{\mathrm{d}P_V^E|_\Lambda}\right] \quad (12.19)$$

so that, with

$$R_\Lambda^V = \ln\frac{\mathrm{d}P_V^E|_\Lambda}{\mathrm{d}P_V^{-E}|_\Lambda}$$

and $f = G(R_\Lambda^V)$, the identity (12.19) is just (12.12). The point is now to obtain for bulk contribution the entropy production in Λ:

$$\bar{S}_\Lambda(\eta_\Lambda) = E\int_{-T}^T \sum_{i,i+e_1\in\Lambda_0}[\eta_t(i)(1-\eta_t(i+e_1))-\eta_t(i+e_1)(1-\eta_t(i))]\,\mathrm{d}N_i^1(t), \quad (12.20)$$

where $N_i^1(t)$ is the number of jumps between i and $i+e_1$ up to time t. This can be computed directly from a standard Girsanov formula. The other terms in R_Λ^V are computed from a Girsanov formula for the nonMarkovian point process $\mathrm{d}P_V^E|_\Lambda$ and for this I first need to identify the intensities (see Brémaud 1981; Lipster and Shiryayev 1978). In order to have a Gibbsian structure (allowing me to pass to (12.13)) these intensities must be the same in the bulk of Λ as they were in the bulk of V. This is easy to verify by considering the conditional expectation of a function g in Λ_0 at time $t+\delta$ given the past history in Λ_0:

$$\mathbb{E}_V^E[g(\eta_{t+\delta}(i), i\in\Lambda_0) \mid \eta_s(i), s\in[-T,t], i\in\Lambda_0]$$
$$= \mathbb{E}_V^E[\mathbb{E}_V^E[g(\eta_{t+\delta}(i), i\in\Lambda_0) \mid \eta_s(i), s\in[-T,t], i\in V_0] \mid \eta_s(i),$$
$$s\in[-T,t], i\in\Lambda_0]. \quad (12.21)$$

The conditional expectation inside is explicit from the Markov process in V_0:

$$\mathbb{E}_V^E[g(\eta_{t+\delta}(i), i\in\Lambda_0) \mid \eta_s(i), s\in[-T,t], i\in V_0]$$
$$= g(\eta_t(i), i\in\Lambda_0) + \delta L_V^E g(\eta_t(i), i\in V_0) + O(\delta^2) \quad (12.22)$$

with L_V^E the Markov generator of the process. Since L_V^E is a sum over all bonds with local rates, we see that the process restricted to Λ has the same rates as the process P_Λ^E, except for the boundary of Λ_0, where a birth and death process is added. As a result the Girsanov formula for R_Λ^V is indeed of the form

$$R_\Lambda^V = \bar{S}_\Lambda - F_\Lambda^V + O(L'T),$$

where the term $O(L'T)$ is uniformly bounded of order of the boundary of Λ_0 (that is L') multiplied with the time-extension T and with unbounded correction term

$$F_\Lambda^V(\eta_\Lambda) = \sum_{i \in \partial \Lambda} \sum_{b_i} \int_{-T}^{T} \left[\ln \frac{\kappa_{b_i}^E(\eta, t)}{\kappa_{b_i}^{-E}(\eta, t)} + \eta_t(i) \ln \frac{\lambda_{b_i}^E(\eta, t)\kappa_{b_i}^{-E}(\eta, t)}{\lambda_{b_i}^{-E}(\eta, t)\kappa_{b_i}^E(\eta, t)} \right] dN_{b_i}(t), \quad (12.23)$$

where the second sum is over all bonds b_i starting at site $i \in \Lambda_0$ with the other end $j \in \Lambda_0^c$ and, with $b_i = \langle ij \rangle$,

$$\lambda_{b_i}^E(\eta, t) \equiv e^{E_j/2} \mathbb{E}_V^E[1 - \eta_t(j) \mid \eta_s(k), k \in \Lambda_0, s \in [-T, t]]$$

and

$$\kappa_{b_i}^E(\eta, t) \equiv e^{E_j/2} \mathbb{E}_V^E[\eta_t(j) \mid \eta_s(k), k \in \Lambda_0, s \in [-T, t]]$$

for $E_j = \pm E$ if $j = i \pm e_1$ and $E_j = 0$ if $j = i \pm e_2$. The expression (12.20) is the entropy production, that is, field times current in Λ, and (12.23) is the boundary term. This establishes (12.13).

References

Balian, R. 1982 *Du microscopique au macroscopique*, Vol. 2. École Polytechnique.

Bonetto, F., Lebowitz, J. L. and Rey-Bellet, L. 2000 Fourier's law: a challenge to theorists. Preprint, mp-arc 00-89.

Brémaud, P. 1981 *Point Processes and Queues, Martingale Dynamics.* Springer.

Bricmont, J. 1996 Science of chaos or chaos in science? In *The Flight from Science and Reason.* Ann. NY Acad. Sci. **775**, 131. (*Physicalia Magazine* **17** (1995), 159.)

den Hollander, F. 2003 Relative entropy for random motion in a random medium. In *Entropy* (ed. A. Greven, G. Keller and G. Warnecke), pp. 215–231. Princeton University Press.

Eckmann, J.-P., Pillet, C.-A. and Rey-Bellet, L. 1999 Non-equilibrium statistical mechanics of anharmonic chains coupled to two heat baths at different temperatures. *Commun. Math. Phys.* **201**, 657–697.

Evans, D. J., Cohen, E. G. D. and Morriss, G. P. 1993 Probability of second law violations in steady flows. *Phys. Rev. Lett.* **71**, 2401–2404.

Gallavotti, G. 1996 Chaotic hypothesis: Onsager reciprocity and fluctuation–dissipation theorem. *J. Statist. Phys.* **84**, 899–926.

Gallavotti, G. 1998 Chaotic dynamics, fluctuations, nonequilibrium ensembles. *Chaos* **8**, 384–392.

Gallavotti, G. 1999 A local fluctuation theorem. *Physica* A **263**, 39–50.

Gallavotti, G. and Cohen, E. G. D. 1995a Dynamical ensembles in nonequilibrium statistical mechanics. *Phys. Rev. Lett.* **74**, 2694–2697.

Gallavotti, G. and Cohen, E. G. D. 1995b Dynamical ensembles in stationary states. *J. Statist. Phys.* **80**, 931–970.

Georgii, H.-O. 2003 Probabilistic aspects of entropy. In *Entropy* (ed. A. Greven, G. Keller and G. Warnecke), pp. 37–54. Princeton University Press.

Goldstein, S. 2003 Boltzmann's approach to statistical mechanics. In *Chance in Physics: Foundations and Perspectives* (ed. D. Dürr). Springer.

Jaynes, E. T. 1992 The Gibbs Paradox. In *Maximum-Entropy and Bayesian Methods* (ed. G. Erickson, P. Neudorfer and C. R. Smith). Kluwer, Dordrecht.

Kurchan, J. 1998 Fluctuation theorem for stochastic dynamics. *J. Phys.* A **31**, 3719–3729.

Lebowitz, J. L. 1999 Microscopic origins of irreversible macroscopic behavior. *Physica* A **263**, 516–527.

Lipster, R. S. and Shiryayev, A. N. 1978 *Statistics of Random Processes. II. Applications.* Springer.

Maes, C. 1999 The fluctuation theorem as a Gibbs property. *J. Statist. Phys.* **95**, 367–392.

Maes, C. and Netocny, K. 2003 Time-reversal and entropy. *J. Statist. Phys.* **110**, 269–310.

Maes, C. and Redig, F. 2000 Positivity of entropy production. *J. Statist. Phys.* **101**, 3–16.

Maes, C., Redig, F. and van Moffaert, A. 2000a On the definition of entropy production via examples. *J. Math. Phys.* **41**, 1528–1554.

Maes, C., Redig, F. and Verschuere, M. 2000b No current without heat. *J. Statist. Phys.* **106**, 569–587.

Maes, C., Redig, F. and Verschuere, M. 2001a Entropy production for interacting particle systems. *Markov Process. Rel. Fields* **7**, 119–134.

Maes, C., Redig, F. and Verschuere, M. 2001b From global to local fluctuation theorems. *Moscow Math. J.* **1**, 421–438.

Lebowitz, J. L. and Spohn, H. 1999 A Gallavotti–Cohen type symmetry in the large deviation functional for stochastic dynamics. *J. Statist. Phys.* **95**, 333–365.

Minlos, R., Roelly, S. and Zessin, H. 2001 Gibbs states on space-time. *Potential Analysis* **13**(4).

Penrose, R. 1989 *The Emperor's New Mind*. Oxford University Press.

Ruelle, D. 1999 Smooth dynamics and new theoretical ideas in nonequilibrium statistical mechanics. *J. Statist. Phys.* **95**, 393–468.

Simon, B. 1993 *The Statistical Mechanics of Lattice Gases*, Vol. 1. Princeton University Press.

Varadhan, S. R. S. 2003 Large deviations and entropy. In *Entropy* (ed. A. Greven, G. Keller and G. Warnecke), pp. 199–214. Princeton University Press.

Chapter Thirteen

Entropy: a Dialogue

Joel L. Lebowitz
Rutgers, The State University of New Jersey

Christian Maes
Instituut voor Theoretische Fysica, K. U. Leuven

A scientist who worked all his life in field theory but was always very curious about entropy, an emotional topic among some of his colleagues, dies and is ushered into a very large room. There, behind a desk, sits an old angel. On the desk and on the shelves lining the walls are many books and instruments, some familiar, some strange. The eye of the scientist (S) is attracted to a shiny sphere with some special aura surrounding it. After some exchange of pleasanteries with the angel (A) he asks what is inside the sphere.

A: The sphere is filled with one heavenly mole (2^{81}) of very small hard balls: they occupy a volume fraction 2^{-18} of the outer ball which has a volume of $1\,\text{m}^3$. These particles move according to Hamiltonian dynamics with elastic collisions and are perfectly shielded from *all* external influences. I started them a very very long time ago in a particular microscopic state X, which I picked at 'random' from a region Ω of the phase space; the phase points in Ω correspond to all the particles being (without overlap) inside a ball with the same origin and one-tenth the radius of the big ball, with a total energy (all kinetic) between $E \pm \Delta E$, E corresponding to average speed of 10^5 m s^{-1} and $\Delta E = 2^{-18} E$. I have watched them evolve since then out of curiosity for how such a classical system would behave.

This is great, thinks S. Here is my chance to find out about entropy but I better check first to see if there is any trick here.

S: How interesting. I assume that your computers, I mean brains, are big enough and accurate enough so that you know the exact position and velocity of each particle at every instant of time.

A: Yes, of course.

S: But then you also know in advance how it will evolve. Why do you bother with this experiment?

A: You see, while I (and other angels of my category) have essentially unlimited memory and therefore can have perfect knowledge of the past and present state of the system, I do not have any special computational abilities and therefore am unable to predict the future any better than mortals are. It is therefore fun to watch these particles evolve in accordance with Newtonian dynamics. In fact, I think of it as an analog computer.

S: OK, so please let us discuss it. According to my understanding, the initial time evolution should have been described very accurately on the mesoscopic level by the Boltzmann equation (BE). This should lead to a spatially uniform state with a Maxwellian distribution of velocities in a relatively short time. Since then, I imagine the system has been in equilibrium with small fluctuations which are quite well described by the linearized BE with Gaussian noise.

A: Absolutely. That is exactly how it evolved on the mesoscopic scale.

S: Fine, but let me ask you: are you really sure that your system is truly isolated?

A: Yes. Just for fun I occasionally exactly reverse all the velocities and let the system evolve for a few years and then reverse the velocities again. The particles all retrace their trajectories perfectly. Actually, doing that is not cost free—decreasing the entropy always requires some effort.

S: So I guess you can make them all return to their initial state, but have you ever seen them all go back spontaneously into the little ball where you started them from?

A: Oh, you mean a Poincaré recurrence? Actually, yes, I have witnessed such an event once but I don't really expect to see it again. I will surely be replaced here by another angel long before then.

S: And just before that return, how did the entropy of this isolated system behave?

A: That was the most interesting part. Suddenly, for no apparent reason, it just kept decreasing until it was back to where it started at the beginning and then after a brief time interval it started increasing again as if nothing special happened.

S: But what about the second law? You have actually seen it violated. So it is not an absolute law. This is exactly what caused the heated lunch-time discussions in our cafeteria.

A: Sure. Isn't that exactly what Maxwell, Boltzmann, and Gibbs clearly said? All you have to do is wait long enough and have your system well isolated. These are both things that come naturally to me—remember, I am older than your universe by quite a bit, which, by the way, is why I am into this classical business. But enough of me, you look puzzled?

ENTROPY: A DIALOGUE

S: Yes, I am puzzled. To help clarify it for me, let me ask you what is the present entropy of this system?

A: Looking at it at this instant, it seems to be in a typical equilibrium configuration, pretty uniform spatially with a Maxwellian distribution of velocities. Hence the macrostate is the equilibrium one. So if you want the value of the entropy, just take the log of the phase space volume appropriate to its energy, and get the value given in the textbooks.

S: This is what has been puzzling me for a long time. According to the textbooks, this is the entropy appropriate for ignorant mortals who have to use some phase space probability density μ. Given such a μ they compute the entropy via $-\int \mu \log \mu \, dX$; this formula, first introduced by Gibbs and later extended by Shannon and others, has been almost synonymous with entropy in our world. You on the other hand know the microscopic state, so your probability density is in effect a delta function—why shouldn't your entropy then be $-\infty$? Or, if you use some discretization to take into account quantum mechanics or whatever, it should then be zero. Why should the volume of the energy surface be relevant for you?

A: Oh, but I was not talking about anybody in particular. Of course I do know, to any precision I care, the exact locations and velocities of the particles. I thought you were asking me about the present macrostate corresponding to the energy and density profile of the system and how many microstates, i.e. phase space volume, fit in it. Obviously, no matter what we really know, this quantity is unambiguously defined, up to relatively negligible terms which depend on the precision with which you define your macrostate, and it gives rise to the standard entropy.

S: What do you mean by standard entropy? Are there different entropies?

A: Well, of course. Not only are there different types of entropy that are useful for different types of problems but within many of these you have still differences depending on what relevant macrovariables you take. There is the measure-theoretic entropy, the topological entropy, the von Neumann entropy, the Gibbs entropy, the H-functional, the Clausius thermodynamic entropy, the Boltzmann entropy, etc.

S: I have always felt that it is unfortunate to have so many quantities with the same name, but what are you talking about in the present context?

A: Here we are speaking in a statistical mechanical context and I was referring to the Boltzmann entropy and also to the thermodynamic equilibrium entropy. You must know that these are only defined once you specify the macrovariables. So, if you give me another set of macrovariables, you get a different Boltzmann entropy.

S: But is this choice completely arbitrary?

A: Yes and no, it depends for what purposes you are going to use this entropy. If you want to use it as a predictive tool for the time evolution, as in the second law, then it is closely linked with the type and level of description you are using. Actually, for the gas in this ball you could usefully compute it from the one-particle distribution function as in the kinetic theory of gases. The Boltzmann equation will then describe the way it changes (except of course in some very special cases, like the ones we discussed earlier). So you see you are free but certain choices make much more sense. In fact, for thermodynamics I would even say you have a rather limited choice.

S: Then what determines a good choice?

A: What you are really asking is what makes a function on the microstate a good macrovariable. One thing that is crucial is to have a function that is additive. The reason is that then most of phase space, say corresponding to a given energy, is occupied by one very large region where these macrovariables are constant. It also helps to use quantities which are conserved locally. This essentially selects the energy, momentum, and particle number.

S: How do you measure these volumes in phase space?

A: You count. It is a bit tricky but you can for all practical purposes also take some nice density over which you integrate.

S: Then what would be an example of having different entropies in your sense?

A: There are many but here is a trivial one. Suppose that I have here red and green particles and that initially, all red particles were on the right side of my ball and all green particles were to the left. Same pressures, same temperatures. Now I let them go and they mix. I get something brown. Did the entropy increase?

S: This I know. Of course the entropy increases. I can even compute by how much.

A: Aha, but now suppose that you were colorblind, you could not see the difference between green and red and you would see something pretty uniformly grey from beginning to end.

S: And the particles can be assumed to be individually each taking exactly the same paths as before?

A: Yes. The microscopic dynamics is identical.

S: Strange. Now I would say there is no increase in entropy. Is that not related to the Gibbs Paradox?

A: Indeed, as is its solution. You see, there are many entropies. In one case you add the extra macrovariable giving you both the color and density-profile, but if you don't care about color you just inspect the overall density profile.

S: There is one thing I am now confused about. If there are these different entropies corresponding to what you and I decide to include as macrovariables, how can the measurement of entropy be objective, yielding the same result for all of us?

A: That is a good question. The solution is that you should think about what is relevant and what is not for the problem you are considering. The possibility of having different entropies is not at all harmful as long as you understand that it corresponds to different situations, e.g. are you interested in properties of the system which depend on the color of the particles or not? Of course, in the former case you better have someone or some instrument which is color sensitive.

S: Do you mean something like contextual?

A: Something like that but don't make it too complicated. Just imagine a colorblind man trying to sort out green from red particles. As for contextuality, you also have it in quantum mechanics. Maybe it is to this that you refer.

This is an opportunity to ask about quantum statistics, feels the scientist. And who knows, perhaps the angel will tell me some quantum secrets.

S: That reminds me. As you know everything, is there still something like indistinguishability for you?

A: No, I can distinguish all particles.

S: That is what I thought, but then how do you get quantum statistics and why do you divide by this $N!$ in front of the Liouville measure?

A: Now it is me not understanding your question.

S: I have seen many textbooks arguing for extensivity of the entropy via quantum mechanics and I have learnt that quantum particles follow a Bose–Einstein or Fermi–Dirac statistics for their occupation numbers. But they must be indistinguishable.

A: Did that not sound a bit strange to you?

S: No, no—it looked pretty convincing.

A: But they did not know about me, it seems. The problem of extensivity of entropy is unchanged in going from classical to quantum physics.

S: Are you saying that this indistinguishability is similar to your example of colorblindness, where you can exchange red for green?

A: It is. You see, classical particles with the same charge, mass, and what have you, similarly don't come equipped with distinguishing labels. They are in fact only distinguished by their dynamic properties. That is the history of their momenta and positions. There is no reason therefore to be mystical about this $N!$; it emerges naturally in classical mechanics just like in quantum mechanics.

S: Perhaps so, but, at any rate, quantum mechanics is much more complicated. Since there is no obvious quantum phase space of microstates, there is also no obvious phase space volume, so tell me, how should one define the quantum Boltzmann entropy?

A: Boltzmann speaks about the connection between the microscopic and macroscopic thermal properties of physical systems. This can be divided into two parts. The first part gives a microscopic formula for computing the entropy $S_B(M)$ of a macroscopic system specified to be in a macrostate M, e.g. one described by the hydrodynamic variables entering the Navier–Stokes equation. This goes over without any deep conceptual problems to quantum mechanics. Given M you can still use Boltzmann's formula $S_B(M) = k \log |\hat{\Gamma}(M)|$, where $|\hat{\Gamma}(M)|$ is now the dimension of the linear subset of the Hilbert space of the system corresponding to M. This makes sense because the variables defining the macrostate, e.g. those specifying (to some appropriate accuracy) the particle number, the momentum, and the energy in each macroscopically small but microscopically large region of the volume occupied by the system, can be chosen to essentially commute and so they form an orthogonal decomposition of the Hilbert space.

S: That's great. I actually remember reading about that in von Neumann's book but had forgotten about it since all the textbooks and papers always use for the entropy the formula $-k \operatorname{Tr}(\hat{\mu} \log \hat{\mu})$, where $\hat{\mu}$ is the density matrix, but that is zero if you are in the pure state Ψ.

A: Yes. This quantum version of the Gibbs entropy, also called the von Neumann entropy, has produced even more confusion than its classical counterpart. But aside from this unnecessary confusion you, I mean your colleagues still down there, do have a real problem when it comes to the second part of Boltzmann's micro–macro connection. Given a point X in the phase space we always have a well-defined M: call it $M(X)$. This defines the Boltzmann entropy $S_B(X) = S_B(M(X))$ for a single macroscopic system like the hard spheres inside this box. As you know, this is important for giving a clear microscopic interpretation to the second law. There is no such relationship between Ψ and M for quantum systems: the wave function of the system does not necessarily specify a unique macrostate. Think of Schrödinger's cat.

S: That is exactly what I was just thinking about. How can I understand this?

The angel shakes his wings and thinks, here is somebody who in fact wants to understand quantum mechanics!

A: Well, to tell you the truth I don't really understand it either. I think that there are other ways to think about quantum mechanics but I am not really the right angel to explain them. Why don't you see my colleague, Angel Psi?

S: Good. I want to see Angel Psi. I have many questions for him.

A: I wish you good luck.

References

Boltzmann, L. 1896, 1898 *Vorlesungen über Gastheorie*, 2 vols. J. A. Barth, Leipzig. This book has been translated into English by S. G. Brush, *Lectures on Gas Theory* (Cambridge University Press, 1964) and it is a treasure for those trying to understand foundational issues.

Bricmont, J. 1996 Science of chaos or chaos in science? In *The Flight from Science and Reason. Ann. NY Acad. Sci.* **775**, 131. This article first appeared in the publication of the Belgian Physical Society, *Physicalia Magazine* **17**, 159, (1995), where it is followed by an exchange between Prigogine and Bricmont.

Gibbs, J. W. 1878 *On the Equilibrium of Heterogeneous Substances*. Connecticut Academy of Sciences. Reprinted in *The Scientific Papers of J. Willard Gibbs* (Dover, New York, 1961). In this monumental piece of work, Gibbs discusses, among many other things, the so-called *subjective* aspect of entropy.

Goldstein, S. 1998 Quantum mechanics without observers. *Phys. Today* (March), p. 42, (April), p. 38. As in the next article, problems with and alternatives to standard quantum mechanics are discussed.

Goldstein, S. and Lebowitz, J. L. 1995 Quantum mechanics. In *The Physical Review: the First Hundred Years* (ed. H. Stroke). AIP Press, New York.

Goldstein, S. and Lebowitz, J. L. 2003 On the (Boltzmann) entropy of nonequilibrium systems. *Physica* D. (In the press and on archive at cond-mat/0304251.)

Jaynes, E. T. 1992 The Gibbs Paradox. In *Maximum-Entropy and Bayesian Methods* (ed. G. Erickson, P. Neudorfer and C. R. Smith). Kluwer, Dordrecht. This article discusses both elements of Gibbs (1878) and Pauli (1973).

Lebowitz, J. L. 1993a Macroscopic laws and microscopic dynamics, time's arrow and Boltzmann's entropy. *Physica* A **194**, 1–97.

Lebowitz, J. L. 1993b Boltzmann's entropy and time's arrow. *Phys. Today* **46**, 32–38.

Lebowitz, J. L. 1999 A century of statistical mechanics: a selective review of two central issues. *Rev. Mod. Phys.* **71**, 346–357.

Lebowitz, J. L. 1994a Microscopic reversibility and macroscopic behavior: physical explanations and mathematical derivations. In *25 Years of Non-Equilibrium Statistical Mechanics, Proc. Sitges Conf., Barcelona, Spain, 1994*. Lecture Notes in Physics (ed. J. J. Brey, J. Marro, J. M. Rubí and M. San Miguel). Springer.

Lebowitz, J. L. 1994b Time's arrow and Boltzmann's entropy. *Physical Origins of Time Asymmetry* (ed. J. Halliwell and W. H. Zurek), pp. 131–146. Cambridge University Press.

Lebowitz, J. L. 1999 Microscopic origins of irreversible macroscopic behavior. *Physica* A **263**, 516–527.

Maxwell, J. C. 1990–95 *Scientific Letters and Papers* (ed. P. M. Harman). Cambridge University Press. In Vol. II, pp. 331–332, Maxwell's demon is introduced. This demon, or, as Maxwell himself wished to call it, Maxwell's valve or finite being, is not unlike the angel in the fable. Maxwell emphasizes that his purpose in invoking this thought experiment was to explain that the second law of thermodynamics has only a *statistical certainty*. See also 'Tait's Thermodynamics,' *Nature* **17**, 257 (1878).

Pauli, W. 1973 *Thermodynamics and the Kinetic Theory of Gases* (ed. C. P. Enz). Pauli Lectures on Physics, Vol. 3. MIT Press, Cambridge, MA. It includes an interesting discussion on the question of extensivity of entropy.

Penrose, R. 1989 *The Emperor's New Mind*. Oxford University Press. Chapter 7 contains a wonderfully clear description of some of the most important aspects of entropy and the second law.

von Neumann, J. 1955 *Mathematical Foundations of Quantum Mechanics*, pp. 398–416. Princeton University Press. (Translated from German edition (1932, Springer) by R. T. Beyer.) Here you will find, in Chapter V, a quantum version and discussion of the Boltzmann entropy.

PART 4
Entropy and Information

Chapter Fourteen

Classical and Quantum Entropies: Dynamics and Information

Fabio Benatti

Università di Trieste

The role of the dynamical entropy of Kolmogorov and Sinai is reviewed in connection with the compressibility of classical information and the instability of classical dynamical systems. Two recent proposals of quantum dynamical entropies are then presented with reference to their applications to quantum information theory and quantum chaos.

14.1 Introduction

In general, states cannot be attributed with certainty to physical systems, but only with some probability; the uncertainty about the actual state is measured by the *Shannon entropy*[1] for classical systems (Billingsley 1965; Walters 1982) and by the *von Neumann entropy* for quantum ones (Ohya and Petz 1993; Wehrl 1978).

If the uncertainty is about predictions concerning the future of physical systems, it can be decreased by gaining information from the time-evolution itself. However, the dynamics may go on producing new information at each successive observation so that forecasting is not made more reliable by knowledge of the past; classically, this kind of uncertainty about the future is measured by the *Kolmogorov–Sinai* (KS) *entropy* (Billingsley 1965; Walters 1982).

The KS entropy[2] is a powerful indicator of unpredictability in classical dynamical systems: it governs the exponential amplification of errors and, as such, it is a signature of chaos (Schuster 1995); it gives the maximal *compression rate* of classical information (Billingsley 1965; Cover and Thomas 1991), and it measures the *algorithmic complexity*[3] of classical trajectories (Alekseev and Yakobson 1981; Brudno 1983).

---

[1] See Section 3.1 of Georgii's contribution to this book.

[2] See Chapter 17 by Keane, Section 16.1.2 of Young's and Section 3.2 of Georgii's contributions to this book.

[3] See Rissanen's contribution to this book.

In extending the notion of dynamical entropy to quantum systems, the technical difficulties presented by noncommutativity are accompanied by the following question of principle: information is gathered from observations, that is, by measuring certain observables. However, since quantum measurement processes act themselves as a source of noise and affect the intrinsic properties of the dynamics, the question arises whether this typical phenomenon is or is not to be accounted for when discussing unpredictability in quantum mechanics.

Several quantum dynamical entropies have been proposed so far (Accardi et al. 1997; Alicki and Fannes 1994; Connes et al. 1987; Voiculescu 1995). Quantum chaos and quantum information provide an ideal testing ground for the different candidates and might shed light on the fundamental issues involved. Against this background we shall present two quantum dynamical entropies (Alicki and Fannes 1994; Connes et al. 1987) and examine their differences and their applications.

14.2 Shannon and von Neumann Entropy

States of classical systems are described by probability measures μ on measure spaces $\mathcal{X}$, the 'volumes' $\mu(E)$ of measurable subsets $E \subset \mathcal{X}$ giving the probabilities for a point $x \in \mathcal{X}$ to belong to E. Realistic measurements cannot sort out points, but can only distinguish between certain cells E_j, $j = 1, 2, \ldots, D$, of size dependent on the accuracy of the instruments used. If $E_i \cap E_j = \emptyset$ when $i \neq j$, and $\bigcup_{j=1}^{D} E_j = \mathcal{X}$, these cells are the *atoms* of a so-called *finite partition* $\mathcal{E}$ of $\mathcal{X}$.

Every partition $\mathcal{E}$ defines a discrete probability distribution $\mu_\mathcal{E} = \{\mu(E_j)\}_{j=1}^{D}$ and the accessible information provided by the corresponding *coarse graining* of $\mathcal{X}$ is measured by the Shannon entropy of $\mu_\mathcal{E}$ (Billingsley 1965; Walters 1982)

$$S(\mu_\mathcal{E}) := \sum_j \eta(\mu(E_j)), \qquad (14.1)$$

where $\eta(x) = -x \log x$ if $x > 0$, while $\eta(0) = 0$.

In standard quantum mechanics, the simplest states of physical systems are vectors $|\psi\rangle$ of some Hilbert space $\mathcal{H}$. The physical observables are linear operators X on $\mathcal{H}$ and the vector states define expectations (mean values) on them: $X \mapsto \langle\psi|X|\psi\rangle$. By means of the trace operation (sum over the diagonal entries of matrices), vector states are also identified with projectors $P_\psi := |\psi\rangle\langle\psi|$ called *pure states*: $\langle\psi|X|\psi\rangle = \text{Tr}(P_\psi X)$.

In this way pure states can be generalized to *mixtures* $\rho = \sum_j \lambda_j |\psi_j\rangle\langle\psi_j|$, with certain weights $\lambda_j > 0$ such that $\sum_j \lambda_j = 1$, also called *density matrices*. They define expectations $X \mapsto \rho(X) := \text{Tr}(\rho X)$ and are positive operators with eigenvalues $0 \leqslant r_\ell$, $\sum_\ell r_\ell = \text{Tr}(\rho) = 1$, forming discrete probability distributions.

The corresponding uncertainty is measured by the von Neumann entropy (Ohya and Petz 1993; Wehrl 1978)

$$S(\rho) := -\operatorname{Tr}(\rho \log \rho) = \sum_{\ell} \eta(r_\ell). \tag{14.2}$$

Therefore, $S(\rho) = 0$ if and only if ρ is a projector, that is, if and only if $\rho^2 = \rho$, which means $r_1 = 1$ and $r_\ell = 0$ otherwise.

Remark 14.1. In quantum mechanics, physical observables (linear operators) form noncommutative algebras. Physical observables of classical systems are functions f on $\mathcal{X}$ and constitute commutative algebras. Similarly to ρ in the quantum case, the measure μ defines the expectation $\omega_\mu(f) = \int_{\mathcal{X}} d\mu(x) f(x)$ on them. Therefore, in the following we will refer to both classical and quantum states as expectations, that is, as positive linear functionals (expectations) on the algebras of observables.

14.2.1 Coding for Classical Memoryless Sources

In classical information theory[4] a source A_D is the component of a transmission channel emitting signals $\{s_j\}_{j=1}^{D}$ with frequencies $\pi = \{p(s_j)\}_{j=1}^{D}$ such that $\sum_{j=1}^{D} p(s_j) = 1$. The state of the source is thus described by the discrete probability distribution π and its Shannon entropy

$$S(\pi) = -\sum_{j=1}^{D} p(s_j) \log p(s_j)$$

measures the average uncertainty about the outcoming signals (Cover and Thomas 1991).

A source A_D is *memoryless* if the signals are statistically independent so that strings $s_{[0,n-1]} := s_0 s_1 \cdots s_{n-1}$ occur with probabilities

$$p(s_{[0,n-1]}) = \prod_{k=0}^{n-1} p(s_k).$$

Then, by the weak law of large numbers, for n large enough, the strings can be divided into a *typical* subset $\mathcal{T}$ with $\simeq e^{nS(\pi)}$ strings for which $p(s_{[0,n-1]}) \simeq e^{-nS(\pi)}$ and its *complementary set* $\mathcal{T}^c$ of negligible measure[5] (*asymptotic equipartition property*) (Cover and Thomas 1991).

The strings of $\mathcal{T}$ can then be listed and one-to-one encoded into the bit-strings of their positions in the list. This requires $\simeq S(\pi)/\log 2$ bits per symbol. Should

---

[4] See Section 3.2 of Georgii's contribution to this book.

[5] See Section 3.2, Theorems 3.4–3.6 of Georgii's and Section 15.2 of Rissanen's contribution to this book.

the entire set of D^n strings be one-to-one encoded, $\log D/\log 2$ bits per symbol would be necessary. However, one-to-one encoding of strings in $\mathcal{T}^c$ is not needed, for they occur with negligible frequency.

Theorem 14.2 (Cover and Thomas 1991). *Asymptotically, long strings of signals from a classical memoryless source A_D with frequency distribution π can be reliably encoded by using only and no less than $S(\pi)/\log 2$ bits per symbol.*

14.2.2 Coding for Quantum Memoryless Sources

Remarkably, undetected eavesdropping during data transmission can be eliminated by encoding the signals s_j into nonorthogonal, normalized vector states $|s_j\rangle$ of some Hilbert space, e.g. photon polarization or spin states. The use of photons to transmit secret keys and the possibility of employing quantum systems to process and retrieve information is the task of *quantum information theory* (Bouwmeester et al. 2000).

Remark 14.3. If the Hilbert space $\mathcal{H}$ of vector states of a quantum system has finite dimension d ($d = 2$ for a spin $1/2$ particle), the system observables form the algebra M_d of $d \times d$ matrices. Unlike classical spins $1/2$, whose only vector states $|\uparrow\rangle$ and $|\downarrow\rangle$ carry one bit of information each, vector states of quantum spins $1/2$ are generic linear combinations $a|\uparrow\rangle + b|\downarrow\rangle$, $a, b \in C$. They are said to carry one *q-bit* of information (Schumacher 1995). Algebraically, one passes from a diagonal 2×2 matrix algebra A_2 to the whole matrix algebra M_2.

Quantum sources M_D emit D vector states $|s_j\rangle$ with frequencies

$$\pi = \{p(s_j)\}_{j=1}^D,$$

so that their states are density matrices

$$\rho_\pi = \sum_{j=1}^D p(s_j)|s_j\rangle\langle s_j|$$

with von Neumann entropies $S(\rho_\pi)$. If the states $|s_j\rangle$ are not orthogonal, the weights $p(s_j)$ are not the eigenvalues r_ℓ^π of ρ_π; from convexity, $0 \leqslant S(\rho_\pi) < S(\pi)$ (Ohya and Petz 1993; Wehrl 1978).

The natural question is thus whether the information coming from a quantum memoryless source can be compressed and, if so, whether it is the quantum entropy, $S(\rho_\pi)$, or the classical one, $S(\pi)$, that plays a role in the process.

Remark 14.4. Compression will be achieved by coding the states from the source into a smaller set of states in such a way that the information contained in the former might be reliably retrieved from the latter.

The degree of reliability can be measured by the so-called *fidelity* parameter (Schumacher 1995): given a state $\rho = \sum_\ell \lambda_\ell |\psi_\ell\rangle\langle\psi_\ell|$ on $\mathcal{H}$ the fidelity of the coding $|\psi_\ell\rangle \mapsto |\phi_\ell\rangle \in \mathcal{H}$ is $F := \sum_\ell \lambda_\ell |\langle\phi_\ell|\psi_\ell\rangle|^2$. Since $|\langle\phi_\ell|\psi_\ell\rangle| < 1$ unless $|\phi_\ell\rangle = |\psi_\ell\rangle$, the difference $1 - F$ measure the reliability of using the vectors ϕ_ℓ to store the information carried by the vectors ψ_ℓ.

A quantum source M_D is memoryless if vectors $|s_{[1,n]}\rangle := |s_1 \otimes s_2 \cdots \otimes s_n\rangle$ are emitted with frequencies

$$p(s_{[1,n]}) = \prod_{k=1}^{n} p(s_k).$$

In such a case, the state of the ensemble of $|s_{[1,n]}\rangle$ is the tensor product state $\rho_{[1,n]}^\otimes := \rho_\pi \otimes \rho_\pi \cdots \otimes \rho_\pi$ with eigenvalues

$$\boldsymbol{r}_{[1,n]} := \prod_{j=1}^{n} r_{s_j}^\pi,$$

respectively eigenvectors

$$|\boldsymbol{r}_{[1,n]}\rangle := \bigotimes_{k=1}^{n} |r_{s_k}^\pi\rangle,$$

that are (tensor) products of eigenvalues r_j^π, respectively eigenvectors $|r_j^\pi\rangle$ of the state ρ_π. The ensemble von Neumann entropy is $S(\rho_{[1,n]}^\otimes) = nS(\rho_\pi)$. Thus, indices $s_1 s_2 \cdots s_n$ occur with probabilities $\boldsymbol{r}_{[1,n]}$ and the asymptotic equipartition property identifies typical and untypical eigenvectors, the role of $S(\pi)$ now being played by $S(\rho_\pi)$. For n large enough, the typical eigenvectors $|\boldsymbol{r}_{[1,n]}\rangle$ span a typical Hilbert space $\mathcal{H}_T$ of dimension $\simeq e^{n\,S(\rho_\pi)}$ carrying most of the probability of the state $\rho_{[1,n]}^\otimes$.

Then, a coding $|s_{[1,n]}\rangle \mapsto |\psi_{[1,n]}\rangle$ can be constructed (Schumacher 1995) from $\mathcal{H}_T$ into the tensor product of $S(\rho_\pi)/\log 2$ copies of two-dimensional Hilbert spaces in such a way that the original $|s_{[1,n]}\rangle$ can be reliably retrieved from the code vectors $|\psi_{[1,n]}\rangle$ with negligible errors, the coding protocol having *fidelity* arbitrarily close to 1.

Theorem 14.5 (Schumacher 1995). *Asymptotically, long strings of signals from a quantum memoryless source M_D with state ρ_π can be reliably encoded by using only and no less than $S(\rho_\pi)/\log 2$ q-bits per symbol.*

14.3 Kolmogorov–Sinai Entropy

Equipping $(\mathcal{X}, \mu)$ with measurable and invertible dynamical maps $T : \mathcal{X} \mapsto \mathcal{X}$, we obtain triples $(\mathcal{X}, \mu, T)$ representing discrete-time, reversible, classical systems, with $\{T^k x\}_{k\in\mathbb{N}}$ denoting the trajectory through $x \in \mathcal{X}$ at $t = 0$.[6]

---

[6] See Section 16.1.2 of Young's contribution to this book.

Given a finite partition $\mathcal{E} = \{E_j\}_{j=1}^{D}$, the trajectories can be symbolically encoded by associating to them the sequences $s(x) = s_0 s_1 \cdots$ of the indices labeling the atoms visited at successive times: $[s(x)](k) = s_k \Leftrightarrow T^k x \in E_{s_k}$.

Remark 14.6. All $s(x)$ with initial segment $s_{[0,n-1]} = s_0 s_1 \cdots s_{n-1}$ are associated with points x belonging to the atoms

$$\bigcap_{k=0}^{n-1} T^{-k}(E_{s_k})$$

of the finer partitions $\mathcal{E}_{[0,n-1]}^{T}$ of $\mathcal{X}$ that are obtained from the evolving partitions

$$T^{-k}(\mathcal{E}) = \{T^{-k}(E_j)\}_{j=1}^{D}.$$

The code-sequences constitute a subset of infinite D-nary sequences Σ_D. Intuitively, the larger the number of different code-sequences, the harder the predictions (Alekseev and Yakobson 1981). Concretely, let S_k be the stochastic variable with values $1, 2, \ldots, D$ signaling the atom visited at time k and consider the set of probabilities

$$p(s_{[0,n-1]}) := \text{Prob}\{S_0 = s_0, S_1 = s_1, \ldots, S_{n-1} = s_{n-1}\}$$

$$= \mu\left(\bigcap_{k=0}^{n-1} T^{-k}(E_{s_k})\right). \tag{14.3}$$

The ensemble of strings $s_{[0,n-1]} := s_0 s_1 \cdots s_{n-1}$ has Shannon entropy

$$H_\mu(\mathcal{E}_{[0,n-1]}^{T}) := -\sum_{s_{[0,n-1]}} p(s_{[0,n-1]}) \log p(s_{[0,n-1]}). \tag{14.4}$$

Because of T-invariance of μ, the S_k form a stationary stochastic process and the *entropy per unit-time*,

$$h_\mu(T, \mathcal{E}) := \lim_{n \to +\infty} \frac{1}{n} H_\mu(\mathcal{E}_{[0,n-1]}),$$

is well defined (Walters 1982).

Definition 14.7 (Walters 1982). The *dynamical entropy* of Kolmogorov and Sinai is the largest entropy per unit-time over finite partitions: $h_\mu(T) := \sup_{\mathcal{E}} h_\mu(T, \mathcal{E})$.

For fixed $\mathcal{E}$ of $\mathcal{X}$, let $\mathcal{E}_{[0,n-1]}(x)$ be the atom of $\mathcal{E}_{[0,n-1]}^{T}$ containing $x \in \mathcal{X}$.

Theorem 14.8 (Shannon–McMillan–Breiman[7] (Mañé 1987)). *If* $(\mathcal{X}, \mu, T)$ *is ergodic, then*

$$\lim_{n \to +\infty} -\frac{1}{n} \log \mu(\mathcal{E}_{[0,n-1]}^{T}(x)) = h_\mu(T, \mathcal{E}), \quad \mu\text{-a.e.}$$

---

[7] See Theorem 3.21 in Georgii's and Theorem 16.4 of Young's contributions to this book.

If $h_\mu(T, \mathcal{E}) > 0$, a fast depletion of $\mathcal{E}^T_{[0,n-1]}(x)$ is expected for it consists of points that stay within a minimal distance from the segment of trajectory $\{T^j x\}_{j=0}^{n-1}$.

14.3.1 KS Entropy and Classical Chaos

Let $\mathcal{X}$ be a compact space with distance $|\cdot|$. *Classical chaos* means *extreme sensitivity to initial conditions* such that errors $|\delta x|$ increase as $|\delta x(n)| \simeq e^{n\lambda}|\delta x|$ with *positive Lyapunov exponent* (Schuster 1995)

$$\lambda := \lim_{n \to +\infty} \lim_{\delta x \to 0} \frac{1}{n} \log \frac{|\delta x(n)|}{|\delta x|}. \tag{14.5}$$

Then, volumes containing points with close-by trajectories are expected to decrease as $\log \mu(\mathcal{E}_{[0,n-1]}(x)) \simeq -n \sum_j \lambda_j^+$, where λ_j^+ are the positive Lyapunov exponents.

Theorem 14.9 (Pesin (Mañé 1987)[8]). *For smooth, ergodic $(\mathcal{X}, \mu, T)$:*

$$h_\mu(T) = \sum_j \lambda_j^+.$$

Example 14.10. A prototype (Schuster 1995) of chaos is the map $T : (x, y) \mapsto (x+y, x+2y) \bmod(1)$ of the unit square into itself (*Arnold Cat Map*). The 2×2 matrix implementing T has an eigenvalue $\lambda > 1$ whence $h_\mu(T) = \log \lambda$.

14.3.2 KS Entropy and Classical Coding

Classical transmission channels are examples of dynamical systems. In particular, stationary sources are described by triples (Σ_D, μ, σ), where Σ_D is the set of infinite D-nary sequences and σ the left shift on them. The so-called cylinders (Walters 1982)

$$E^{[s_m, s_{m+1}, \ldots, s_n]}_{[m, m+1, \ldots, n]} := \{s : s[\ell] = s_\ell, \ \ell = m, m+1, \ldots, n\} \tag{14.6}$$

generate the algebra of measurable subsets. The elementary cylinders $E^s_{\{0\}}$ are atoms of a partition $\mathcal{E}$ of Σ_D and $\sigma(E^s_{\{0\}}) = E^s_{\{1\}}$. Thus, the cylinders in (14.6) are, as in Remark 14.6, atoms of finer partitions $\mathcal{E}^\sigma_{[0,n-1]}$. The shift-invariant measure μ corresponds to assigning them volumes

$$p(s_{[0,n-1]}) := \mu(E^{[s_0, \ldots, s_{n-1}]}_{[0, \ldots, n-1]})$$

that depend only on $s_0, s_1, \ldots, s_{n-1}$, but not on the sites $0, 1, \ldots, n-1$. If

$$p(s_{[0,n-1]}) \neq \prod_{k=1}^{n} p(s_k),$$

---

[8] See Theorem 16.7 of Young's contribution to this book.

the classical source is not memoryless and Theorem 14.2 does not help. However, Theorem 14.8 can be used (Billingsley 1965; Cover and Thomas 1991) to divide length-n strings $s_{[0,n-1]}$ into typical and untypical sets in such a way that, for large n, nearly $e^{n\,h_\mu(\sigma)}$ strings $s_{[0,n-1]}$ have probabilities $e^{-n\,h_\mu(\sigma)}$.

Theorem 14.11 (Billingsley 1965; Cover and Thomas 1991). *Asymptotically, long strings of signals from a stationary, ergodic classical source (Σ_D, μ, σ) can be reliably encoded by using only and no less than $h_\mu(\sigma)/\log 2$ bits per symbol.*

Example 14.12. A stationary, ergodic Markov source (Walters 1982) is described by a probability distribution $\pi = \{p_j\}_{j=1}^D$ and a transition matrix $[p_{ij}]$, $i, j = 1, 2, \ldots, D$, such that

$$p(s_{[0,n-1]}) = p_{s_{n-1}s_{n-2}} \cdots p_{s_1 s_0} p_{s_0} \quad \text{with } p_{ij} \geqslant 0, \ \sum_{i=1}^D p_{ij} = 1$$

and π as the only eigenvector with eigenvalue 1. Then,

$$h_\mu(\sigma) = -\sum_{ij} p_j p_{ij} \log p_{ij}.$$

If $p_{ij} = p_i$ for all j, the source is memoryless, the Markov chain becomes a Bernoulli shift (independent coin tossing) and

$$h_\mu(\sigma) = -\sum_{i=1}^D p_j \log p_i = S(\pi).$$

14.3.3 KS Entropy and Algorithmic Complexity

If $(\mathcal{X}, \mu, T)$ behaves irregularly, one expects difficulties in reproducing its trajectories via a computer. Indeed, the existence of a simple algorithm able to output them would contradict their expected complexity.[9]

According to Kolmogorov, the *algorithmic complexity* (Alekseev and Yakobson 1981; Cover and Thomas 1991; Ford et al. 1991; Li and Vitanyi 1997) of a string $s_{[0,n-1]}$ can be consistently defined as the length $\ell(p)$ of the shortest binary program p that, run on a universal computer U (Turing machine), outputs $s_{[0,n-1]}$, $U(p) = s_{[0,n-1]}$, and halts: $K(s_{[0,n-1]}) := \inf\{\ell(p) : U(p) = s_{[0,n-1]}\}$. Kolmogorov proved $K(s_{[0,n-1]})$ to be essentially machine independent.

Remark 14.13. Complexity is not a computable quantity[10], but estimates are equally helpful (Ford et al. 1991). A sequence $s_{[0,n-1]}$ of 0's and 1's can always

---

[9] See Section 15.2 of Rissanen's contribution to this book.
[10] See Section 15.2 of Rissanen's contribution to this book.

be reproduced by copying it literally, whence $K(s_{[0,n-1]}) \leqslant n + 2\log_2 n + c$, where c is a constant referring to the length of the printing instructions, and $2\log_2 n$ bits provide the computer with the length of the string.

Of course, the bound in the above remark can be very loose; for instance, $s_{[0,n-1]}$ may consist of all 1s, in which case the program *print 1 n times* requires only $2\log_2 n + c$ bits. In the latter case, the regularity of the string has allowed the information necessary to reproduce it to be compressed. Thus one is led to call a sequence random if it can be reproduced only by copying it, that is, when $K(s_{[0,n-1]})$ scales as n (Ford et al. 1991; Li and Vitanyi 1997).

Trajectories $\{T^j x\}_{j \in \mathbb{N}}$ correspond to infinitely long sequences; in such a case, their complexity $K_x(T)$ is measured by the largest *complexity per symbol* (Alekseev and Yakobson 1981) of the associated symbolic sequences $s^{\mathcal{E}}(x)$ with respect to a coarse-graining $\mathcal{E}$.

Definition 14.14 (Alekseev and Yakobson 1981).

$$K_x(T) := \sup_{\mathcal{E}} \limsup_{n \to +\infty} \frac{1}{n} K(s^{\mathcal{E}}_{[0,n-1]}(x)).$$

Remarkably, the theorem below confirms the expectation that irregular time-evolutions have complex trajectories.

Theorem 14.15 (Brudno 1983). *If $(\mathcal{X}, \mu, T)$ is ergodic, $K_x(T) = h_\mu(T)$ μ-a.e.*

Example 14.16. Most trajectories of Example 14.10 are complex: $K_{(x,y)}(T) = \log \lambda$.

14.4 Quantum Dynamical Entropies

Discrete-time quantum dynamical systems are specified by triples $(\mathcal{M}, \omega, \Theta)$, where $\mathcal{M}$ is an algebra of operators, ω a Θ-invariant state, and $\Theta : \mathcal{M} \mapsto \mathcal{M}$ an automorphism, that is an invertible linear map on $\mathcal{M}$ that preserves the algebraic relations: $\Theta(m_1 m_2) = \Theta(m_1)\Theta(m_2)$ for all $m_{1,2} \in \mathcal{M}$ (Benatti 1993; Ohya and Petz 1993).

Example 14.17. The *Koopman formulation* of classical mechanics turns classical dynamical systems $(\mathcal{X}, \mu, T)$ into algebraic triples $(\mathcal{A}_{\mathcal{X}}, \omega_\mu, \Theta_T)$, where $\mathcal{A}_{\mathcal{X}}$ is the commutative von Neumann algebra $L^\infty_\mu(\mathcal{X})$ of (μ-essentially bounded) functions f on $\mathcal{X}$, ω_μ is the state defined by $\omega_\mu(f) = \int_{\mathcal{X}} d\mu(x) f(x)$ (see Remark 14.1) and Θ_T is the automorphism of $\mathcal{A}_{\mathcal{X}}$ such that $f(x) \mapsto \Theta_T(f)(x) = f(Tx)$ (Benatti 1993).

Example 14.18. The simplest quantum systems are d-level systems (the theoretical modelling of racemization describes optically active molecules as 2-level systems like spin $1/2$ particles). The physical observables are $d \times d$ matrices in M_d and the unitary evolution $U = \exp(iH)$ is generated by a Hamiltonian $H \in M_d$ with discrete spectrum of energies: $H|e_\ell\rangle = e_\ell|e_\ell\rangle$. Then, the time-1 Heisenberg evolution is an automorphism mapping $M \in M_d$ into

$$\Theta(M) := U M U^{-1} = \sum_{j,k=1}^{d} \langle e_j|M|e_k\rangle e^{i(e_j-e_k)}, \quad M \in M_d.$$

The states are density matrices $\rho \in M_d$ that, if they commute with H, are Θ-invariant.

Example 14.19. Finite tensor products of matrix algebras M_d can be embedded into the infinite tensor product $M^N := \otimes^N M_d$ as

$$M_{[0,n]} := \bigotimes_{k=0}^{n} (M_d)_k \otimes 1_{[n+1},$$

where $1_{[n+1}$ denotes the tensor product of infinitely many identity matrices to the right of the nth site. M_d^N is called a *quantum spin half-chain* (Fannes et al. 1992).

The shift σ on M_d^N acts as follows: $\sigma : M_{[0,n]} \mapsto M_{[1,n+1]}$ and translation-invariant states ω are defined by local density matrices $\rho_{[0,n]} \in M_{[0,n]}$ such that $\rho_{[0,n]} \upharpoonright M_{[0,n-1]} = \rho_{[0,n-1]}$, where $\rho \upharpoonright M$ denotes the restriction of the state ρ to a subalgebra M. Moreover, $\rho_{[0,n]} = \rho_{[s,n+s]}$ for all $s \in N$. Typical choices of $\rho_{[0,n]}$ that generalize the notion of Markov sources in Example 14.12 are called *quantum Markov chains* (Accardi 1981; Fannes et al. 1992).

Example 14.20. Given a family of unitary operators e_j, $j \in Z$, such that

$$e_j = e_j^*, \qquad e_j^2 = 1, \qquad e_i e_j = e_j e_i (-1)^{g(|i-j|)}, \tag{14.7}$$

where $g : N \mapsto \{0, 1\}$, the *localized words* $w_{s_{[m,n]}} = e_{s_m} e_{s_{m+1}} \cdots e_{s_n}$, with $s_m < s_{m+1} < \cdots < s_n$, generate an algebra $\mathcal{M}_0$. Then, $\mathcal{M}_0$ is equipped with the shift automorphism $\sigma(e_i) = e_{i+1}$ and with the σ-invariant state ω such that $\omega(1) = 1$ and $\omega(w_{s_{[m,n]}}) = 0$ unless $s_{[m,n]} = \emptyset$, in which case $w_\emptyset = 1$. Via the so-called *GNS construction* based on ω (Ohya and Petz 1993), $\mathcal{M}_0$ is completed to a von Neumann algebra $\mathcal{M}$. These quantum dynamical systems $(\mathcal{M}, \omega, \sigma)$ are called *bit-streams* (Alicki and Narnhofer 1995; Golodets and Størmer 1998) and their properties strongly depend on the function g ruling the commutation relations among operators at different sites.

Quantum spin half-chains (M_d^N, ω, σ) provide the appropriate description of quantum sources. A recent result (Petz and Mosonyi 2001) extends Theorem 14.5 to (completely) ergodic stationary quantum sources with memory described by local density matrices

$$\rho_{[0,n-1]} = \sum_{s_{[0,n-1]}} p(s_{[0,n-1]}) \rho(s_{[0,n-1]}).$$

It uses the notion of *mean entropy*

$$s(\omega) := \lim_{n\to\infty} \frac{1}{n} S(\rho \restriction M_{[0,n-1]}). \tag{14.8}$$

Remark 14.21. The von Neumann entropy over tensor products satisfies (Ohya and Petz 1993; Wehrl 1978)

$$|S(\rho \restriction M_1) - S(\rho \restriction M_2)| \leqslant S(\rho \restriction M_1 \otimes M_2) \leqslant S(\rho \restriction M_1) + S(\rho \restriction M_2). \tag{14.9}$$

Then, subadditivity and σ-invariance of ω guarantee the existence of (14.8) (Ohya and Petz 1993; Wehrl 1978).

It turns out (Petz and Mosonyi 2001) that $s(\omega)$ can be used to single out high-probability subspaces that allow one to code the local states $\rho(s_{[0,n-1]})$ into other density matrices $\tilde{\rho}(s_{[0,n-1]})$ on $M_{[0,n-1]}$ with a close to 1 fidelity parameter (see Remark 14.4)

$$F := \sum_{s_{[0,n-1]}} p(s_{[0,n-1]}) \operatorname{Tr}(\rho(s_{[0,n-1]}) \tilde{\rho}(s_{[0,n-1]})). \tag{14.10}$$

When $\omega = \bigotimes_{k\in\mathbb{Z}} (\rho_\pi)_k$, one recovers Theorem 14.5: $s(\omega) = S(\rho_\pi)$.

Theorem 14.22 (Petz and Mosonyi 2001). *Asymptotically, long strings of signals from stationary, completely ergodic quantum sources $(\mathcal{M}_N, \omega, \sigma)$ can be reliably encoded by using only and no less than $s(\omega)/\log 2$ q-bits per signal.*

Classical sources can also be described by algebraic triples $(A_D^N, \omega_\mu, \sigma)$: cylinders as in (14.6) correspond to local commutative subalgebras $A_{[0,n]}$. The *quasi-local* commutative C^* algebra A_D^N is a *classical spin half-chain*. Once equipped with the shift $\sigma : A_{\{n\}} \mapsto A_{\{n+1\}}$, the probabilities $p(s_{[0,n]})$ define a σ-invariant global state ω_μ over A_D^N with local restrictions $\rho_{[0,n]} := \omega_\mu \restriction A_{[0,n]}$. Then, $h_\mu(\sigma)$ equals the mean entropy density $s(\omega_\mu)$ defined as in (14.8) (for classical memoryless sources, ω_μ is a tensor product state, $\omega_\mu = \bigotimes_{k\in\mathbb{Z}} (\rho_\mu)_k$, whence $s(\omega_\mu) = S(\rho_\mu)$).

Remark 14.23. For stationary classical sources the dynamical entropy coincides with the mean entropy. One would like any quantum dynamical entropy to yield the mean entropy $s(\omega)$ in the case of a stationary quantum source (M_d^N, ω, σ). However, an approach like the one that follows fails (Benatti 1993):

(i) let $M \subset \mathcal{M}$ be any finite-dimensional subalgebra and $M_{[0,n-1]}^\Theta$ the algebra generated by the subalgebras $\{\Theta^j(M)\}_{j=0}^{n-1}$;

(ii) calculate

$$h_\omega(\Theta, M) := \lim_n \frac{1}{n} S(\omega \restriction M_{[0,n-1]}^\Theta);$$

and

(iii) define the dynamical entropy of $(\mathcal{M}, \omega, \Theta)$ to be

$$h_\omega(\Theta) := \sup_{M \subset \mathcal{M}} (\Theta, M).$$

Due to noncommutativity, $M^\Theta_{[0,1]}$ may already be infinite dimensional, while the nontensorial form of $M^\Theta_{[0,n-1]}$ does not allow the use of subadditivity as in Remark 14.21. Thus $h_\omega(\Theta, M)$ is ill-defined.

Following too closely the classical construction does not help; there must be a point where quantum mechanics takes a different turn. A hint as to how to proceed comes from considering *partitions of unit* and *decompositions of states* that from an entropic point of view are classically indistinguishable.

14.4.1 Partitions of Unit and Decompositions of States

According to Example 14.17, the characteristic functions e_j of the atoms E_j of a finite partition $\mathcal{E}$ of a classical dynamical system $(\mathcal{X}, \mu, T)$ are orthogonal projections $e_j \in \mathcal{A}_\mathcal{X}$ such that $e_j e_k = \delta_{jk} e_j$ and $\sum_{j=1}^D e_j = 1$. Thus, partitions of $\mathcal{X}$ provide partitions of unit in $\mathcal{A}_\mathcal{X}$. They also provide *decompositions* of ω_μ on $\mathcal{A}_\mathcal{X}$ into linear convex combinations of other states $\hat{\omega}^j_\mathcal{E}$:

$$\omega_\mu = \sum_j \omega_\mu(e_j) \hat{\omega}^j_\mathcal{E},$$

$$\mathcal{A}_\mathcal{X} \ni f \mapsto \hat{\omega}^j_\mathcal{E}(f) = \frac{\omega_\mu(e_j f)}{\omega_\mu(e_j)} = \frac{1}{\mu(E_j)} \int_{E_j} \mathrm{d}\mu(x) f(x). \quad (14.11)$$

Remark 14.24. With the atoms E_j of $\mathcal{E}$ interpreted as orthogonal projections e_j, the partition $\mathcal{E}$ is identified with the commutative subalgebra $A_\mathcal{E} \subset \mathcal{A}_\mathcal{X}$ generated by them. Also, the atoms $\bigcap_{k=0}^{n-1} T^{-k}(E_{s_k})$ of the finer partitions $\mathcal{E}^T_{[0,n-1]}$ are identified with products $e_{s_{[0,n-1]}} := \prod_{k=1}^{n-1} \Theta^k_T(e_{s_k})$ of evolving characteristic functions. These are projections generating commutative subalgebras $A^{\Theta_T}_{[0,n-1]}$.

14.4.2 CNT Entropy: Decompositions of States

The *Connes–Narnhofer–Thirring* (CNT) *entropy* of a quantum dynamical system $(\mathcal{M}, \omega, \Theta)$ is constructed by means of decompositions of ω (Connes et al. 1987).

Let $S(\rho; \sigma) := \mathrm{Tr}(\sigma(\log \sigma - \log \rho))$ denote the *relative entropy* of two density matrices ρ and σ. Convexity arguments show that $S(\rho; \sigma) > 0$ and $S(\rho; \sigma) = 0$ if and only if $\rho = \sigma$ (Ohya and Petz 1993). Given a density matrix ρ and a subalgebra $N \subseteq M_d$, the *entropy of N with respect to ρ* is defined by

(again $\rho \upharpoonright N$ means restricting the state ρ to act on the subalgebra N)

$$H_\rho(N) := \sup_{\rho=\sum_j \lambda_j \rho_j} \sum_j \lambda_j S(\rho_j \upharpoonright N; \rho \upharpoonright N)$$
$$= S(\rho \upharpoonright N) - \sum_j \lambda_j S(\rho_j \upharpoonright N). \tag{14.12}$$

Given a generic quantum dynamical system $(\mathcal{M}, \omega, \Theta)$ and a finite-dimensional subalgebra $N \subseteq \mathcal{M}$, the problem that the algebra

$$N_{[0,n-1]}^\Theta := \bigvee_{k=0}^{n-1} \Theta^k(N)$$

generated by the evolving N may become infinite dimensional is avoided by introducing the so-called *n-subalgebra functionals* that are based on convex decompositions of ω labelled by multi-indices $s = s_0 s_1 \cdots s_{n-1}$: $\omega = \sum_s \lambda_s \omega_s$. To each subalgebra $\Theta^j(N)$ there corresponds an index s_j taking values in a set I_j and the supremum in (14.12) is replaced by (with $\eta(x) = -x \log x$)

$$H_\omega(N, \Theta(N), \ldots, \Theta^{n-1}(N))$$
$$:= \sup_{\omega=\sum_s \lambda_s \omega_s} \left\{ \sum_s \eta(\lambda_s) - \sum_{k=0}^{n-1} \sum_{s_k} \eta(\lambda_{s_k}^k) \right.$$
$$\left. + \sum_{k=0}^{n-1} \sum_{s_k} \lambda_{s_k}^k S(\omega \upharpoonright \Theta^k(N); \hat{\omega}_{s_k}^k \upharpoonright \Theta^k(N)) \right\}, \tag{14.13}$$

where the states

$$\hat{\omega}_{s_k}^k := \sum_{s, s_k \text{ fixed}} \frac{\lambda_s \omega_s}{\lambda_{s_k}^k}, \qquad \lambda_{s_k}^k := \sum_{s, s_k \text{ fixed}} \lambda_s,$$

provide decompositions $\omega = \sum_{s_k} \lambda_{s_k}^k \hat{\omega}_{s_k}^k$.

Definition 14.25 (Connes et al. 1987). The CNT entropy of $(\mathcal{M}, \omega, \Theta)$ is

$$h_\omega^{\text{CNT}}(\Theta) := \sup_N h_\omega^{\text{CNT}}(\Theta, N), \tag{14.14}$$

where

$$h_\omega^{\text{CNT}}(\theta, N) := \lim_{n \to +\infty} \frac{1}{n} H_\omega(N, \Theta(N), \ldots, \Theta^{n-1}(N)). \tag{14.15}$$

Remark 14.26. From (14.12) it follows that $0 \leqslant H_\omega(N) \leqslant S(\rho \upharpoonright N)$. The functionals in (14.13), $H_\omega^{[0,n-1]}(N)$ for short, are bounded by $nH_\omega(N)$, for the first line in (14.13) is nonpositive. Moreover, they are Θ-*invariant* and *subadditive* (Connes et al. 1987):

$$H_\omega^{[0,n-1]}(N) \leqslant H_\omega^{[0,p-1]}(N) + H_\omega^{[p,n-1]}(N)$$

and

$$H_\omega^{[p,n-1]}(N) = H_\omega^{[0,n-p-1]}(N).$$

Then, the limit in (14.15) exists.

Classical Case. Let $A_\mathcal{E} \subset A_\mathcal{X}$ be the finite commutative matrix algebra associated with a finite partition $\mathcal{E}$ of $\mathcal{X}$. Since $\omega_\mu(e_j) = \mu(E_j)$, the Shannon entropy of $\mu_\mathcal{E}$ equals the von Neumann entropy $S(\omega_\mu \upharpoonright A_\mathcal{E})$ of the state ω_μ restricted to $A_\mathcal{E}$. By inserting in (14.12) a decomposition of ω_μ as in (14.11) obtained from the partition of unit $\{e_j\}$ induced by $\mathcal{E}$, it follows that $S(\hat{\omega}_\mu^j \upharpoonright A_\mathcal{E}) = 0$ and thus that $H_{\omega_\mu}(A_\mathcal{E}) = S(\omega_\mu \upharpoonright A_\mathcal{E})$. Moreover, the projections $e_{s_{[0,n-1]}}$ of Remark 14.24 and the assumed Θ_T-invariance of ω_μ give

$$\left. \begin{array}{l} \lambda_{s_{[0,n-1]}} = \omega_\mu(e_{s_{[0,n-1]}}) = p(s_{[0,n-1]}), \\[4pt] \lambda_{s_k}^k = p(s_k), \quad \hat{\omega}_{s_k}^k(\Theta^k(f)) = \dfrac{\omega_\mu(e_{s_k} f)}{\omega_\mu(e_{s_k})}. \end{array} \right\} \quad (14.16)$$

Then, $H_{\omega_\mu}^{[0,n-1]}(A_D) = H_\mu(\mathcal{E}_{[0,n-1]}^T)$ (compare (14.4)) and $h_{\omega_\mu}^{\text{CNT}}(\Theta_T, A_D) = h_\mu(T, \mathcal{E})$.

Remark 14.27. Decompositions of states in a purely noncommutative context correspond to $y_j \in \mathcal{M}$ such that $\sum_j y_j^* y_j = 1$ (Benatti 1993), with y^* the adjoint of y. If ρ is a state on $\mathcal{M} = M_d$, then one can decompose ρ as $\rho = \sum_j \sqrt{\rho} y_j^* y_j \sqrt{\rho}$. Notice that $\rho = \sum_j \rho y_j^* y_j$ is not a decomposition because the operators $\rho y_j^* y_j$ are not positive and, once normalized, cannot give rise to density matrices. Operators of the form $y_j \rho y_j^*$ are alright, but they do not decompose ρ, $\rho \neq \sum_j y_j \rho y_j^*$ unless the y_j commute with ρ.

14.4.3 AF Entropy: Partitions of Unit

Given a quantum dynamical system $(\mathcal{M}, \omega, \Theta)$ let $\mathcal{M}_0 \subseteq \mathcal{M}$ be a Θ-invariant subalgebra. The *Alicki–Fannes* (AF) *entropy* (Alicki and Fannes 1994) is based on partitions of unit, namely, on sets $\mathcal{Y} = \{y_j\}_{j=1}^D \in \mathcal{M}_0$ such that $\sum_j y_j^* y_j = 1$. Let us introduce

- the time-evolving partitions of unit $\Theta^k(\mathcal{Y}) := \{\Theta^k(y_j)\}_{j=1}^D$;

- the refined partitions

$$\mathcal{Y}_{[0,n-1]}^\Theta = \{\Theta^{n-1}(y_{j_{n-1}}) \Theta^{n-2}(y_{j_{n-2}}) \cdots \Theta(y_{j_1}) y_{j_0}\};$$

- the $D^n \times D^n$ density matrices $\rho[\mathcal{Y}_{[0,n-1]}^\Theta]$ with matrix elements

$$\rho[\mathcal{Y}_{[0,n-1]}^\Theta]i,j = \omega(y_{j_0}^*\Theta(y_{j_1}^*)\cdots\Theta^{n-1}(y_{j_{n-1}}^*y_{i_{n-1}})\cdots\Theta(y_{i_1})y_{i_0}), \tag{14.17}$$

and von Neumann entropy $S(\rho[\mathcal{Y}_{[0,n-1]}^\Theta])$.

Definition 14.28 (Alicki and Fannes 1994). The AF entropy of $(\mathcal{M}, \omega, \Theta)$ is

$$h_{\omega,\mathcal{M}_0}^{\mathrm{AF}}(\Theta) := \sup_{\mathcal{Y}\in\mathcal{M}_0} \limsup_{n\to+\infty} \frac{1}{n} S(\rho[\mathcal{Y}_{[0,n-1]}^\Theta]). \tag{14.18}$$

Remark 14.29. The density matrices $\rho[\mathcal{Y}_{[0,n-1]}^\Theta]$ form a family of states as in Example 14.19 and define a global state $\omega_\mathcal{Y}$ on the quantum spin half-chain $\mathcal{M}_D^N$. In this sense, partitions of unit provide symbolic models of $(\mathcal{M}, \omega, \theta)$. However, because of noncommutativity $\omega_\mathcal{Y}$ is not shift-invariant, whence the lim sup.

Classical Case. Step functions form a Θ_T-invariant subalgebra $\mathcal{A}_0 \subset \mathcal{A}_\mathcal{X}$. Let $\mathcal{E} = \{e_j\}_{j=1}^D$ be the partition of unit corresponding to a partition of $\mathcal{X}$. Then, because of commutativity and orthogonality,

$$\omega_\mu(e_{j_0}\cdots\Theta_T^{n-1}(e_{j_{n-1}}e_{i_{n-1}})\cdots e_{i_0}) = \prod_{k=0}^{n-1}\delta_{j_k i_k}\omega_\mu(e_{s_{[0,n-1]}}). \tag{14.19}$$

Thus $\rho[\mathcal{E}_{[0,n-1]}]$ is a diagonal matrix with eigenvalues $p(s_{[0,n-1]})$ as in (14.3), whence $h_{\omega_\mu,\mathcal{A}_0}^{\mathrm{AF}}(\Theta_T) \geqslant h_\mu(T,\mathcal{E})$. In fact (Alicki et al. 1996a), $h_{\omega_\mu,\mathcal{A}_0}^{\mathrm{AF}}(\Theta_T) = h_\mu(T)$.

Remark 14.30. The CNT entropy is based on decompositions of the invariant state ω, the AF entropy on partitions of unit that extract information about the state by constructing the matrix (14.17) thereby modifying ω and interfering with the dynamics (compare Remark 14.27). These effects disappear in the commutative case, but are to be kept under control in the quantum case if the AF entropy is chosen as a measure of the unpredictability of quantum dynamics. Indeed, the simple extraction of information via partitions might work as an extrinsic source of randomness that has nothing to do with the intrinsic features of the dynamics.

14.5 Quantum Dynamical Entropies: Perspectives

Both the AF and CNT entropy coincide with the KS entropy on classical dynamical systems. They also coincide on finite-dimensional quantum systems as in Example 14.18, where $h_\omega^{\mathrm{AF}}(\Theta) = h_\omega^{\mathrm{CNT}}(\Theta) = 0$. This can be seen as follows.

CNT entropy. The functionals in (14.13) are *monotone under embeddings*,

$$N_i \subseteq M_j \implies H_\omega(N_1, N_2, \ldots, N_n) \leqslant H_\omega(M_1, M_2, \ldots, M_n), \tag{14.20}$$

and *invariant under repetitions* of subalgebras (Connes et al. 1987). Thus, $\Theta^k(N) \subseteq \Theta^k(M_d) = M_d$ implies

$$H_\omega^{[0,n-1]}(N) \leqslant H_\omega(M_d, M_d, \ldots) \leqslant H_\omega(M_d) \leqslant \log d \quad \text{and} \quad \lim_n \frac{1}{n}$$

yields the result.

AF entropy. Let $\mathcal{Y} = \{y_j\}_{j=1}^D \in M_d$ be a partition of unit and $\{|j\rangle\}_{j=1}^D$ be an orthonormal basis in the D-dimensional Hilbert space $\mathbf{C}^D$. Let $\rho \in M_d$ be a density matrix. The states ρ on M_d and $\rho^{\mathcal{Y}} = \sum_{i,j=1}^D y_j \rho y_i^* \otimes |j\rangle\langle i|$ on $M_d \otimes M_D$ are such that $S(\rho^{\mathcal{Y}}) = S(\rho)$ (Fannes and Tuyls 2000). Notice the following restrictions:

$$\rho^{\mathcal{Y}} \upharpoonright M_D = \rho[\mathcal{Y}] \quad \text{and} \quad \rho^{\mathcal{Y}} \upharpoonright M_d = \sum_{j=1}^D y_j \rho y_j^* =: \Pi_{\mathcal{Y}}[\rho].$$

Then, using (14.9) we get

$$S(\rho[\mathcal{Y}]) \leqslant S(\rho) + S(\Pi_{\mathcal{Y}}[\rho]). \tag{14.21}$$

Both ρ and $\Pi_{\mathcal{Y}}[\rho]$ are states on M_d; therefore, their von Neumann entropies are bounded by $\log d$ and again $\lim_n (1/n)$ yields the result.

Remark 14.31. It is no surprise that d-level systems as in Example 14.18 exhibit zero quantum dynamical entropy, for their evolution is quasiperiodic due to discrete energy spectrum, hence highly regular. It is in infinite-dimensional systems that one has to look for complex behaviours.

Moving from finite to infinite quantum systems, not only do we find nonzero entropy, but also different CNT and AF entropies.

Quantum Markov Chains. The dynamical systems (M_d^N, ω, σ) in Example 14.19 describe quantum sources with memory. By analogy with the classical case (see the discussion in Section 14.4), one expects the dynamical entropy to coincide with the mean entropy. Indeed one has (Alicki and Fannes 1994; Park 1994)

$$h_\omega^{\text{CNT}}(\sigma) = s(\omega), \qquad h_{\omega, \mathcal{M}_0}^{\text{AF}}(\sigma) = s(\omega) + \log d. \tag{14.22}$$

Remark 14.32. The partitions of unit used to calculate $h_{\omega, \mathcal{M}_0}^{\text{AF}}(\sigma)$ come from the algebra $\mathcal{M}_0$ of strictly local operators. The dependence of the result on different choices of $\mathcal{M}_0$ is taken care of in Fannes and Tuyls (2000). The extra term $\log d$ is a direct effect of the fact observed in Remark 14.30 that partitions of unit do perturb the state.

Bit-Streams. The dynamical systems in Example 14.20 are more interesting. For $g_0 = (0, 0, 0, \ldots)$ the various bits e_j all commute and $(\mathcal{M}, \omega, \sigma)$ is an algebraic version of a classical Bernoulli shift, while for $g_1 = (1, 1, 1, \ldots)$

different e_j's anticommute. Instead, take a *sufficiently irregular* g_3 such that for any localized words $w_{s_{[m,n]}}$, there exists a countable set $J([m, n])$ of integers such that for all $p, q \in J([m, n])$ the shifted words $w_{s_{[m+p,n+p]}}$ and $w_{s_{[m+q,n+q]}}$ anticommute. Then, one has (Alicki and Narnhofer 1995)

$$h_\omega^{\text{CNT}}(\sigma) = \begin{cases} \log 2 & g = g_0, \\ \frac{1}{2}\log 2 & g = g_1, \\ 0 & g = g_3, \end{cases} \tag{14.23}$$

$$h_{\omega,\mathcal{M}_0}^{\text{AF}}(\sigma) = \log 2 \quad \text{independently of } g. \tag{14.24}$$

Remark 14.33. For g_0 the result is expected as the CNT and AF entropy both reduce to the KS entropy in the commutative context, in this case a balanced toin-cossing (see Example 14.12). The departure in the other two cases indicates that the CNT entropy, being based on decompositions which do not perturb the evolving system, is sensitive to the degree of algebraic independence of increasingly separated observables. If correlations persist as in the case of the irregular g_3, then it vanishes.

14.5.1 Quantum Dynamical Entropies and Quantum Chaos

Quantum chaos comprises a variety of peculiar behaviours exhibited by quantum systems with chaotic classical limit (Casati and Chirikov 1995) despite the fact that their time-evolution is possibly quasiperiodic.

A typical example is a quantized version of the Arnold Cat Map (Ford et al. 1991), where the Hilbert space is N-dimensional and Example 14.10 emerges when $N \to \infty$. We know that, no matter how large N, *Quantum Cat Maps* have zero dynamical entropy. The reason is that the ultimate discrete coarse-graining imposed by $N^{-1} > 0$ implies a regular time-behaviour as in Example 14.18 and forbids going to finer and finer details of the dynamics as in (14.5) where $|\delta x| \to 0$ before $1/n \to 0$.

Indeed, if we identify quantum chaos with $h_\omega^{\text{CNT}} > 0$ or $h_\omega^{\text{AF}} > 0$, then there is no room for *chaos* in finite-dimensional Hilbert spaces. However, if in (14.15) and (14.18) we consider the dependence on N of the arguments of the limits, we discover that by letting classical mechanics ($N \to \infty$) emerge first, then $\lim_n 1/n \lim_N \neq 0$ (Benatti et al. 2003).

It is thus important to study the *semi-classical behaviour* when $N \to +\infty$ of the functionals before the temporal limit $1/n$ is taken. It is in fact expected that quantum systems with chaotic classical limit should approximately behave as their chaotic classical counterparts up to times of the order of $\log N$ called *breaking-times* (Alicki et al. 1996b; Casati and Chirikov 1995).

14.5.2 Dynamical Entropies and Quantum Information

Because of their definition, the AF entropy singles out optimal partitions of unit and the CNT entropy optimal decompositions of states. Both correspond to optimal sets of operators (see Remark 14.27). In view of (14.22), it remains to be investigated whether and how they might be put in correspondence with the high probability subspace used for encoding quantum sources with high fidelity as in Theorem 14.22.

In the case of the CNT entropy, the need for a better understanding of the possible connections with compression of information is enhanced by the fact that the 1-algebra functional in (14.12) is what appears in quantum information theory as *maximal accessible information* (Benatti 1996), and optimal decompositions as in Remark 14.27 seem to have an important role to play in connection with *quantum entanglement* (Benatti and Narnhofer 2001).

14.5.3 Dynamical Entropies and Quantum Randomness

There are two ways in which quantum mechanics challenges the notion of complexity presented in Section 14.3.3.

The first is that applying algorithmic considerations to the computation of eigenvalues and eigenvectors of the quantized Arnold Cat Map of Section 14.5.1, one does not find any complexity for any value of N (Ford et al. 1991). This occurs in spite of the *Correspondence Principle* that teaches us that classical mechanics emerges from quantum mechanics in the classical limit (Ford et al. 1991) and that the classical Arnold Cat Map is complex (compare Example 14.16).

As the quantum dynamical entropies are expected to behave almost classically up to the $\log N$ breaking-time, the question arises of whether there is no violation of the Correspondence Principle, but simply a manifestation of the noninterchangeability of classical, $\lim_N$, and temporal, $\lim_n (1/n)$, limits.

The second challenge is related to the possible use of quantum dynamical entropies to quantify the degree of complexity of sequences of quantum signals. In particular, in view of (14.23), as bit-streams may have zero h^{CNT} and nonzero h^{AF}, does this indicate that there is no unique notion of *quantum algorithmic complexity*, but (at least) two, depending on whether we account for perturbations due to observations or not?

References

Accardi, L. 1981 *Phys. Rep.* **77**, 169.
Accardi, L., Ohya, M. and Watanabe, N. 1997 *Open Systems Infor. Dynamics* **4**, 71.
Alekseev, V. M. and Yakobson, M. V. 1981 *Phys. Rep.* **75**, 287.
Alicki, R. and Fannes, M. 1994 *Lett. Math. Phys.* **32**, 75.
Alicki, R. and Narnhofer, H. 1995 *Lett. Math. Phys.* **33**, 241.

Alicki, R., Andries, J., Fannes, M. and Tuyls, P. 1996a *Rev. Math. Phys.* **8**, 167.
Alicki, R., Makowiec, D. and Miklaszewski, W. 1996b *Phys. Rev. Lett.* **77**, 838.
Benatti, F. 1993 *Deterministic Chaos in Infinite Quantum Systems*. Sissa Notes in Physics. Springer.
Benatti, F. 1996 *J. Math. Phys.* **37**, 5244.
Benatti, F. and Narnhofer, H. 2001 *Phys. Rev.* A **63**, 042306.
Benatti, F., Cappellini, V., De Cock, M., Fannes, M. and Vanpeteghem, D. 2003 *Classical Limit of Quantized Toral Automorphisms and Dynamical Entropies*. (In preparation.)
Billingsley, P. 1965 *Ergodic Theory and Information*. Wiley.
Bouwmeester, D., Ekert, A. and Zeilinger, A. 2000 *The Physics of Quantum Information*. Springer.
Brudno, A. A. 1983 *Trans. Moscow Math. Soc.* **44**, 127.
Casati, G. and Chirikov, B. 1995 *Quantum Chaos. Between Order and Disorder*. Cambridge University Press.
Connes, A., Narnhofer, H. and Thirring, W. 1987 *Commun. Math. Phys.* **112**, 691.
Cover, T. M. and Thomas, J. A. 1991 *Elements of Information Theory*. Wiley.
Fannes, M. and Tuyls, P. 2000 *Inf. Dim. Analysis, Quantum Prob. Rel. Top.* **2**, 511.
Fannes, M., Nachtergaele, B. and Werner, R. F. 1992 *Commun. Math. Phys.* **144**, 443.
Ford, J., Mantica, G. and Ristow, G. H. 1991 *Physica* D **50**, 493.
Golodets, V. Ya. and Størmer, E. 1998 *Ergod. Theory Dynam. Sys.* **18**, 859.
Li, M. and Vitanyi, P. 1997 *An Introduction to Kolmogorov Complexity and Its Application*. Springer.
Lindblad, G. 1988 Dynamical entropy for quantum systems. In *Quantum Probability and Applications II*. Lecture Notes in Mathematics, vol. 1303, p. 183.
Mañé, R. 1987 *Ergodic Theory and Differentiable Dynamics*. Ergebnisse der Mathematik und ihrer Grenzgebiete, 3. Folge, Vol. 8. Springer.
Ohya, M., and Petz, D. 1993 *Quantum Entropy and Its Use*. Springer.
Park, Y. M. 1994 *Lett. Math. Phys.* **36**, 63.
Petz, D. and Mosonyi, M. 2001 Stationary quantum source coding. *J. Math. Phys.* **42**, 4857.
Schumacher, B. 1995 *Phys. Rev.* A **51**, 2738.
Schuster, H. G. 1995 *Deterministic Chaos*, 3rd edn. VCH, Weinheim.
Voiculescu, D. 1995 *Commun. Math. Phys.* **170**, 249.
Walters, P. 1982 *An Introduction to Ergodic Theory*. Graduate Text in Mathematics, vol. 79. Springer.
Wehrl, A. 1978 *Rev. Mod. Phys.* **50**, 221.

Chapter Fifteen

Complexity and Information in Data

J. Rissanen
Helsinki Institute for Information Technology

In this expository chapter we discuss some of the recent results in statistical modeling based on information and coding theoretic ideas. The central theme is a formalization of the intuitive ideas of the complexity of a given data string, given a parametric class of probability models, and the useful information that can be extracted from the string with the model class. Such an approach to modeling has its roots in the algorithmic theory of information and Kolmogorov sufficient statistics decomposition, which we outline below after a discussion of the fundamentals of coding and modeling.

Briefly, for a parametric class of probability models there is a universal parameter-free model as a density function, whose negative logarithm, evaluated at the data string given, is defined to be its (stochastic) *complexity*. It breaks up into two parts, the first representing the code length needed to describe the model in a set of 'distinguishable' models from the given amount of data. This model incorporates all the information that can be extracted from the data with the given class of models, and its code length is defined to be the *information* in the data. The second part in the complexity represents the code length for noninformative noise. Such a decomposition provides a powerful justification for the *minimum description length* (MDL) principle for model selection, which is to search for the model or model class with the minimum stochastic complexity.

The universal model can be found as the solution to a minmax problem, which is to find a probability model such that it permits encoding of data (within the means available, i.e. the parametric model class) with the shortest mean code length, where the mean is taken with respect to the worst-case data-generating model lying inside or outside of the parametric class.

We illustrate these ideas with the linear quadratic denoising problem, for which the universal model together with the complexity and the information can be calculated in a particularly nice form. This allows a decomposition of the data into the information-bearing part, the 'law,' and the noise, which gets defined as the part in the data that cannot be compressed with the model class considered. The result is a superior algorithm to separate and remove the noise from the data.

15.1 Introduction

In intuitive terms, the objective of statistical modeling is to extract information from a given set of data. Clearly, what is meant by 'information' is not the familiar ideas of Shannon information nor Fisher information but something that has to do with the machinery we think has generated the data. Giving a formal definition of the 'information' involved and a measure for it turns out to be an intricate issue. In fact, we cannot even do it in isolation but we will have to introduce another related notion, the *complexity* of the data (Rissanen 1996). To give an idea of these notions consider a binary string of length n, such as $x^n = 00101110\ldots 1$. We may ask, what are the *complexity* and the (useful) *information* in this string, relative to models belonging to, say, the Bernoulli class, Markov class, etc., the class of all programs in a universal computer language or all computable distributions. It would be hopeless to try to formalize these notions in an absolute sense; both of them are in the eye of the beholder, as it were. The model classes are necessary even to express the properties we wish to learn and obtain 'information' about.

Complexity may be formalized in terms of the least number of bits with which the string can be described or encoded, when a suitable form of description that depends on the model class selected is agreed upon. For instance, if we permit the model class of all programs, then the so-called algorithmic or Kolmogorov complexity of an alternating string like $0101\ldots 01$ is low, because we only need to describe the rule that the string is alternating starting with 0 and its length. However, if we take the Bernoulli class, we can only count the number of 1s and 0s and encode the string as one of the strings with about equal number of 0s and 1s, which makes the string look like a purely random string with maximum complexity.

The (useful) *information* refers to the regular features in the string or its broad properties, as seen through the class of models agreed upon. Hence their shortest description should always be less than the complexity, which means that the alternating string should have little information relative to the class of all programs. In fact, the information equals the complexity, leaving nothing unexplained as 'noise.' The amount of information relative to the Bernoulli class is also small, because the property learned is just that the frequencies of the occurrences of zeros and ones are equal. This does not help in reducing the length of describing the string, which leaves the entire string as unexplained noise. However, even when a string has maximum complexity, such as a string obtained by tossing a fair coin, the information should be small, relative to both classes of models. This is because, again, such a string has few learnable properties.

The problem, then, is to find a way to quantify the strength of the properties constraining the data string at hand, so that we can measure the amount of information that can be learned from the data with a model class selected. This can be made precise in the algorithmic theory of complexity, although it leads to insurmountable noncomputability problems and precludes a practical use of both the algorithmic complexity and the associated Kolmogorov sufficient statistics with its measure of information. Because of the conceptual value of

15.2 Basics of Coding

In coding we want to transmit or store sequences of elements of a finite set $A = \{a_1, \ldots, a_m\}$ in terms of binary symbols 0 and 1. The set A is called the *alphabet* and its elements are called *symbols*, which can be of any kind, often numerals. The sequences are called *messages*, or often just data when the symbols are numerals. A code, then, is a one-to-one function $C : A \to B^*$ taking each symbol in the alphabet into a finite binary string, called the *codeword*. It is extended to sequences $x = x_1, \ldots, x_n$

$$C : A^* \to B^*$$

by the operation of *concatenation*: $C(xx_{n+1}) = C(x)C(x_{n+1})$, where xy denotes the string obtained when symbol or string y is appended to the end of the string x. We want the extended code, also written as C, to be not only invertible but also such that the codewords $C(a_i)$ can be separated and recognized in the code string $C(x)$ without a comma. This implies an important restriction on the codes, the so-called prefix property, which states that *no codeword is a prefix of another*. This requirement, making a code a *prefix code*, implies the important Kraft inequality

$$\sum_{a \in A} 2^{-n_a} \leqslant 1, \tag{15.1}$$

where $n_a = |C(a)|$ denotes the length of the codeword $C(a)$.

To prove this, consider the leaves of any complete binary tree, i.e. a tree where each node has either two sons or none. Any such tree can be obtained by starting with a two-leaf tree and splitting successively the leaf nodes until the tree is obtained. For the two-leaf tree the Kraft inequality holds with equality: $2^{-1} + 2^{-1} = 1$. Splitting a node w of length $|w|$, the equality $2^{-|w0|} + 2^{-|w1|} = 2^{-|w|}$ holds for the son nodes, just as for the probabilities of a Bernoulli process, which by an easy induction implies that the Kraft inequality holds with equality for any complete binary tree. The codewords of a prefix code are leaves of a tree, which may be completed. Hence, the claim holds. It is easily extended even to countable alphabets, where the sum becomes the limit of strictly increasing finite sums bounded from above by unity.

The codeword lengths of a prefix code, then, define a probability distribution by $P(a) = 2^{-n_a}/K$, where K denotes the sum on the left-hand side of the Kraft inequality. Even the converse is true in essence. Indeed, assume conversely that a set of integers $n_1, \ldots, n_m$ satisfies the Kraft inequality. We shall describe a prefix code with the lengths of the codewords $C(a_i)$ given by these integers. Let $\hat{n}$ be the maximum of these integers. Write each term as

$$2^{-n_i} = 2^{r_i} \times 2^{-\hat{n}},$$

where $r_i = \hat{n} - n_i$, and

$$\sum_i 2^{-\hat{n}} 2^{r_i} \leqslant 1,$$

by the assumption. This is a sum of the leaf probabilities $2^{-\hat{n}}$ over a subset of the leaves $0\ldots 0, 0\ldots 01, \ldots, 1\ldots 1$ of a balanced tree of depth $\hat{n}$. Hence, we get the required prefix code by partitioning this subset into m segments of $2^{r_1}, 2^{r_2}, \ldots$ adjacent leaves such that each segment has a common mother node of length n_i, which we take as the codeword.

An important practical data-coding problem results when we ask how to design a prefix code when the symbols are taken independently with probabilities $P(a)$, and we want the code to have as short a mean length $L(P) = \sum_i P(a_i) n(a_i)$,

$$L^*(P) = \min_{n(a_i)} L(P),$$

as possible, where the lengths $n(a_i)$ are required to satisfy the Kraft inequality. This may also be regarded as a measure of the *mean complexity* of the alphabet A together with the distribution P. Indeed, intuitively, we consider an object 'complex' if it takes a large number of binary digits to describe (encode) it.

The optimal lengths and a corresponding prefix code can be found by Huffman's algorithm, but far more important is the following remarkable property.

Theorem 15.1. *For any prefix code,*

(i)
$$L(P) \geqslant -\sum_i P(a_i) \log_2 P(a_i) \equiv H(P),$$

(ii)
$$L(P) = H(P) \quad \Longleftrightarrow \quad n(a_i) \equiv -\log_2 P(a_i),$$

where $0 \log_2 0 = 0$.

Proof. We give a very simple proof, usually credited to Shannon, of this fundamental theorem. Letting log stand for the binary logarithm and ln for the natural one, we have

$$\sum P(a_i) \log \frac{2^{-n(a_i)}}{P(a_i)} - (\log e) \sum P(a_i) \ln \frac{2^{-n(a_i)}}{P(a_i)}$$
$$\leqslant (\log e) \sum P(a_i) \left[\frac{2^{-n(a_i)}}{P(a_i)} - 1 \right]$$
$$= (\log e) \left[\sum 2^{-n(a_i)} - 1 \right] \leqslant 0.$$

The first inequality follows from $\ln x \leqslant x - 1$. □

The lower bound $H(P)$ of the mean code lengths in the theorem is the famous *entropy*, which may be taken as a measure of the *ideal* mean complexity of A in light of P. Indeed, $H(P) \leqslant L^*(P) \leqslant H(P)+1$, and for large alphabets the entropy is close to $L^*(P)$. The ideal code length, $-\log P(a_i)$, the *Shannon–Wiener information*, may be taken as the *complexity* of the element a_i, relative to the given distribution P. The role of the distribution P is that of a *model* for the elements in the set A in that it describes a property of them, which, in turn, restricts the elements in a collective sense rather than individually. Entropy, then, measures the strength of such a restriction in an inverse manner. This is the meaning of entropy. One should remember that it is defined for a given distribution, whose choice depends on the circumstances and clearly deprives any absolute meaning from the entropy.

Shannon's theorem with the same proof can be stated in the more general form

$$\min_q E_p \log \frac{p(X)}{q(X)} = 0, \qquad (15.2)$$

the minimum reached for $q = p$, where q and p are either probability mass functions or density functions. The ratio $E_p \log p(X)/q(X) = D(p\|q)$ is the important Kullback–Leibler distance between the two distributions. Notice that $D(p\|q) \neq D(q\|p)$ in general.

15.3 Kolmogorov Sufficient Statistics

We give a very brief account of the Kolmogorov complexity and his sufficient statistics decomposition relevant to the theme in this chapter and refer the reader to the comprehensive text by Li and Vitanyi (1993). Let U be a computer that can execute programs $p_U(x^n)$ delivering the desired binary string $x^n = x_1, \ldots, x_n$ as the output. In other words, when the program $p_U(x^n)$ is fed into the computer, after a while the machine prints out x^n and stops. We may then view the computer as defining a many-to-one decoding map that recovers the binary string x^n from any of its programs. In the terminology above, a program $p_U(x^n)$ is a codeword of length $|p_U(x^n)|$ for the string x^n. Clearly, there are a countable number of programs for each string, for we can add any number of instructions canceling each other to create the same string. Notice that when the machine stops it will not start by itself, which means that no program that generates x^n and stops can be a prefix of a longer program that does the same. Hence, if we place the set of all programs with the said property in a binary tree, where they appear as leaves, the countably infinite 'code' tree satisfies the Kraft inequality.

The *Kolmogorov complexity* of a string x^n, relative to a universal computer U, is defined as

$$K_U(x^n) = \min_{p_U(x^n)} |p_U(x^n)|.$$

In words, it is the length of the shortest programs in the language of U that generate the string.

The set of all programs for U may be ordered, first by the length and then the programs of the same length alphabetically. Hence, each program p has an index $i(p)$ in this list. Importantly, this set has a generating grammar, which can be programmed in another universal computer's language with the effect that each universal computer can execute another computer's programs by use of this translation or compiler program, as it is also called. Hence, in the list of the programs for U there is in some place a shortest program $p_U(V)$ that is capable of translating all the programs of another universal computer V. But this then means that

$$K_V(x^n) \leqslant K_U(x^n) + |p_V(U)| = K_U(x^n) + C_U,$$

where C_U does not depend on the string x^n. By exchanging the roles of U and V we see that the dependence of the complexity of long strings x^n on the particular universal computer used is diluted. We cannot, of course, claim that the dependence has been eliminated, because the constant C_U can be as large as we wish by picking a 'bad' universal computer. However, by recognizing that for a reasonably long string x^n the shortest program in a universal computer will have to capture essentially all the regular features in the string, we can safely regard such a U as fixed, and the Kolmogorov complexity $K_U(x^n)$, indeed, provides us a virtually absolute measure of the string's complexity.

Although the dependence of the complexity on the universal computer is not too serious a handicap, there is another more serious one, namely, the shortest programs, cannot be found in any 'mechanical' way. More precisely, there is no program such that, when given the data string x^n, the computer will calculate the complexity $K_U(x^n)$ as a binary integer. The usual proof of this important theorem due to Kolmogorov appeals to the undecidability of the so-called halting problem. We give here another proof following Nohre (1994).

Suppose to the contrary that a program q_U with the said property exists. We can then write a program p_U, using q_U as a subroutine, which finds the shortest string $z^n = p_U(\lambda)$ from the empty string λ such that

$$K_U(z^n) > |p_U|. \qquad (15.3)$$

In essence, the program p_U examines the strings x^t, sorted alphabetically by nondecreasing length, computes $K_U(x^t)$ with q_U and checks if (15.3) holds. It is clear that such a shortest string exists, because $K_U(x^t)$ has no upper bound and $|p_U|$ has some fixed finite length. But by the definition of Kolmogorov complexity $K_U(z^n) \leqslant |p_U|$, which contradicts (15.3). Hence, no program q_U with the assumed property exists.

Despite the negative content of the proved result, it is, or should be, of utmost importance in statistical inference, because it clearly sets a limit for what can be meaningfully asked. Any statistical effort trying to search for the 'true' underlying data-generating distribution by systematic (mechanical) means is hopeless, even if we, by the 'truth', mean only the best model. Therefore, human intuition

COMPLEXITY AND INFORMATION IN DATA 305

and intelligence in this endeavor are indispensable. It is also clear in this light that, while many physical phenomena are simple, because their data admit laws (or, as Einstein put it, God is good), to find the laws is inherently difficult! It has generally taken geniuses to discover even some of the simplest laws in physics. In the algorithmic theory most strings are maximally complex, although we cannot prove a single one to be so, which raises the intriguing question of whether most data strings arising in the physical world are also maximally complex or nearly so, i.e. obeying *no* laws. The complexity, of course, is defined in terms of the least number of bits it takes to write down the data.

Although we have identified the shortest program for a string x with its 'ideal' model, because it will have to capture all the regular features in the string, it does not quite correspond to intuition. In fact, we associate with a model only the regular features rather than the entire string-generating machinery. We outline next the construction due to Kolmogorov by which the desired idea of a model is captured in the algorithmic theory. We refer to Cover and Thomas (1991) for a more comprehensive account including the history of the work left unpublished by Kolmogorov.

The Kolmogorov complexity is immediately extended to the conditional complexity $K(x \mid y)$ as the length of the shortest program that generates string x from another string y and causes the computer to stop. One can then show that the joint complexity $K(x, y)$ of x and y breaks up as

$$K(x, y) \cong K(x) + K(y \mid x) \tag{15.4}$$
$$\cong K(y) + K(x \mid y), \tag{15.5}$$

where without formalization by $\cong$ we just mean 'almost equality,' say equality up to a constant not depending on the length of the strings x and y. We now take y to be a program that describes the 'summarizing properties' of string $x = x^n$. A 'summarizing property' of data may be formalized as a finite set of strings A where the data belong to, together with other sequences sharing this property. Hence, the property A need not specify the sequence completely. We may now think of programs consisting of two parts, where the first part describes optimally such a set A with the number of bits given by the Kolmogorov complexity $K(A)$, and the second part merely describes x^n in A with about $\log |A|$ bits, $|A|$ denoting the number of elements in A. Notice that the set A itself is such that it can be described by a computer program; in fact, since it is taken here as finite we surely can describe it at least by listing all its elements, but possibly by a shorter program. A simple example is the set of all binary strings of length n with m ones in each. Clearly, we do not need to list all its elements. The sequence x^n then gets described in $K(A) + \log |A|$ bits. We now consider sets A for which

$$K(A) + \log |A| \cong K(x^n). \tag{15.6}$$

For instance, the singleton set $A = \{x^n\}$ certainly satisfies this, because the second term is zero. Clearly, this set represents all the summarizing properties of x^n, because it represents *all* the properties; $K(A)$ is too large and A too small. If we then consider an increasing sequence of sets A while requiring (15.6) to hold, there will be a largest one that includes all the strings that share the summarizing properties of x^n. We might then say that such a largest set A has 'extracted' all the relevant properties of x^n and nothing else. But this in view of (15.6) means that we should ask for a set $\hat{A}$ for which (15.6) holds and for which $K(\hat{A})$ is minimal. Such a set $\hat{A}$, or its defining program, is called Kolmogorov's *minimal sufficient statistic* for the description of x^n.

It follows from (15.5) and (15.6) that both

$$K(x^n \mid \hat{A}) \cong \log|\hat{A}| \quad \text{and} \quad K(\hat{A} \mid x^n) \cong 0.$$

The latter just expresses the obvious: that the summarizing properties of x^n can be computed from x^n with little cost, i.e. are determined by x^n, and that the bits describing $\hat{A}$ are then the 'interesting' bits in the program (code) for x^n. The former implies that the rest, about $\log|\hat{A}|$ in number, are noninformative noise bits, i.e. x^n given $\hat{A}$ is random. We define $K(\hat{A})$ to be the Kolmogorov *information* in the string x^n.

15.4 Complexity

We begin with a brief discussion of a *model*. We consider sets X and Y and their Cartesian product $X \times Y$ together with the extension $X^n \times Y^n$ to sets of strings of length n. Perhaps the most common type of model is defined by a function $F : X \to Y$, which is extended to sequences with the same notation $F : X^n \to Y^n$, together with an error function $\delta(y, F(x))$. We consider here the logarithmic error functions $\delta(y, F(x)) = -\log p(y \mid \hat{x})$, where $\hat{x} = F(x)$ and $p(y \mid \hat{x})$ is a conditional density function. This is extended to sequences by independence $p(y^n \mid x^n)$. Many of the usual error functions, above all the quadratic one, can be represented as the negative logarithm of conditional density functions. Of particular interest to us are the parametric models, $\mathcal{M}_\gamma = \{p(y^n \mid x^n; \gamma, \theta) : \theta \in \Omega\}$, where γ is a structure index, such as the pair of orders p, q in ARMA models, and $\theta = \theta_1, \ldots, \theta_k$ ranges over some, usually compact, subset Ω of the k-dimensional Euclidean space, k depending on γ, such as $k = p + q$ in the ARMA models. (An ARMA model is a linear input/output model of type $y_t + \sum_{i=1}^p a_i y_{t-i} = x_t + \sum_{i=1}^q b_i x_{t-i}$ for $t = 1, 2, \ldots$.) Put $\mathcal{M} = \bigcup_\gamma \mathcal{M}_\gamma$, an example of which is the set of all ARMA models or for curve fitting the set of all polynomials. The parameters θ often include parameters in the function F and others in the conditional probability or density function, such as the variance. Finally, an important special case is the one where the data sequence x^n is absent, in which case we write x^n rather than y^n for the single data sequence, and $p(x^n; \gamma, \theta)$ for the models. The theory is similar for both types of data. An example is prediction, where we wish to predict each 'future' value x_{t+1} from the observed past values $x^t = x_1, \ldots, x_t$ for $t = 1, 2, \ldots$. We also drop the requirement that a real code length must be integer valued, and hence the negative logarithm of any probability or density is regarded as an ideal code length. By a harmless abuse of notation even the word 'ideal' is often dropped. To summarize, a probability distribution, a model, and a code can be identified.

As indicated in the Introduction the idea with complexity is to obtain the shortest (ideal) code length with which the data sequence x^n can be encoded when the codes are somehow designed with the models in the classes $\mathcal{M}_\gamma = \{p(x^n; \gamma, \theta)\}$ and $\mathcal{M} = \bigcup_\gamma \mathcal{M}_\gamma$, respectively. In order to avoid the computability problem in the algorithmic theory, the sense in which the code length is shortest will have to be probabilistic rather than literally shortest for every sequence.

The code we are looking for must be designed from the model class $\mathcal{M}_\gamma$ given, rather than from a fixed distribution as the case was with Shannon, (15.2). Hence, we should look for a density function $p(x^n; \gamma)$ that is universal for the class. A first attempt could be to consider the best possible code length $\min_\theta \log 1/p(x^n; \gamma, \theta)$. The minimizing parameter is the maximum likelihood estimate $\hat{\theta}(x^n)$, and the data sequence x^n could be encoded optimally with the code $p(\cdot; \gamma, \hat{\theta}(x^n))$. However, we cannot define $-\log p(x^n; \gamma, \hat{\theta}(x^n))$ as the complexity of the data sequence x^n, relative to the class $\mathcal{M}_\gamma$, because $p(x^n; \gamma, \hat{\theta}(x^n))$ defines no distribution. In fact, its integral over the data sequences is greater than unity. Clearly, the code defined by $p(\cdot; \gamma, \hat{\theta}(x^n))$ is tailored to the data sequence x^n, as it were, and we could not construct it from the knowledge of the model class $\mathcal{M}_\gamma$ alone. Recall that a code or a model must be defined without reference to any particular data sequence.

Nevertheless, we may take $-\log p(x^n; \gamma, \hat{\theta}(x^n))$ as the ideal target for the universal density function, and we generalize Shannon's problem (15.2) as follows,

$$\min_q \max_{g \in G} E_g \log \frac{p(X^n; \gamma, \hat{\theta}(X^n))}{q(X^n)}, \qquad (15.7)$$

where G is a class larger than $\mathcal{M}_\gamma$. The capitalized X^n refer to random variables whose outcomes are the sequences x^n. Unlike in Barron et al. (1998), where g was restricted to the model class $\mathcal{M}_\gamma$, the idea here is to model the situation where the data have properties that cannot be captured by the model class considered, which is done by considering data-generating density functions g that lie outside the class $\mathcal{M}_\gamma$. In fact, we can let G consist of all density functions such that $G = \{g : E_g \log(g(X^n)/p(X^n; \gamma, \hat{\theta}(X^n))) < \infty\}$. This excludes the singular distributions, which clearly do not restrict the data in any manner and hence do not specify any properties in them that we wish to model.

Provided the set of parameters Ω is such that the integral

$$C_n(\gamma) = \int_{\hat{\theta}(y^n) \in \Omega} p(y^n; \gamma, \hat{\theta}(y^n)) \, dy^n \qquad (15.8)$$

is finite, the solution to (15.7), i.e. the minimizing q, is given by the *normalized maximum likelihood* (NML) density function

$$\hat{p}(x^n; \gamma) = \frac{p(x^n; \gamma, \hat{\theta}(x^n))}{C_n(\gamma)}, \qquad (15.9)$$

and the minmax value is given by $\log C_n(\gamma)$. Notice, that the NML distribution is defined in terms of the models in the class $\mathcal{M}_\gamma$ so that it can be computed at least to an approximation. For many of the usual classes it can be done very accurately, for we have (Rissanen 1996)

$$L(x^n; \mathcal{M}_\gamma) = -\log \hat{p}(x^n; \gamma)$$
$$= -\log p(x^n; \gamma, \hat{\theta}(x^n)) + \log C_n(\gamma), \qquad (15.10)$$

$$\log C_n(\gamma) = \frac{k}{2} \log \frac{n}{2\pi} + \log \int_\Omega |I(\theta)|^{1/2}\, d\theta + o(1), \qquad (15.11)$$

where $I(\theta)$ is the Fisher information matrix

$$I(\theta) = \left\{ -E_\theta \frac{\partial^2 \log p(X; \gamma, \theta)}{\partial \theta_i \partial \theta_j} \right\}.$$

Here E_θ denotes the expectation with respect to $p(x; \gamma, \theta)$. We define $L(x^n; \mathcal{M}_\gamma)$ to be the (stochastic) *complexity* of the data sequence x^n, given the class $\mathcal{M}_\gamma$. We give an example of these formulas for the Bernoulli class in the next section.

We mention that for many model classes the status of the NML distribution as defining the complexity and universal model can be strengthened, essentially in the manner that the worst-case minmax bound $\log C_n(\gamma)$ is actually the norm, in that it also holds for almost all data-generating distributions that fall within the class $\mathcal{M}_\gamma$ (Rissanen 1986).

15.5 Information

The definition of the useful information in data is more intricate than that of the complexity, and we can only sketch the procedure. If we generate a data sequence y^n of a fixed length either with a model $p(y^n; \theta)$ or with another $p(y^n; \theta')$, we can tell with the ML estimate $\hat{\theta}(y^n)$ which one generated the data with a small probability of making a mistake, if the two parameters are well apart. In fact, if we let n grow and the distance defined by the square root of $(\theta - \theta')^T I(\theta)(\theta - \theta')$ exceeds $O(1/\sqrt{n})$ per parameter, where $I(\theta)$ is the Fisher information matrix and the superscript 'T' denotes transposition, the probability of making a mistake goes to zero, but if the distance stays less than this bound, the probability does not go to zero as n grows. We may then regard two models as virtually indistinguishable from the data y^n, for large n, if their distance is smaller than the given critical bound. Now, if we partition the bounded parameter space Ω into hypercubes of side length $O(1/\sqrt{n})$ in the metric given, then, quite remarkably, the number of such hypercubes is $C_n(\gamma)$ (Myung et al. 2000) (see also Balasubramanian 1997). The centers of the hypercubes may then be regarded as the models that can be distinguished from a growing data sequence with diminishing probability of making a mistake. A description or encoding of the center of the hypercube that contains the parameter $\hat{\theta}(x^n)$ defining the optimal model requires $\log C_n(\gamma)$ bits. This code length we define to be the *information* in the data sequence x^n that can be extracted with the given model class $\mathcal{M}_\gamma$. It was called the *geometric complexity* in Myung et al. (2000) (see also Balasubramanian 1997), and prior to that *model cost*, suggesting a penalty for not knowing the data-generating distribution. Our term information for the same model cost emphasizes the positive aspect of this notion, namely, as the amount of useful, and, in fact, learnable 'information' one can gain from the data with the model class considered.

Let $A_n(\theta)$ denote the hypercube where $\hat{\theta}(x^n)$ falls. Then, the sequences y^n such that $\hat{\theta}(y^n) \in A_n(\theta)$ do not add any more information to what can be extracted from x^n. They all require about the same code length, namely, $\log 1/p(x^n; \gamma, \hat{\theta}(x^n))$, and because of this they cannot be further compressed with the given model class, except by an insignificant amount. But, clearly, neither can the common code length $\log C_n(\gamma)$ for the distinguishable models be reduced. Much as in the Kolmogorov sufficient statistics decomposition (15.6), the complexity in (15.10)—the shortest code length for x^n, relative to the class $\mathcal{M}_\gamma$—breaks up into the code length for the useful information specifying the set $X_n(\theta) = \{y^n : \hat{\theta}(y^n) \in A_n(\theta)\}$ and the code length for the noninformative sequences in this set, the 'volume' of which is $1/p(x^n; \gamma, \hat{\theta}(x^n))$. The set $X_n(\theta)$, then, plays the role of the finite set A in (15.6).

In some cases it is possible to establish a one-to-one correspondence between the set of distinguishable models whose number is $C_n(\gamma)$ and the sequences $\hat{x}^n = \hat{x}^n(\theta)$ specified by these models. We then obtain a decomposition of the data sequence as $x^n = \hat{x}^n + e^n$, where e^n represents noninformative and incompressible 'noise.' It corresponds to one of the strings in the set $X_n(\theta)$ specified by the distinguishable model determined by $\hat{x}^n(\theta)$. This happens, for instance, in the linear quadratic regression problems (Rissanen 2001); we briefly discuss the resulting denoising algorithm in the next section.

The complexity and the information can be defined for the larger class of models $\mathcal{M} = \bigcup_\gamma \mathcal{M}_\gamma$ in just the same way. We calculate the NML distribution

$$\hat{p}(x^n; \mathcal{M}) = \frac{\hat{p}(x^n; \hat{\gamma}(x^n))}{\mathcal{C}_n}, \qquad (15.12)$$

$$\mathcal{C}_n = \sum_{y^n} \hat{p}(y^n; \hat{\gamma}(y^n)), \qquad (15.13)$$

where $\hat{\gamma}(y^n)$ denotes the structure parameter maximizing $\hat{p}(y^n; \gamma)$. Since $\mathcal{C}_n$ is just a constant, the search for $\hat{\gamma}(x^n)$ is tantamount to the *minimum description length* (MDL) principle (Rissanen 1978, 1989) for model selection,

$$\min_{\gamma} \log \frac{1}{\hat{p}(x^n; \gamma)} = \min_{\gamma} \left\{ \log \frac{1}{p(x^n; \gamma, \hat{\theta}(x^n))} + \log C_n(\gamma) \right\}, \quad (15.14)$$

which we now see as the principle that seeks to separate the useful information from noise. Notice in particular that so long as we compare models within the family $\mathcal{M}$, there is no need to add a code length for γ to the criterion.

We conclude this section with a derivation of the complexity and the information for the Bernoulli class.

Example 15.2. The Bernoulli class $\mathcal{B}$ is obtained by extending the probability distribution on $\{0, 1\}$, defined by $P(x = 0) = p$ as the parameter, to sequences by independence. The NML distribution is given by

$$\hat{P}(x^n) = P(x^n; \hat{p}(x^n)) \bigg/ \sum_m \binom{n}{m} \left(\frac{m}{n}\right)^m \left(\frac{n-m}{n}\right)^{n-m}, \quad (15.15)$$

where

$$\hat{p}(x^n) = \frac{n_0(x^n)}{n} \quad \text{and} \quad P(x^n; \hat{p}(x^n)) = \hat{p}(x^n)^{n_0(x^n)}(1 - \hat{p}(x^n))^{n-n_0(x^n)}.$$

Write this in the form

$$\hat{P}(x^n) = \left(1 \bigg/ \binom{n}{n_0(x^n)}\right) \times \pi_n(\hat{p}(x^n)), \quad (15.16)$$

where

$$\pi_n(\hat{p}(x^n)) = \binom{n}{n_0(x^n)} P(x^n; \hat{p}(x^n)) \bigg/ \sum_m \binom{n}{m} \left(\frac{m}{n}\right)^m \left(\frac{n-m}{n}\right)^{n-m}. \quad (15.17)$$

In order to evaluate the resulting ideal code length $-\ln \hat{P}(x^n)$, we use the important and ubiquitous Stirling approximation formula in the form refined by Robbins,

$$\ln n! = (n + \tfrac{1}{2}) \ln n - n + \ln \sqrt{2\pi} + R(n), \quad (15.18)$$

where

$$\frac{1}{12(n+1)} \leqslant R(n) \leqslant \frac{1}{12n}.$$

This permits us to evaluate the terms in the sum in equation (15.17) to a sufficient accuracy for us as

$$\ln \binom{n}{m} \cong nH\left(\frac{m}{n}\right) \ln 2 - \tfrac{1}{2} \ln \left[\frac{m}{n} \frac{n-m}{n}\right] - \tfrac{1}{2} \ln n - \ln \sqrt{2\pi},$$

where $H(p)$ is the binary entropy at p. This gives

$$\binom{n}{m} \left(\frac{m}{n}\right)^m \left(\frac{n-m}{n}\right)^{n-m} \cong \frac{1}{\sqrt{2\pi n}} \left[\frac{m}{n} \frac{n-m}{n}\right]^{-1/2}.$$

COMPLEXITY AND INFORMATION IN DATA   311

Recognizing the sum in the denominator of equation (15.17) (with step length $1/n$ rather than $1/\sqrt{n}$) as an approximation of a Riemann integral, we get it approximately as

$$\sqrt{n/(2\pi)} \int_0^1 \frac{1}{\sqrt{p(1-p)}} \, dp = \sqrt{n\pi/2},$$

where the integral of the square root of the Fisher information

$$J(p) = \frac{1}{p(1-p)}$$

is a Dirichlet integral with the value π. Finally, we get the complexity as

$$-\log \hat{P}(x^n) = nH(n_0(x^n)/n) + \tfrac{1}{2} \log \tfrac{1}{2} n\pi + o(1), \qquad (15.19)$$

in which the first term is the code length for the noninformative part in the complexity and the rest is the information. The added term $o(1)$, which goes to zero at the rate $O(1/n)$ as $n \to \infty$, takes care of the errors made by the application of Stirling's formula.

15.6 Denoising with Wavelets

As an application of the above theory, we briefly discuss the denoising problem (Rissanen 2000). Wavelet transforms provide an orthonormal basis for the data string y^n, thus

$$y^n = Wc^n, \qquad (15.20)$$
$$c^n = W'y^n, \qquad (15.21)$$

where both y^n and the coefficients c^n may be viewed as column vectors. The wavelet equations define the orthonormal matrix W, which of course can be huge, without the need to even write it down. If we define the column vector $\hat{c}^n$ by retaining, say, only k coefficients in c^n with their indices in the set $\gamma = \{i_1, \ldots, i_k\}$ while setting the rest to zero, we have

$$\hat{y}^n = W\hat{c}^n, \qquad (15.22)$$
$$e^n = y^n - \hat{y}^n. \qquad (15.23)$$

Because of orthonormality of W, $\hat{y}^n$ is the orthogonal projection of y^n onto the subspace spanned by the $k = k_\gamma$ columns of W, corresponding to the nonzero coefficients in $\hat{c}^n$ and indices in γ. For the same reason, Parseval's equality holds $\sum_i \hat{y}_i^2 = \sum_i \hat{c}_i^2$. In the formalism above, we have a class of Gaussian models $\mathcal{M}_\gamma = \{p(y^n; \gamma, \hat{y}^n, \tau)\}$, where p is the normal density function of mean $\hat{y}^n$ and the $n \times n$ diagonal covariance matrix with τ on the diagonal.

The NML density function $\hat{f}(y^n; \gamma, \tau_0, R)$ can be found in an exact form (Rissanen 2000), except that the range of the integration Ω will involve two hyperparameters, a lower bound for τ and an upper bound for $(\hat{c}^n, \hat{c}^n) = \sum_i \hat{c}_i^2$. These, however, influence the optimization of γ in an essential manner. To overcome the problem we eliminate these parameters by computing $\hat{f}(y^n; \gamma, \hat{\tau}_0(y^n), \hat{R}(y^n))$ and its normalization, which requires four new parameters. But these just amount to an additive constant, which does not affect the desired optimization of the exact and perfectly symmetric criterion:

$$\min_\gamma \left\{ (n - k_\gamma) \ln \frac{(c^n, c^n) - (\hat{c}^n, \hat{c}^n)}{n - k_\gamma} + k \ln \frac{(\hat{c}^n, \hat{c}^n)}{k_\gamma} - \ln \frac{k_\gamma}{n - k_\gamma} \right\}. \quad (15.24)$$

This still looks like a formidable optimization problem requiring a search over all subsets of the n columns of W. The search is actually simple since the optimum subcollection of columns of W, determined by γ, consists either of k largest squared coefficients in c^n or, intriguingly, of k smallest squared coefficients, for some value of k. The result is a superior denoising algorithm, where noise gets defined as that part in the data sequence that cannot be compressed with the given model class rather than by the high-frequency part as in the traditional risk-based schemes. A dramatic numerical example of speech data illustrating the difference is given in Rissanen (2000).

References

Barron, A. R., Rissanen, J. and Yu, B. 1998 The MDL principle in modeling and coding. *IEEE Trans. Information Theory* (Special Issue to commemorate 50 years of Information Theory) **IT-44**, 2743–2760.

Balasubramanian, V. 1997 Statistical inference, Occam's razor and statistical mechanics on the space of probability distributions. *Neural Computation* **9**, 349–268.

Cover, T. M. and Thomas, J. A. 1991 *Elements of Information Theory*. Wiley.

Li, M. and Vitanyi, P. 1993 *An Introduction to Kolmogorov Complexity and Its Applications*. Springer.

Myung, I. J., Balasubramanian, V. and Pitt, M. A. 2000 Counting probability distributions: differential geometry and model selection. *Proc. Natl Acad. Sci.* **97**, 11 170–11 175.

Nohre, R. 1994 Some topics in descriptive complexity. PhD thesis, Linkoping University, Linkoping, Sweden.

Rissanen, J. 1978 Modeling by shortest data description. *Automatica* **14**, 465–471

Rissanen, J. 1986 Stochastic complexity and modeling. *Ann. Stat.* **14**, 1080–1100

Rissanen, J. 1989 *Stochastic Complexity in Statistical Inquiry*. World Scientific.

Rissanen, J. 1996 Fisher information and stochastic complexity. *IEEE Trans. Information Theory* **11-42**, 40–47.

Rissanen, J. 2000 MDL denoising. *IEEE Trans. Information Theory* **IT-46**, 2537–2543.

Rissanen, J. 2001 Strong optimality of normalized ML models as universal codes and information in data. *IEEE Trans. Information Theory* **IT-47**, 1712–1717.

Chapter Sixteen

Entropy in Dynamical Systems

Lai-Sang Young
Courant Institute of Mathematical Sciences

In this chapter, the word *entropy* is used exclusively to refer to the entropy of a dynamical system. It measures the *rate of increase in dynamical complexity* as the system evolves with *time*. This is not to be confused with other notions of entropy connected with spatial complexity.

I will attempt to give a brief survey of the role of entropy in dynamical systems and especially in smooth ergodic theory. The topics are chosen to give a flavor of this invariant; they are also somewhat biased toward my own interests. This article is aimed at nonexperts. After a review of some basic definitions and ideas, I will focus on one main topic, namely, the relation of entropy to Lyapunov exponents and dimension, which I will discuss in some depth. Several other interpretations of entropy, including its relations to volume growth, periodic orbits and horseshoes, large deviations and rates of escape, are treated briefly.

16.1 Background

16.1.1 Dynamical Systems

Roughly speaking, a dynamical system is a deterministic process which evolves with time. A *discrete-time* system is generated by the iteration of a map from a space to itself, i.e. $f : X \to X$. For $x \in X$, the set

$$\{x, fx, f^2x = f(fx), \ldots, f^n x = f(f^{n-1}x), \ldots\}$$

is called the *orbit* or forward orbit of x. If f is invertible, then the full orbit of x is given by $\{f^n x, n \in \mathbb{Z}\}$. A *continuous-time* dynamical system is one generated by a one-parameter group (or semi-group) of maps $f^t : X \to X$, $t \in \mathbb{R}$ (respectively, $\mathbb{R}^+$). While many of the results of this paper have both discrete- and continuous-time versions, we will, for simplicity of exposition, restrict ourselves to the case of discrete time.

One of the goals of dynamical systems is to understand the asymptotic or large-time behavior of maps and flows, and entropy is a measure of the rate of increase in complexity in the orbit structures of f^n as n tends to infinity.

This research is partially supported by a grant from the NSF.

We have not imposed any conditions on X or f so far. In general one assumes that X has certain structures which are preserved by f. For example, *ergodic theory* is concerned with the study of measure-preserving transformations of probability spaces. Topological, geometric, and analytic methods are used in other approaches. In this chapter, entropy will be discussed from all these viewpoints.

16.1.2 Topological and Metric Entropies

We begin with topological entropy because it is simpler, even though chronologically, metric entropy was defined first. Topological entropy was first introduced by Adler et al. (1965). The following definition is due to Bowen and Dinaburg. A good reference for this section is Walters (1982), which contains many of the early references.

Definition 16.1. Let $f : X \to X$ be a continuous map of a compact metric space X. For $\varepsilon > 0$ and $n \in \mathbb{Z}^+$, we say $E \subset X$ is an (n, ε)-separated set if for every $x, y \in E$, there exists i, $0 \leq i < n$, such that $d(f^i x, f^i y) > \varepsilon$. Then the topological entropy of f, denoted by $h_{\text{top}}(f)$, is defined to be

$$h_{\text{top}}(f) = \lim_{\varepsilon \to 0} \left\{ \limsup_{n \to \infty} \frac{1}{n} \log N(n, \varepsilon) \right\},$$

where $N(n, \varepsilon)$ is the maximum cardinality of all (n, ε)-separated sets.

Roughly speaking, we would like to equate 'higher entropy' with 'more orbits,' but since the number of orbits is usually infinite, we need to fix a finite 'resolution,' i.e. a scale below which we are unable to tell points apart. Suppose we do not distinguish between points that are less than ε apart. Then $N(n, \varepsilon)$ represents the number of distinguishable orbits of length n, and if this number grows like $\sim e^{nh}$, then h is the topological entropy. Another way of counting the number of distinguishable n-orbits is to use (n, ε)-*spanning* sets, i.e. sets E with the property that for every $x \in X$ there exists $y \in E$ such that $d(f^i x, f^i y) < \varepsilon$ for all $i < n$; $N(n, \varepsilon)$ is then taken to be the minimum cardinality of these sets. The original definition in Adler et al. (1965) uses open covers, but it conveys essentially the same idea.

Facts 16.2. It follows immediately from Definition 16.1 that

(i) $h_{\text{top}}(f) \in [0, \infty]$, and

(ii) if f is a differentiable map of a compact d-dimensional manifold, then

$$h_{\text{top}}(f) \leq d \log \|Df\|.$$

Metric or *measure-theoretic* entropy for a transformation was introduced by Kolmogorov and Sinai in 1959; the ideas go back to Shannon's information theory. Let $(X, \mathcal{B}, \mu)$ be a probability space, i.e. X is a set, $\mathcal{B}$ a σ-algebra of subsets of X, and μ a probability measure on $(X, \mathcal{B})$. Let $\alpha = \{A_1, \ldots, A_k\}$ be a finite partition of X. Then the *entropy of the partition* α, written $H(\alpha)$, is defined to be

$$H(\alpha) := H(\mu(A_1), \ldots, \mu(A_k)), \quad \text{where } H(p_1, \ldots, p_k) = -\sum p_i \log p_i.$$

If β is another finite partition, then the *conditional entropy* of α given β, written $H(\alpha \mid \beta)$, is defined to be

$$H(\alpha \mid \beta) = \sum_{B \in \beta} \mu(B) H(\mu(A_1 \mid B), \ldots, \mu(A_k \mid B)).$$

These definitions have the following interpretations. Letting i be the α-*address* of $x \in X$ if $x \in A_i$, $H(\alpha)$ measures the amount of uncertainty, on average, as one attempts to predict the α-address of a randomly chosen point x, while $H(\alpha \mid \beta)$ represents the uncertainty in predicting the α-address of a point given its β-address.

To this 'static' situation, we introduce a map $f : X \to X$, which we assume is a measure-preserving transformation, i.e. for all $A \in \mathcal{B}$, we have $f^{-1}A \in \mathcal{B}$ and $\mu(A) = \mu(f^{-1}A)$. Observe that $f^{-i}\alpha = \{f^{-i}A_1, \ldots, f^{-i}A_k\}$ is also a partition, and the *join* of α and β, written $\alpha \vee \beta$, is the partition $\{A \cap B : A \in \alpha, B \in \beta\}$. In particular, the elements of $\bigvee_{i=0}^{n-1} f^{-i}\alpha$ are sets of the form $\{x : x \in A_{i_0}, fx \in A_{i_1}, \ldots, f^{n-1}x \in A_{i_{n-1}}\}$ for some $(i_0, i_1, \ldots, i_{n-1})$.

Definition 16.3. The metric entropy of f, written $h_\mu(f)$, is defined as follows:

$$h_\mu(f, \alpha) := \lim_{n \to \infty} \frac{1}{n} H\left(\bigvee_0^{n-1} f^{-i}\alpha\right) = \lim_{n \to \infty} H\left(f^{-n}\alpha \bigm| \bigvee_0^{n-1} f^{-i}\alpha\right);$$

$$h_\mu(f) := \sup_\alpha h_\mu(f, \alpha).$$

It is an easy exercise to show that the two limits in the first displayed formula above are equal. The first is the average uncertainty *per iteration* in guessing the α-addresses of a typical n-orbit, while the second is the uncertainty in guessing the α-address of $f^n x$ given the α-addresses of $x, fx, \ldots, f^{n-1}x$. Instead of taking the supremum over all partitions, the following facts make it easier to compute the quantity in the second line: $h_\mu(f) = h_\mu(f, \alpha)$ if α is a generator; it can also be realized as $\lim h_\mu(f, \alpha_n)$, where α_n is an increasingly refined sequence of partitions such that $\bigvee \alpha_n$ partitions X into points.

The following theorem gives yet another interpretation of metric entropy.

Theorem 16.4 (the Shannon–McMillan–Breiman Theorem). *Let f and α be as above, and write $h = h_\mu(f, \alpha)$. Assume for simplicity that (f, μ) is ergodic. Then given $\varepsilon > 0$, there exists N such that the following holds for all $n \geq N$:*

(i) $\exists Y_n \subset X$ with $\mu(Y_n) > 1 - \varepsilon$ such that Y_n is the union of $\sim e^{(h \pm \varepsilon)n}$ atoms of $\bigvee_0^{n-1} f^{-i}\alpha$ each having μ-measure $\sim e^{-(h \pm \varepsilon)n}$;

(ii)
$$-\frac{1}{n} \log \mu\left(\left(\bigvee_0^{n-1} f^{-i}\alpha\right)(x)\right) \to h \quad \text{a.e. and in } L^1,$$

where $\alpha(x)$ is the element of α containing x.

16.2 Summary

Both h_μ and h_{top} measure the exponential rates of growth of n-orbits: h_μ counts the number of *typical* n-orbits, while h_{top} counts *all* distinguishable n-orbits.

We illustrate these ideas with the following coin-tossing example. Consider $\sigma : \Pi_0^\infty \{H, T\} \to \Pi_0^\infty \{H, T\}$, where each element of $\Pi_0^\infty \{H, T\}$ represents the outcome of an infinite series of trials and σ is the shift operator. Since the total number of possible outcomes in n trials is 2^n, $h_{\text{top}}(\sigma) = \log 2$. If $P(H) = p$, $P(T) = 1 - p$, then $h_\mu(\sigma) = -p \log p - (1-p) \log(1-p)$. In particular, if the coin is biased, then the number of *typical* outcomes in n trials is $\sim e^{nh}$ for some $h < \log 2$.

From the discussion above, it is clear that $h_\mu \leq h_{\text{top}}$. We, in fact, have the following *variational principle*.

Theorem 16.5. *Let f be a continuous map of a compact metric space. Then*

$$h_{\text{top}}(f) = \sup_\mu h_\mu(f),$$

where the supremum is taken over all f-invariant Borel probability measures μ.

The idea of expressing entropy in terms of local information has already appeared in Theorem 16.4(ii). Here is another version of the same idea, which is more convenient for certain purposes. In the context of continuous maps of compact metric spaces, for $x \in X$, $n \in \mathbb{Z}^+$ and $\varepsilon > 0$, define

$$B(x, n, \varepsilon) := \{y \in X : d(f^i x, f^i y) < \varepsilon, \ 0 \leq i < n\}.$$

Theorem 16.6 (Brin and Katok 1983). *Assume (f, μ) is ergodic. Then for μ-a.e. x,*

$$h_\mu(f) = \lim_{\varepsilon \to 0} \left\{ \limsup_{n \to \infty} -\frac{1}{n} \log \mu B(x, n, \varepsilon) \right\}.$$

Thus, metric entropy can also be interpreted as being the rate of loss of information on nearby orbits. This leads naturally to its relation to Lyapunov exponents, which is the topic of the next section.

16.3 Entropy, Lyapunov Exponents, and Dimension

In this section, we focus on a set of ideas in which entropy plays a central role. Recall the definition of *Lyapunov exponents*. Let $f : M \to M$ be a differentiable map of a Riemannian manifold. For $x \in M$ and $v \in T_x M$, let

$$\lambda(x, v) = \lim_{n \to \infty} \frac{1}{n} \log \|DT_x^n(v)\|$$

if this limit exists. A theorem due to Oseledec states that if μ is an f-invariant probability on M, then $\lambda(x, v)$ is well defined for μ-a.e. x and every vector v. Let $\{\lambda_i(x)\}$ be the set of all growth rates starting from x. These functions are called the Lyapunov exponents of (f, μ). If f is a diffeomorphism, i.e. if it is invertible and its inverse is also differentiable, then the tangent space $T_x M$ at μ-a.e. x is the direct sum of subspaces $E_i(x)$, where $v \in E_i(x)$ has growth rate $\lambda_i(x)$. Thus the multiplicity of $\lambda_i(x)$ is the dimension of $E_i(x)$. We remark that since $Df(E_i(x)) = E_i(fx)$, the functions $x \mapsto \lambda_i(x)$ and $\dim E_i(x)$ are constant μ-a.e. if (f, μ) is ergodic.

Setting for this section. We consider (f, μ), where f is a C^2 diffeomorphism of a compact Riemannian manifold M and μ is an f-invariant Borel probability measure. For simplicity, we assume that (f, μ) is ergodic. (This assumption is not needed, but without it the results below are a little more cumbersome to state.) Let $\lambda_1 > \lambda_2 > \cdots > \lambda_r$ denote the distinct Lyapunov exponents of (f, μ), and let E_i be as above.

We have before us two ways of measuring dynamical complexity: metric entropy, which measures the growth in randomness or number of 'typical' orbits, and Lyapunov exponents, which measure the rates at which nearby orbits diverge. The first is a purely probabilistic concept, while the second is primarily geometric. We now compare these two invariants. Write $a^+ = \max(a, 0)$.

Theorem 16.7 (Pesin's Formula (Pesin 1977)). *If μ is equivalent to the Riemannian volume on M, then*

$$h_\mu(f) = \sum \lambda_i^+ \dim E_i.$$

Theorem 16.8 (Ruelle's Inequality (Ruelle 1978b)). *Without any assumptions on μ, we have*

$$h_\mu(f) \leq \sum \lambda_i^+ \dim E_i.$$

To illustrate what is going on, consider the following two examples. Assume for simplicity that both maps depicted below are affine on each of the shaded rectangles, and that their images are as shown. The first map is called the 'baker's transformation.' Lebesgue measure is preserved, and with μ equal to Lebesgue measure, we have $h_\mu(f) = \log 2 = \lambda_1$, the positive Lyapunov exponent. For the second map, we are interested only in those points that remain in the

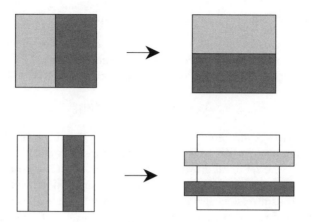

Figure 16.1.

shaded vertical strips in all (forward and backward) times. It is easy to see that these points form a Cantor set Λ. Lebesgue measure is irrelevant here, for Λ has Lebesgue measure zero. Instead, we equip f with a measure μ which makes (f, μ) isomorphic to the $(\frac{1}{2}, \frac{1}{2})$-Bernoulli process. More precisely, μ is supported on Λ, and if $\alpha = \{A_1, A_2\}$ is the partition of Λ into points in the left and right vertical strips, then every element of $\bigvee_0^{n-1} f^{-i}\alpha$ has μ-measure $1/2^n$. With respect to this measure, $h_\mu(f) = \log 2 < \lambda_1$.

Theorems 16.7 and 16.8 suggest that entropy is created by the exponential divergence of nearby orbits. In a conservative system, i.e. in the setting of Theorem 16.7, all the expansion goes back into the system to make entropy, leading to the equality in Pesin's entropy formula. A strict inequality occurs when some of the expansion is 'wasted' due to 'leakage' or dissipation from the system.

The reader may have noticed that the results above involve only positive Lyapunov exponents. Indeed, there is a complete characterization of when the entropy formula in Theorem 16.7 holds. We first state the result, with explanations to follow.

Theorem 16.9 (Ledrappier 1984; Ledrappier and Strelcyn 1982; Ledrappier and Young 1985a). *For (f, μ) with $\lambda_1 > 0$, Pesin's entropy formula holds if and only if μ is an SRB measure.*

Definition 16.10. An f-invariant Borel probability measure μ is called a *Sinai–Ruelle–Bowen measure* or *SRB measure* if f has a positive Lyapunov exponent μ-a.e. and μ has absolutely continuous conditional measures on unstable manifolds.

We elaborate on what it means for a measure to have 'absolutely continuous conditional measures on unstable manifolds.' For $x \in M$, the local unstable set

ENTROPY IN DYNAMICAL SYSTEMS

of size δ at x is defined to be

$$W_\delta^u(x) := \left\{ y \in M : d(x, y) < \delta \text{ and } \limsup_{n \to \infty} \frac{1}{n} \log d(f^{-n}x, f^{-n}y) < 0 \right\}.$$

It is a standard fact from nonuniform hyperbolic theory that if f has a positive Lyapunov exponent μ-a.e., then at μ-a.e. x there exists $\delta(x) > 0$ such that $W_{\delta(x)}^u(x)$ is an embedded disc tangent to the subspace at x corresponding to positive Lyapunov exponents. These discs, called *local unstable manifolds*, can be chosen so that $f(W_{\delta(x)}^u(x)) \supset W_{\delta(fx)}^u(fx)$. They inherit from M Riemannian structures, which induce on them Riemannian measures. We call μ an SRB measure if the conditional measures of μ on $\{W_\delta^u\}$ are absolutely continuous with respect to the Riemannian measures on them.

SRB measures are important for the following reasons.

(i) They are more general than invariant measures that are equivalent to Lebesgue (they are allowed to be singular in stable directions).

(ii) They can live on attractors (which cannot support invariant measures equivalent to Lebesgue).

(iii) Under suitable conditions they are 'physical,' meaning they reflect the properties of positive Lebesgue measure sets.

We can state this more precisely as follows.

Fact 16.11 (Pugh and Shub 1989). If μ is an ergodic SRB measure with no zero Lyapunov exponents, then there is a positive Lebesgue measure set V with the property that for every continuous function $\varphi : M \to \mathbb{R}$,

$$\frac{1}{n} \sum_{i=0}^{n-1} \varphi(f^i x) \to \int \varphi \, d\mu$$

for Lebesgue-a.e. $x \in V$.

Theorems 16.7–16.9 and Definition 16.10 are generalizations of ideas first developed for Anosov and Axiom A systems by Sinai, Ruelle, and Bowen. See Sinai (1972), Ruelle (1978a), and Bowen (1975), and the references therein. For more information on SRB measures, we refer the reader to the expository article by Young (2003).

Next we investigate the discrepancy between entropy and the sum of positive Lyapunov exponents and show that it can be expressed in terms of fractal dimension. Let ν be a probability measure on a metric space, and let $B(x, r)$ denote the ball of radius r centered at x.

Definition 16.12. We say that the *dimension* of ν, written $\dim(\nu)$, is well defined and equal to α if for ν-a.e. x

$$\lim_{\varepsilon \to 0} \frac{\log \nu(B(x, \varepsilon))}{\log \varepsilon} = \alpha.$$

The existence of $\dim(\nu)$ means that locally ν has a scaling property. This is clearly not true for all measures.

Theorem 16.13 (Ledrappier and Young 1985b). *Corresponding to every $\lambda_i \neq 0$, there is a number δ_i with $0 \leqslant \delta_i \leqslant \dim E_i$ such that*

(a) $h_\mu(f) = \sum \lambda_i^+ \delta_i = -\sum \lambda_i^- \delta_i$;

(b) $\dim(\mu \mid W^u) = \sum_{\lambda_i > 0} \delta_i$, $\dim(\mu \mid W^s) = \sum_{\lambda_i < 0} \delta_i$.

The numbers δ_i have the interpretation of being the *partial dimensions* of μ in the directions of the subspaces E_i. Here $\mu \mid W^u$ refers to the conditional measures of μ on unstable manifolds. The fact that these conditional measures have well-defined dimensions is part of the assertion of the theorem.

Observe that the entropy formula in Theorem 16.13 can be thought of as a refinement of Theorems 16.7 and 16.8. When μ is equivalent to the Riemannian volume on M, $\delta_i = \dim E_i$ and the formula in Theorem 16.13(a) is Pesin's Formula. In general, we have $\delta_i \leqslant \dim E_i$. Plugged into (a) above, this gives Ruelle's Inequality. We may think of $\dim(\mu \mid W^u)$ as a measure of the dissipativeness of (f, μ) in forward time.

I would like to include a very brief sketch of a proof of Theorem 16.13, using it as an excuse to introduce the notion of *entropy along invariant foliations*.

Sketch of Proof. I. The 'conformal' case. By 'conformal,' we refer here to situations where all the Lyapunov exponents are equal to λ for some $\lambda > 0$. This means, in particular, that f cannot be invertible, but let us not be bothered by that. In fact, let us pretend that, locally, f is a dilation by e^λ. Let $B(x, n, \varepsilon)$ be as defined toward the end of Section 16.1. Then

$$B(x, n, \varepsilon) \sim B(x, \varepsilon e^{-\lambda n}).$$

By Theorem 16.6,

$$\mu B(x, n, \varepsilon) \sim e^{-nh},$$

where $h = h_\mu(f)$. Comparing the expressions above and setting $r = e^{-\lambda n}$, we obtain

$$\mu B(x, r) \sim r^{h/\lambda},$$

which proves that $\dim(\mu)$ exists and equals h/λ.

II. The general picture. Let $\lambda_1 > \cdots > \lambda_u$ be our positive Lyapunov exponents. A more refined version of the unstable manifolds theorem (see the paragraph following Definition 16.10) says that for each $i \leqslant u$, there exists at μ-a.e. x an immersed submanifold $W^i(x)$ passing through x and tangent to $E_1(x) \oplus \cdots \oplus E_i(x)$. These manifolds are invariant, i.e. $fW^i(x) = W^i(fx)$, and the leaves of W^i are contained in those of W^{i+1}. Our strategy here is to work our way up this hierarchy of W^is, dealing with one exponent at a time.

For each i, we introduce the notion of (metric) entropy along W^i, written h_i. Intuitively, this number is a measure of the randomness of f along the leaves of W^i; it ignores what happens in transverse directions. Technically, it is defined as in Definition 16.3, using (infinite) partitions whose elements are contained in the leaves of W^i. (Observe that the conditional version of $h_\mu(f, \alpha)$ in Definition 16.3 continues to make sense.) We prove that for each i there exists δ_i such that

(i) $h_1 = \delta_1 \lambda_1$;

(ii) $h_i - h_{i-1} = \delta_i \lambda_i$ for $i = 2, \ldots, u$; and

(iii) $h_u = h_\mu(f)$.

The proof of (i) is similar to that in the 'conformal' case since it involves only one exponent. To give an idea of why (ii) is true, consider the action of f on the leaves of W^i, and pretend somehow that a quotient dynamical system can be defined by collapsing the leaves of W^{i-1} inside W^i. This 'quotient' dynamical system has exactly one Lyapunov exponent, namely, λ_i. It behaves as though it leaves invariant a measure with dimension δ_i and has entropy $h_i - h_{i-1}$. A fair amount of technical work is needed to make this precise, but once properly done, it is again the single exponent principle at work. Summing the equations in (ii) over i, we obtain $h_u = \sum_{i=1}^{u} \delta_i \lambda_i$. Step (iii) says that zero and negative exponents do not contribute to entropy. This completes the outline of the proof in Ledrappier and Young (1985b). □

16.3.1 Random Dynamical Systems

We close this section with a brief discussion of dynamical systems subjected to random noise. Let ν be a probability measure of Diff(M), the space of diffeomorphisms of a compact manifold M. Consider one- or two-sided sequences of diffeomorphisms

$$\ldots, f_{-1}, f_0, f_1, f_2, \ldots$$

chosen independently with law ν. This set-up applies to stochastic differential equations

$$d\xi_t = X_0 \, dt + \sum_i X_i \circ dB_t^i,$$

where the X_i are time-independent vector fields and the f_i are time-one-maps of the stochastic flow (defined using the Stratonovitch integral).

The random dynamical system above defines a Markov process on M with $P(E \mid x) = \nu\{f, f(x) \in E\}$. Let μ be the marginal of a stationary measure of this process, i.e. $\mu = \int f_* \mu \, d\nu(f)$. As individual realizations of this process, our random maps also have a system of invariant measures $\{\mu_{\bar{f}}\}$ defined for

$$\nu^{\mathbb{Z}}\text{-a.e. } \bar{f} = (f_i)_{i=-\infty}^{\infty}.$$

These measures are invariant in the sense that

$$(f_0)_*\mu_{\bar f} = \mu_{\sigma \bar f},$$

where σ is the shift operator; they are related to μ by

$$\mu = \int \mu_{\bar f}\, d\nu^{\mathbb{Z}}.$$

Slightly extending ideas for a single diffeomorphism, it is easy to see that Lyapunov exponents and metric entropy are well defined for almost every sequence $\bar f$ and are nonrandom. We continue to use the notation $\lambda_1 > \cdots > \lambda_r$ and h.

Theorem 16.14 (Ledrappier and Young 1988). *Assume that μ has a density w.r.t. Lebesgue measure. Then*

$$h = \sum \lambda_i^+ \dim E_i.$$

We remark that, unlike Theorem 16.7, no assumptions are made here on the individual maps f_i, i.e. the above formula derives its validity entirely from the randomness of the noise. Moreover, μ has a density if the transition probabilities $P(\cdot \mid x)$ do. In the light of Theorem 16.9, when $\lambda_1 > 0$, the measures $\mu_{\bar f}$ may be regarded as *random SRB measures*.

16.4 Other Interpretations of Entropy

16.4.1 Entropy and Volume Growth

Let $f : M \to M$ be a differentiable map of a compact m-dimensional C^∞ Riemannian manifold. In this subsection, $h(f)$ is the topological entropy of f. As we will see, there is a strong relation between $h(f)$ and the rates of growth of areas or volumes of f^n-images of embedded discs. To make these ideas precise, we need the following definitions.

For $\ell, k \geq 1$, let $\Sigma(\ell, k)$ be the set of C^k mappings $\sigma : Q^\ell \to M$, where Q^ℓ is the ℓ-dimensional unit cube. Let $\omega(\sigma)$ be the ℓ-dimensional volume of the image of σ in M counted with multiplicity, i.e. if σ is not one-to-one, and the image of one part coincides with that from another part, then we will count the set as many times as it is covered. For $n = 1, 2, \ldots, k \leq \infty$, and $\ell \leq m$, let

$$V_{\ell,k}(f) = \sup_{\sigma \in \Sigma(\ell,k)} \limsup_{n \to \infty} \frac{1}{n} \log \omega(f^n \circ \sigma),$$

$$V(f) = \max_\ell V_{\ell,\infty}(f),$$

and

$$R(f) = \lim_{n \to \infty} \frac{1}{n} \log \max_{x \in M} \|Df^n(x)\|.$$

Theorem 16.15.

(i) If f is $C^{1+\varepsilon}$, $\varepsilon > 0$, then $h(f) \leqslant V(f)$ (Newhouse 1988).

(ii) For $f \in C^k, k = 1, \ldots, \infty$,

$$V_{\ell,k}(f) \leqslant h(f) + \frac{2\ell}{k} R(f)$$

(Yomdin 1987).

In particular, for $f \in C^\infty$, (i) and (ii) together imply $h(f) = V(f)$

For (i), Newhouse showed that $V(f)$ as a volume growth rate is in fact attained by a large family of discs of a suitable dimension. The factor $(2\ell/k)R(f)$ in (ii) is a correction term for pathologies that may (and do) occur in low differentiability.

Ideas related to (ii) are also used to resolve a version of the *Entropy Conjecture*. Let $S_\ell(f)$, $\ell = 0, 1, \ldots, m$, denote the logarithm of the spectral radius of $f_* : H_\ell(M, \mathbb{R}) \to H_\ell(M, \mathbb{R})$, where $H_\ell(M, \mathbb{R})$ is the ℓth homology group of M, and let $S(f) = \max_\ell S_\ell(f)$.

Theorem 16.16 (Yomdin 1987). *For $f \in C^\infty$, $S(f) \leqslant h(f)$.*

Intuitively, $S(f)$ measures the complexity of f on a global, topological level; it tells us which handles wrap around which ones and how many times. These crossings create 'horseshoes' (see below) which contribute to entropy. This explains why $S(f)$ is a lower bound for $h(f)$. The reason why $h(f)$ can be strictly larger is that entropy can also be created locally, say, inside a disc, without involving any action on homology.

16.4.2 Growth of Periodic Points and Horseshoes

Let $A = (a_{ij})$ be an $s \times s$ matrix with $a_{ij} = 0$ or 1, and let

$$\Sigma_A := \left\{ x = (x_i) \in \prod_{i=-\infty}^{\infty} \{1, 2, \ldots, s\} : a_{x_i x_{i+1}} = 1 \; \forall i \right\}.$$

Then $\sigma : \Sigma_A \to \Sigma_A$, where $(\sigma x)_i = (x)_{i+1}$ is called a *shift of finite type*. If $a_{ij} = 1$ for all i, j, then σ is also called the *full shift* on s symbols, written $\sigma : \Sigma_s \to \Sigma_s$. The following is a straightforward exercise.

Fact 16.17. *Let $\sigma : \Sigma_A \to \Sigma_A$ be a shift of finite type. Then*

$$h_{\text{top}}(\sigma) = \lim_{n \to \infty} \frac{1}{n} \log(\# \text{ fixed points of } \sigma^n).$$

This relation does not hold in general (examples of maps with infinitely many fixed points are easily constructed), but it is the basis for a number of results in smooth dynamical systems. We mention some of them.

A diffeomorphism $f : M \to M$ is said to be *uniformly hyperbolic* on a compact invariant set $\Lambda \subset M$ if, at every $x \in \Lambda$, the tangent space splits into $E^u(x) \oplus E^s(x)$, $Df \mid E^u$ is expanding, $Df \mid E^s$ is contracting, and the splitting is Df-invariant.

Theorem 16.18.

(i) *Let f be uniformly hyperbolic on a compact invariant set Λ which is maximal in some neighborhood U, i.e. $\Lambda = \bigcap_{n \in \mathbb{Z}} f^n U$. Then*

$$h_{\text{top}}(f|_\Lambda) = \lim_{n \to \infty} \frac{1}{n} \log(\# \text{ fixed points of } f^n|_\Lambda)$$

(Bowen 1975).

(ii) *Let $f : M \to M$ be an arbitrary C^2 diffeomorphism and let μ be an invariant Borel probability measure with no zero Lyapunov exponents. Then*

$$h_\mu(f) \leq \liminf_{n \to \infty} \frac{1}{n} \log(\# \text{ fixed points of } f^n).$$

If $\dim(M) = 2$, this inequality is also valid if $h_\mu(f)$ is replaced by $h_{\text{top}}(f)$ (Katok 1980).

As an abstract geometric condition, uniform hyperbolicity was introduced in the 1960s by Smale in his theory of Axiom A (Smale 1967). It is a relatively strong condition, much stronger, for example, than having an invariant measure with no zero Lyapunov exponents. The result in (i) is proved by setting up a correspondence between the orbits of $f|_\Lambda$ and those of a shift of finite type, and proving that the correspondence is nearly one-to-one. This is done using a special partition called a Markov partition. For a reference, see Bowen (1975).

Outside of the Axiom A category, few diffeomorphisms admit (finite) Markov partitions. Instead of attempting to code all orbits, a more modest goal is to prove the presence of certain structures. Results of the type in (ii) are proved by showing that the map, or some iterate of it, has a *horseshoe*. Horseshoes are building blocks of hyperbolic behavior. We refer the reader to Smale (1967) for his geometric description. Suffice it to say here that he models a folding-type mechanism, and the presence of a horseshoe (or a 'generalized horseshoe') implies the existence of a uniformly hyperbolic invariant set on which the map is topologically conjugate to a shift of finite type. Our next result says that under certain conditions, entropy is essentially carried by horseshoes.

Theorem 16.19 (Katok 1980). *Let $f : M \to M$ be a C^2 diffeomorphism of a compact manifold, and let μ be an invariant Borel probability measure with no zero Lyapunov exponents. If $h_\mu(f) > 0$, then given $\varepsilon > 0$ there exist $N \in \mathbb{Z}^+$ and $\Lambda \subset M$ such that*

(i) $f^N(\Lambda) = \Lambda$, $f^N \mid \Lambda$ is uniformly hyperbolic,

(ii) $f^N \mid \Lambda$ is topologically conjugate to $\sigma : \Sigma_s \to \Sigma_s$ for some s, and

(iii) $\dfrac{1}{N} \log h_{\text{top}}(f^N \mid \Lambda) > \dfrac{1}{N} \log h_\mu(f) - \varepsilon.$

If $\dim(M) = 2$, the statement above is also valid if $h_\mu(f)$ is replaced by $h_{\text{top}}(f)$.

For more information on the relations between topological entropy and various growth properties of a map or flow, see Katok and Hasselblatt (1995).

16.4.3 Large Deviations and Rates of Escape

We first state some weak results that hold quite generally. This is followed by stronger results in more specialized situations. The material in this subsection is taken largely from Young (1990), which contains some related references.

Let $f : X \to X$ be a continuous map of a compact metric space, and let m be a Borel probability measure on X. We think of m as a reference measure, and assume that there is an invariant probability measure μ such that for all continuous functions $\varphi : X \to \mathbb{R}$, the following holds for m-a.e. x:

$$\frac{1}{n} S_n \varphi(x) := \frac{1}{n} \sum_{i=0}^{n-1} \varphi(f^i x) \to \int \varphi \, d\mu := \bar{\varphi} \quad \text{as } n \to \infty.$$

For $\delta > 0$, we define

$$E_{n,\delta} = \left\{ x \in X : \left| \frac{1}{n} S_n \varphi(x) - \bar{\varphi} \right| > \delta \right\}$$

and ask how fast $m E_{n,\delta}$ decreases to zero as $n \to \infty$.

Following standard large-deviation theory, we introduce a dynamical version of relative entropy. Let ν be a probability measure on X. We let $h_m(f; \nu)$ be the essential supremum with respect to ν of the function

$$h_m(f, x) = \lim_{\varepsilon \to 0} \limsup_{n \to \infty} -\frac{1}{n} \log m B(x, n, \varepsilon),$$

where $B(x, n, \varepsilon)$ is as defined in Section 16.1. (See also Theorem 16.6.) Let $\mathcal{M}$ denote the set of all f-invariant Borel probability measures on X, and let $\mathcal{M}_e$ be the set of ergodic invariant measures. The following is straightforward.

Proposition 16.20. *For every $\delta > 0$,*

$$\limsup_{n \to \infty} \frac{1}{n} \log m E_{n,\delta} \geq \sup \left\{ h_\nu(f) - h_m(f; \nu) : \nu \in \mathcal{M}_e, \left| \int \varphi \, d\nu - \bar{\varphi} \right| > \delta \right\}.$$

Without further conditions on the dynamical system, a reasonable upper bound cannot be expected. More can be said, on the other hand, for systems with better statistical properties, such as those satisfying Axiom A or the uniformly hyperbolic condition (see Section 16.4.2). A version of the following result was first proved by Orey and Pelikan. For $\nu \in \mathcal{M}_e$, let λ_ν be the sum of the positive Lyapunov exponents of (f, ν) counted with multiplicity.

Theorem 16.21. *Let $f \mid \Lambda$ be a uniformly hyperbolic attractor, and let $U \supset \Lambda$ be its basin of attraction. Let m be the (normalized) Riemannian measure on U, and let μ be the SRB measure on the attractor (see Section 16.3). Then $(1/n)S_n\varphi$ satisfies a large-deviation principle with rate function*

$$k(s) = -\sup\left\{h_\nu(f) - \lambda_\nu : \nu \in \mathcal{M}_e, \int \varphi\, d\mu = s\right\}.$$

A similar set of ideas applies to *rates of escape* problems. Consider a differentiable map $f : M \to M$ of a manifold with a compact invariant set Λ (which is not necessarily attracting). Let U be a compact neighborhood of Λ and assume that Λ is the maximal invariant set in $\bar{U}$, the closure of U. Define

$$E_{n,U} = \{x \in \bar{U} : f^i x \in \bar{U},\ 0 \leqslant i < n\}.$$

In the theorem below, m is Lebesgue measure, and $\mathcal{M}$ is the set of invariant measures in $\bar{U}$. For $\nu \in \mathcal{M}$, let λ_ν be as before, averaging over ergodic components if ν is not ergodic. The result for the uniformly hyperbolic case was first proved in Bowen (1975).

Theorem 16.22. *In general,*

$$\liminf_{n\to\infty} \frac{1}{n} \log m E_{n,U} \geqslant \sup\{h_\nu(f) - \lambda_\nu : \nu \in \mathcal{M}\}.$$

If Λ is uniformly hyperbolic or partially uniformly hyperbolic (see Young (1990) for a definition), then the limit above exists and is equal to the right side.

The reader may recall that $h_\nu(f) \leqslant \lambda_\nu$ is Ruelle's Inequality (Theorem 16.8). The quantity on the right side of Theorem 16.22 is called *topological pressure*. There is a variational principle similar to that in Theorem 16.5 but with a potential (see Ruelle (1978a) or Bowen (1975) for the theory of equilibrium states of uniformly hyperbolic systems).

We finish with the following interpretations. The numbers λ_ν, $\nu \in \mathcal{M}$, describe the forces that push a point away from Λ. For example, if Λ consists of a single fixed point of saddle type, then the sum of the logarithms of its eigenvalues of modulus greater than 1 is precisely the rate of escape from a neighborhood of Λ. The entropies of the system, $h_\nu(f)$, represent, in some ways, the forces that keep a point in U, so that $h_\nu(f) - \lambda_\nu$ gives the net escape rate. One cannot, however, expect equality in general in Theorem 16.22, the reason being that not all parts of an invariant set are 'seen' by invariant measures. For a simple example, consider the time-one-map of the 'figure 8' flow, where $\Lambda \subset \mathbb{R}^2$ consists of a saddle fixed point p together with its separatrices, both of which are homoclinic orbits. If $|\det Df(p)| < 1$, then Λ is an attractor, from whose neighborhood there is no escape. For its unique invariant measure $\mu = \delta_p$, however, $0 = h_\nu < \lambda_\nu$.

References

Adler, R., Konheim, A. and McAndrew, M. 1965 Topological entropy. *Trans. Amer. Math. Soc.* **114**, 309–319.

Bowen, R. 1975 *Equilibrium States and the Ergodic Theory of Anosov Diffeomorphisms*. Springer Lecture Notes in Mathematics, vol. 470.

Brin, M. and Katok, A. 1983 *On Local Entropy*. Geometric Dynamics, Springer Lecture Notes, vol. 1007, pp. 30–38.

Katok, A. 1980 Lyapunov exponents, entropy and periodic orbits for diffeomorphisms. *Publ. Math. IHES* **51**, 137–174.

Katok, A. and Hasselblatt, B. 1995 *Introduction to the Modern Theory of Dynamical Systems*. Cambridge University Press.

Ledrappier, F. 1984 Proprietes ergodiques des mesures de Sinai. *Publ. Math. IHES* **59**, 163–188.

Ledrappier, F. and Strelcyn, J.-M. 1982 A proof of the estimation from below in Pesin entropy formula. *Ergod. Theory Dynam. Sys.* **2**, 203–219.

Ledrappier, F. and Young, L.-S. 1985a The metric entropy of diffeomorphisms. I. *Ann. Math.* **122**, 509–539.

Ledrappier, F. and Young, L.-S. 1985b The metric entropy of diffeomorphisms. II. *Ann. Math.* **122**, 540–574.

Ledrappier, F. and Young, L.-S. 1988 Entropy formula for random transformations. *Prob. Theor. Rel. Fields* **80**, 217–240.

Newhouse, S. 1988 Entropy and volume. *Ergod. Theory Dynam. Sys.* **8**, 283–299.

Pesin, Ya. 1977 Characteristic Lyapunov exponents and smooth ergodic theory. *Russ. Math. Surv.* **32**, 55–114.

Pugh, C. and Shub, M. 1989 Ergodic attractors. *Trans. AMS* **312**, 1–54.

Ruelle, D. 1978a *Thermodynamic Formalism*. Addison Wesley, Reading, MA.

Ruelle, D. 1978b An inequality of the entropy of differentiable maps. *Bol. Sc. Bra. Mat.* **98**, 83–87

Sinai, Ya. G. 1972 Gibbs measures in ergodic theory. *Russ. Math. Surv.* **27**, 21–69.

Smale, S. 1967 Differentiable dynamical systems. *Bull. Am. Math. Soc.* **73**, 747–817.

Walters, P. 1982 *An Introduction to Ergodic Theory*. Graduate Texts in Mathematics, Vol. 79. Springer.

Yomdin, Y. 1987 Volume growth and entropy. *Israel J. Math.* **57**, 285–300.

Young, L.-S. 1990 Some large deviation results for dynamical systems. *Trans. Am. Math. Soc.* **318**, 525–543.

Young, L.-S. 2003 What are SRB measures, and which dynamical systems have them? *J. Stat. Phys.* **108**, 733–754.

Chapter Seventeen

Entropy in Ergodic Theory

Michael Keane

Wesleyan University

This is an expository contribution devoted to the description in elementary terms of the role which the concept of entropy played in the initial development of ergodic theory. To also give a flavor of current activity, an interesting unsolved problem has been included which can easily be understood by nonspecialists. The solution of this problem would be of great interest.

The introduction by Kolmogorov in the 1950s of Shannon's concept of entropy with the aim of furthering the classification theory of measure-preserving transformations has led in the past half century to significant advances. In the following elementary exposition we attempt to touch on the main developments of the initial period for nonspecialists. To give a glimpse into the future of this lively and exciting discipline we conclude with an equally elementary treatment of an interesting open problem in ergodic theory.

The central objects of study in classical measure-theoretic ergodic theory are the invertible measure-preserving transformations acting on Lebesgue spaces. The classification problem consists of determining which of these objects are isomorphic, or more generally which are homomorphic images of other such objects. Here we mean by Lebesgue space a probability space which is isomorphic as a measure space to the unit interval with Lebesgue measure on the Lebesgue measurable sets; the reason for this restriction is to eliminate pathology arising from the possible difference in measure-theoretic structures themselves having nothing to do with the transformations, which are considered to be the objects of primary interest. It should be clear from Chapters 3 and 16 or the references Petersen (1989) and Walters (1982) how these objects are defined and what it means for them to be isomorphic or homomorphic. Although much has been accomplished during the past half century, the classification problem remains largely unsolved.

Let us consider a classical example from the 1950s. If we are given a probability vector p having a finite number of entries, then we can construct the so-called Bernoulli scheme based on the vector p, which we denote by BS(p). It is the measure-preserving transformation given by the left shift on a bilateral sequence space over a finite alphabet with the same number of elements as entries of the vector p; this sequence space is provided with the probability measure given by placing the probability measure p on each of the coordinates and requiring that the coordinates be mutually independent. If now $p = (1/2, 1/2)$ and $q = (1/3, 1/3, 1/3)$, then an open problem from that time was whether $BS(p)$ and $BS(q)$ were isomorphic.

Some knowledge of structure of measure-preserving transformations was available at that time. Ergodicity was one of the key concepts, corresponding to irreducibility in other similar studies, and two isomorphic transformations certainly both had to be ergodic or both not, but any Bernoulli scheme $BS(p)$ is easily seen to be ergodic. (It is often still difficult to determine whether a given transformation is ergodic; the ongoing investigations of ergodicity of different billiard systems provide a paradigm for the degree of difficulty.) Other invariants were given by the concepts of k-mixing (Bernoulli schemes are k-mixing for all k), and more generally the spectrum of a measure-preserving transformation, which is simply the transformation seen as an abstract operator on the Hilbert space of square-integrable complex-valued functions on the probability space. (Bernoulli schemes all have the same spectrum in this sense.) There are still today some tough unsolved problems around concerning these concepts—for example, it is not yet known whether transformations exist which are 2-mixing but not 3-mixing, or whether transformations with simple Lebesgue spectrum exist—but these ideas were not applicable to Bernoulli schemes.

Kolmogorov saw that a new, numerical invariant for measure-preserving transformations could be introduced using Shannon's entropy; he called this invariant the mean entropy of a measure-preserving transformation. The idea is very simple; any partition of a probability space possesses a numerical Shannon entropy given by the sum of $-r \log r$ over the probabilities r of the different elements of the partition, corresponding to a measure of the amount of information contained in the knowledge of which partition element a random point of the space lies in. If we now specify not only the partition element of a random point of the space but also of its images under the measure-preserving transformation up to a finite time T, then the average rate of information obtained per unit time should exist in the long run, as T tends to infinity. This is easy to show technically, and depends only on a basic subadditivity property of the Shannon entropy known previously and well understood. But which partition of the space should be taken? Kolmogorov proposed taking that partition which gave the maximal rate, and in a joint article with Sinai it was shown that any *generating* partition would attain this maximal rate. A generating partition is one which has the special property that if one knows for each iterate of a point (both backwards and forwards) the partition element to which the iterate belongs, then one can determine the point itself.

This theorem was relevant to Bernoulli schemes. The partition of the sequence space according to the values taken by the central coordinate in $BS(p)$ is an independent partition, in the sense that it is independent of all its translations by the shift transformation, so that its mean entropy is easily seen to be simply the Shannon entropy of the probability vector p, and moreover this partition is clearly a generator, since giving all the coordinates of a point in the sequence space amounts to giving the point itself. Hence in the example given above, the mean entropies were $\log 2$ and $\log 3$, respectively, and it was thus shown that these two measure-preserving transformations were not isomorphic.

The second success due to entropy came shortly thereafter. In the definition of mean entropy sketched above one considers all finite partitions, or, equivalently, all partitions having finite entropy. It is not obvious that a transformation having finite mean entropy does have a generator of finite entropy. First, Rokhlin showed in 1965 in a difficult article that this is indeed true, and subsequently Krieger showed in 1970 that a transformation with finite mean entropy possesses a finite generator with the least possible number of elements. Both of these articles are difficult, and recently more elementary proofs have been published. The philosophical content of these results is that if we wish to consider arbitrary measure-preserving transformations of finite entropy, then it suffices to consider shift transformations on symbolic spaces; the probability measure is then not necessarily the independent one.

The third success of entropy in the classification problem was accomplished around 1970 by Ornstein, who showed that mean entropy is a complete invariant for Bernoulli schemes: two Bernoulli schemes with the same mean entropy are isomorphic. This celebrated result was the beginning of an intensive effort resulting in many further interesting ideas: a general way to prove isomorphism of given transformations to Bernoulli schemes, examples of processes which were spectrally like Bernoulli schemes but not isomorphic to them, finitary isomorphisms which yielded simple intuitive proofs using coding techniques, investigation of other types of identifications such as orbit equivalence, relative isomorphism theory, more general group actions, and many others.

In conclusion, we can safely say that the classification of measure-preserving transformations is well under way, with many interesting positive results, a steady continuation of development, and a small set of central problems which are considered to be the most important at present in the field; among these are the so-called weak Pinsker conjecture, stating that every measure-preserving transformation of positive entropy is isomorphic to the product of a Bernoulli scheme with an ergodic process of small entropy, and the question of whether every measure-preserving transformation with finite entropy has a so-called smooth model, i.e. a representation as a differentiable map on a compact finite-dimensional manifold preserving a smooth probability measure. It is not clear how large a role entropy will play in further developments, but as a basic measurement tool and as a tool for understanding the quantitative asymptotic nature of iterates of measure-preserving transformations it has certainly proved its value.

The reader who is interested in becoming more familiar with the ideas discussed above can consult Petersen (1989) and Walters (1982), both of which contain clear and elementary expositions of the basic development.

This concludes our elementary exposition for nonspecialists of the main developments of the initial period. Subsequently, there has been considerable development, and it would be impossible to do justice to all of the contributions and contributors in the space remaining for this elementary article. However, to give a small taste of current activity in ergodic theory, I would like to single out one very interesting current problem, which is yet unsolved and whose solution would be of global interest. I have chosen this problem because it can also be easily understood by nonspecialists, in the spirit of this book.

The problem which I wish to discuss concerns a specific transformation which I shall call the *binomial transformation*; others also refer to it as the Pascal transformation, but I prefer the descriptive name. Perhaps this is also a suitable place for a disclaimer: the binomial transformation is not an invention of mine and I am not sure who first discovered it, although I believe it was Vershik (1974). Other, more recent, references are Adams and Petersen (1998) and Petersen and Schmidt (1997). It is also a transformation of entropy zero, so that the main body of the theory of entropy up to the present does not give much information about its structure. In general, many of the transformations under consideration at present are transformations of zero entropy, and a number of the open problems in ergodic theory can be reduced to questions involving zero entropy transformations. Unfortunately, this does not help much in many situations.

It is helpful to begin with an informal description of the measure-preserving transformation we shall consider, which is a Lebesgue measure-preserving mapping from the unit interval to itself, essentially invertible. The definition follows a general procedure known to ergodicists as 'cutting and stacking.' At the first step, the unit interval is divided into two equal halves:

This yields two 'stacks' of height one. At each of the following steps, the stacks present are each divided into equal halves, and the right half of each stack is placed on top of the left half of the following stack. This yields for steps two and three the pictures

and

and in general, at step n there are $n+1$ stacks, which we number from left to right $0, 1, \ldots, n$, the stack numbered k having height equal to the binomial coefficient $\binom{n}{k}$. The transformation is now defined by sending each point of the unit interval to the point immediately above it in its stack. It is clear that if the point considered is not on the top level of one of the stacks, then this persists in the following steps and the definition is consistent. Also, the only points of the unit interval which can belong to the top of one of the stacks at each step lie at the ends of the intervals, so that the transformation is defined except for countably many points, and clearly preserves Lebesgue measure.

Our next task is to formalize this definition. We set

$$X := \{x \in [0,1] : x \text{ not dyadic rational}\}$$

and abuse terminology slightly by also calling intersections of intervals with X intervals. (This is actually not an abuse if we consider X as a totally ordered set and intervals as being order intervals.) If I is an interval with $I \subseteq X$, we denote by $l(I)$ and $r(I)$ the left and right halves of I; as we shall only consider intervals with dyadic rational 'endpoints' (which then do not belong to I), the 'midpoint' is also dyadic rational and belongs neither to $l(I)$ nor to $r(I)$. Now we have enough notation to give formal definitions.

(i) $B_{0,0}^{(0)} := X$.

(ii) Inductively, for $n \geq 0$, $0 \leq k \leq n+1$, and $0 \leq i \leq \binom{n+1}{k}$

$$B_{k,i}^{(n+1)} := l(B_{k,i}^{(n)}) \qquad \text{for } 0 \leq i < \binom{n}{k},$$

$$B_{k,i}^{(n+1)} := r(B_{k-1,i-\binom{n}{k}}^{(n)}) \qquad \text{for } \binom{n}{k} \leq i < \binom{n+1}{k}.$$

(iii) T_n maps $B_{k,i}^{(n)}$ to $B_{k,i+1}^{(n)}$ linearly and in order-preserving fashion, for each $0 \leq i < \binom{n}{k} - 1$; on the last interval of each stack T_n is not defined.

(iv) $T : X \to X$ is the unique mapping which extends the mappings T_n, $n \geq 0$.

It is now an easy exercise, which we omit, to prove that T is an invertible Lebesgue measure-preserving transformation from X to X.

This concludes the formal definition of the binomial transformation. In order to prove that T is ergodic, we use the following version of the Lebesgue density theorem. We denote the Lebesgue measure of a measurable subset A of the unit interval by $|A|$, and the dyadic interval of the unit interval which contains x by $I_n(x)$.

Lemma 17.1. *Let A be a measurable subset of X. Then, for almost every $x \in A$,*

$$\lim_{n \to \infty} \frac{|A \cap I_n(x)|}{|I_n(x)|} = 1.$$

In this form, the expression under the limit is a martingale with respect to the dyadic filtration, so that the lemma follows from the martingale convergence theorem.

Lemma 17.2. *For almost every $(x, y) \in X \times X$, there exist infinitely many n such that $I_n(x)$ and $I_n(y)$ belong to the same stack.*

Proof. Let $x \in X$ have the dyadic expansion

$$x = \sum_{m=1}^{\infty} x_m 2^{-m}$$

with $x_m \in \{0, 1\}$ for each $m \geq 1$. It is elementary to show by induction that at step n, x belongs to the kth stack ($0 \leq k \leq n$) if and only if

$$k = \#\{m : 1 \leq m \leq n,\ x_m = 1\} = \sum_{m=1}^{n} x_m.$$

Since, under Lebesgue measure, the x_m are independent and identically distributed,

$$\sum_{m=1}^{n} (x_m - y_m)$$

is a centered random walk on the integers which visits 0 infinitely often with probability one, as such random walks are recurrent. $\square$

Theorem 17.3. *T is ergodic.*

Proof. Suppose that $A \subset X$ is measurable with $TA = A$. Then, if I and J are intervals belonging to the same stack at any fixed step, we must have

$$|I \cap A| = |J \cap A|,$$

as I is mapped linearly to J by some power of T. If now $x \in A$ and $y \in A^c$ are such that

$$\frac{|A \cap I_n(x)|}{|I_n(x)|} > \frac{1}{2}$$

and

$$\frac{|A^c \cap I_n(y)|}{|I_n(y)|} > \frac{1}{2},$$

then it follows that $I_n(x)$ and $I_n(y)$ cannot belong to the same stack at step n. Thus the lemmas above imply that $|A \times A^c| = 0$ and T is ergodic. $\square$

The interesting problem which is still open is whether the transformation T is weakly mixing. In the references given it is shown that T cannot have rational eigenvalues, i.e. T is totally ergodic, but it seems to be difficult to show that T cannot have irrational eigenvalues. We even believe that the binomial transformation is strongly mixing.

We conclude by sketching a second proof of ergodicity, shown to us by Bernard Host, which is essentially different from the first proof and which yields more information. It works in the following fashion. Suppose that μ is a probability measure on X invariant under the transformation T. Then μ assigns to each interval in a stack the same measure, which depends only on the step and on the stack number. Let us denote the measure of an interval in the kth stack at level n, under μ, by $a_k^{(n)}$. This is a collection of nonnegative real numbers with $a_0^{(0)} = 1$ and satisfying the equality relations

$$a_k^{(n)} = a_k^{(n+1)} + a_{k+1}^{(n+1)}$$

for $n \geqslant 0$ and $0 \leqslant k \leqslant n$. Such a collection is called a *Bernstein* collection; it is an easy exercise in analysis to show that, given any such Bernstein collection, there exists a probability measure θ on the unit interval such that

$$a_k^{(n)} = \int p^k (1-p)^{n-k} \theta(\mathrm{d}p)$$

for $n \geqslant 0$ and $0 \leqslant k \leqslant n$. Hence we have identified all T-invariant probability measures on X as mixtures of Bernoulli measures giving probability p to 0 and $1-p$ to 1 in the dyadic expansion of an element $x \in X$; these Bernoulli measures are exactly the ones which are ergodic under T, and Lebesgue measure corresponds to the value $p = 1/2$. This is a beautiful proof!

References

Adams, T. M. and Petersen, K. E. 1998 Binomial-coefficient multiples of irrationals. *Monatsh. Math.* **125**, 269–278.

Petersen, K. 1989 *Ergodic Theory* (corrected reprint of the 1983 original). Cambridge Studies in Advanced Mathematics, vol. 2. Cambridge University Press.

Petersen, K. and Schmidt, K. 1997 Symmetric Gibbs measures. *Trans. Am. Math. Soc.* **349**, 2775–2811.

Vershik, A. M. 1974 A description of invariant measures for actions of certain infinite-dimensional groups (in Russian). *Dokl. Akad. Nauk SSSR* **218**, 749–752. (English translation *Sov. Math. Dokl.* **15**, 1396–1400 (1975).)

Walters, P. 1982 *An Introduction to Ergodic Theory*. Graduate Texts in Mathematics, Vol. 79. Springer.

Combined References

Accardi, L. 1981 *Phys. Rep.* **77**, 169.
Accardi, L., Ohya, M. and Watanabe, N. 1997 *Open Systems Infor. Dynamics* **4**, 71.
Adams, T. M. and Petersen, K. E. 1998 Binomial-coefficient multiples of irrationals. *Monatsh. Math.* **125**, 269–278.
Adler, R., Konheim, A. and McAndrew, M. 1965 Topological entropy. *Trans. Amer. Math. Soc.* **114**, 309–319.
Alekseev, V. M. and Yakobson, M. V. 1981 *Phys. Rep.* **75**, 287.
Alicki, R. and Fannes, M. 1994 *Lett. Math. Phys.* **32**, 75.
Alicki, R. and Narnhofer, H. 1995 *Lett. Math. Phys.* **33**, 241.
Alicki, R., Andries, J., Fannes, M. and Tuyls, P. 1996a *Rev. Math. Phys.* **8**, 167.
Alicki, R., Makowiec, D. and Miklaszewski, W. 1996b *Phys. Rev. Lett.* **77**, 838.
Alt, H. W. 1998 An evolution principle and the existence of entropy. First order systems. *Continuum Mech. Thermodyn.* **10**, 61–79.
Au, J. and Wehr, L. R. 1997 Light scattering experiment and extended thermodynamics. *Continuum Mech. Thermodyn.* **9**, 155–164.
Au, J., Müller, I. and Ruggeri, T. 2000 Temperature jumps at the boundary of a rarefied gas. *Continuum Mech. Thermodyn.* **12**, 19–30.
Au, J. D., Torrilhon, M. and Weiss, W. 2001 The shock tube study in extended thermodynamics. *Phys. Fluids* **13**, 2423–2432.
Baillon, J.-B., Clément, Ph., Greven, A. and den Hollander, F. 1993 A variational approach to branching random walk in random environment. *Ann. Prob.* **21**, 290–317.
Baillon, J.-B., Clément, Ph., Greven, A. and den Hollander, F. 1994 On a variational problem for an infinite particle system in a random medium. I. The global growth rate. *J. Reine Angew. Math.* **454**, 181–217.
Balasubramanian, V. 1997 Statistical inference, Occam's razor and statistical mechanics on the space of probability distributions. *Neural Computation* **9**, 349–268.
Balian, R. 1982 *Du microscopique au macroscopique*, Vol. 2. École Polytechnique.
Barbera, E. 1999 On the principle of minimal entropy production for Navier–Stokes–Fourier fluids. *Continuum Mech. Thermodyn.* **11**, 327–330.
Barbera, E., Müller, I. and Sugiyama, M. 1999 On the temperature of a rarefied gas in non-equilibrium. *Meccanica* **34**, 103–113.
Barron, A. R. 1986 Entropy and the central limit theorem. *Ann. Prob.* **14**, 336–342.
Barron, A. R., Rissanen, J. and Yu, B. 1998 The MDL principle in modeling and coding. *IEEE Trans. Information Theory* (Special Issue to commemorate 50 years of Information Theory) **IT-44**, 2743–2760.
Benatti, F. 1993 *Deterministic Chaos in Infinite Quantum Systems*. Sissa Notes in Physics. Springer.

Benatti, F. 1996 *J. Math. Phys.* **37**, 5244.

Benatti, F. and Narnhofer, H. 2001 *Phys. Rev.* A **63**, 042306.

Benatti, F., Cappellini, V., De Cock, M., Fannes, M. and Vanpeteghem, D. 2003 *Classical Limit of Quantized Toral Automorphisms and Dynamical Entropies*. (In preparation.)

Bernstein, B. 1960 Proof of Carathéodory's local theorem and its global application to thermostatics. *J. Math. Phys.* **1**, 222–224.

Bhatnagar, P. L., Gross, E. P. and Krook, M. 1954 A model for collision processes in gases. *Phys. Rev.* **94**, 511.

Bianchini, S. and Bressan, A. 2003 Vanishing viscosity solutions of nonlinear hyperbolic systems. *Ann. Math.* (In the press.)

Billingsley, P. 1965 *Ergodic Theory and Information*. Wiley.

Bird, G. 1994 *Molecular Gas Dynamics and the Direct Simulation of Gas Flows*. Clarendon Press, Oxford.

Boillat, G. and Ruggeri, T. 1997 Moment equations in the kinetic theory of gases and wave velocities. *Continuum Mech. Thermodyn.* **9**, 205–212.

Boillat, G. and Ruggeri, T. 1999 Relativistic gas: moment equations and maximum wave velocity. *J. Math. Phys.* **40**.

Boltzmann, L. 1896, 1898 *Vorlesungen über Gastheorie*, 2 vols. Leipzig: Barth. (English transl. by S. G. Brush, *Lectures on Gas Theory* (Cambridge University Press, 1964).)

Bonetto, F., Lebowitz, J. L. and Rey-Bellet, L. 2000 Fourier's law: a challenge to theorists. Preprint, mp-arc 00-89.

Bouwmeester, D., Ekert, A. and Zeilinger, A. 2000 *The Physics of Quantum Information*. Springer.

Bowen, R. 1975 *Equilibrium States and the Ergodic Theory of Anosov Diffeomorphisms*. Springer Lecture Notes in Mathematics, vol. 470.

Boyling, J. G. 1972 An axiomatic approach to classical thermodynamics. *Proc. R. Soc. Lond.* A **329**, 35–70.

Brémaud, P. 1981 *Point Processes and Queues, Martingale Dynamics*. Springer.

Bressan, A. 1992 Global solutions of systems of conservation laws by wave-front tracking. *J. Math. Analysis Appl.* **170**, 414–432.

Bressan, A. 2000 *Hyperbolic Systems of Conservation Laws*. Oxford University Press.

Bressan, A., Liu, T.-P. and Yang, T. 1999 L^1 stability estimates for $n \times n$ conservation laws. *Arch. Ration. Mech. Analysis* **149**, 1–22.

Bricmont, J. 1996 Science of chaos or chaos in science? In *The Flight from Science and Reason. Ann. NY Acad. Sci.* **775**, 131.

Bricmont, J. 1996 Science of chaos or chaos in science? In *The Flight from Science and Reason. Ann. NY Acad. Sci.* **775**, 131. (*Physicalia Magazine* **17** (1995), 159.)

Brin, M. and Katok, A. 1983 *On Local Entropy*. Geometric Dynamics, Springer Lecture Notes, vol. 1007, pp. 30–38.

Brudno, A. A. 1983 *Trans. Moscow Math. Soc.* **44**, 127.

Buchdahl, H. A. 1966 *The Concepts of Classical Thermodynamics*. Cambridge University Press.

Callen, H. B. 1960 *Thermodynamics*. Wiley, New York.

Capocaccia, D., Cassandro, M. and Olivieri, E. 1974 A study of metastability in the Ising model. *Commun. Math. Phys.* **39**, 185–205.

Carathéodory, C. 1909 Untersuchung über die Grundlagen der Thermodynamik. *Math. Annalen* **67**, 355–386.

Carathéodory, C. 1925 Über die Bestimmung der Energie und der absoluten Temperatur mit Hilfe von reversiblen Prozessen. Sitzungsberichte der preußischen Akademie der Wissenschaften, Math. Phys. Klasse, pp. 39–47.

Carathéodory, C. 1955 *Gesammelte Mathematische Schriften*, Band II. Beck'sche Verlagsbuchhandlung, München.

Carnot, S. 1824 *Réflexions sur la puissance motrice du feu et sur les machines propres a développer cette puissance*. Chez Bachelier, Libraire, Paris.

Casati, G. and Chirikov, B. 1995 *Quantum Chaos. Between Order and Disorder*. Cambridge University Press.

Cassandro, M., Galves, A., Olivieri, E. and Vares, M. E. 1984 Metastable behavior of stochastic dynamics: a pathwise approach. *J. Statist. Phys.* **35**, 603–634.

Catoni, O. and Cerf, R. 1995/97 The exit path of a Markov chain with rare transitions. *ESAIM Prob. Statist.* **1**, 95–144 (electronic).

Cattaneo, C. 1948 Sulla conduzione del calore. *Atti Sem. Mat. Fis. Univ. Modena* **3**, 83–101.

Chen, G.-Q., Frid, H. and Li, Y. 2002 Uniqueness and stability of Riemann solutions with large oscillation in gas dynamics. *Commun. Math. Phys.* **228**, 201–217.

Chernoff, H. 1956 Large sample theory: parametric case. *Ann. Math. Statist.* **23**, 493–507.

Clark, N. A. 1975 Inelastic light scattering from density fluctuations in dilute gases. The kinetic-hydrodynamic transition in a monatomic gas. *Phys. Rev.* A **12**.

Clausius, R. 1854 Über eine veränderte Form des zweiten Hauptsatzes der mechanischen Wärmetheorie. *Poggendroff's Annln Phys.* **93**, 481–506.

Clausius, R. 1864 Ueber die Concentration von Wärme- und Lichtstrahlen und die Gränze Ihre Wirkung. In *Abhandlungen über die Mechanischen Wärmetheorie*, Vol. 1, pp. 322–361. Vieweg & Sohn, Braunschweig.

Clausius, R. 1865 Ueber verschiedene für die Anwendung bequeme Formen der Haubtgleichungen der mechanischen Wärmetheorie. *Vierteljahrschrift der naturforschenden Gesellschaft (Zürich)* **10**, 1–59. (Also in Clausius (1867), pp. 1–56, and translated in Kestin (1976), pp. 162–193.)

Clausius, R. 1867 *Abhandlungungen über die Mechanische Wärmetheorie*, Vol. 2. Vieweg & Sohn, Braunschweig.

Clausius, R. 1876 *Die Mechanische Wärmetheorie*. Vieweg & Sohn, Braunschweig.

Clausius, R. 1887 *Die Mechanische Wärmetheorie*, Vol. 1, 3rd edn. Vieweg & Sohn, Braunschweig.

Coleman, B. D. and Noll, W. 1963 The thermodynamics of elastic materials with heat conduction and viscosity. *Arch. Ration. Mech. Analysis* **13**, 167–178.

Coleman, B. D. and Mizel, V. J. 1964 Existence of caloric equations of state in thermodynamics. *J. Chem. Phys.* **40**, 1116.

Connes, A., Narnhofer, H. and Thirring, W. 1987 *Commun. Math. Phys.* **112**, 691.

Cooper, J. L. B. 1967 The foundations of thermodynamics. *J. Math. Analysis Appl.* **17**, 172–193.

Cover, T. M. and Thomas, J. A. 1991 *Elements of Information Theory*. Wiley.

Cramér, H. 1938 Sur un noveau théorème—limites de la théorie des probabilités. *Actualités Scientifiques et Industrielles* **736**, 5–23. *Colloque consacré à la théorie des probabilités*, Vol. 3. Hermann, Paris

Csiszár, I. 1975 *I*-divergence geometry of probability distributions and minimization problems. *Ann. Prob.* **3**, 146–158.

Csiszár, I. 1984 Sanov property, generalized I-projection and a conditional limit theorem. *Ann. Prob.* **12**, 768–793.

Csiszár, I. and Körner, J. 1981 *Information Theory, Coding Theorems for Discrete Memoryless Systems*. Akdaémiai Kiadó, Budapest.

Dafermos, C. M. 1979 The second law of thermodynamics and stability. *Arch. Ration. Mech. Analysis* **70**, 167–179.

Dafermos, C. M. 1989 Admissible wave fans in nonlinear hyperbolic systems. *Arch. Ration. Mech. Analysis* **106**, 243–260.

Dafermos, C. M. 2000 *Hyperbolic Conservation Laws in Continuum Physics*. Springer.

Day, W. A. 1972 *The Thermodynamics of Simple Materials with Fading Memory*. Springer Tracts in Natural Philosophy, Vol. 22. Springer.

de Groot, S. R. and Mazur, P. 1963 *Non-Equilibrium Thermodynamics*. North-Holland, Amsterdam.

Dembo, A. and Zeitouni, O. 1993 *Large Deviation Techniques and Applications*. Jones and Bartlett, Boston, MA, and London.

den Hollander, F. 2000 *Large Deviations*. Fields Institute Monographs 14. American Mathematical Society, Providence, RI.

Denbigh, K. 1989 The many faces of irreversibility. *Br. J. Phil. Sci.* **40**, 501–518.

Denbigh, K. 1990 How subjective is entropy. In *Maxwell's Demon, Entropy, Information, Computing* (ed. H. S. Leff and A. F. Rex), pp. 109–115. Princeton University Press.

Deuschel, J.-D., Stroock, D. W. and Zessin, H. 1991 Microcanonical distributions for lattice gases. *Commun. Math. Phys.* **139**, 83–101.

DiPerna, R. J. 1979 Uniqueness of solutions to hyperbolic conservation laws. *Indiana Univ. Math. J.* **28**, 137–188.

DiPerna, R. J. 1983 Convergence of approximate solutions to conservation laws. *Arch. Ration. Mech. Analysis* **82**, 27–70.

Duhem, P. 1891 *Hydrodynamique, Elasticité, Acoustique* (2 vols). Paris, Hermann.

Duistermaat, J. J. 1968 Energy and entropy as real morphisms for addition and order. *Synthese* **18**, 327–393.

Dupuis, P. and Ellis, R. S. 1997 *A Weak Convergence Approach to the Theory of Large Deviations*. Wiley Series in Probability and Statistics (Probability and Statistics Section). Wiley-Interscience.

Earman, J. 1974 An attempt to add a little direction to 'the problem of the direction of time'. *Phil. Sci.* **41**, 15–47.

Eckart, C. 1940a The thermodynamics of irreversible processes. I. The simple fluid. *Phys. Rev.* **58**, 267–269.

Eckart, C. 1940b The thermodynamics of irreversible processes. II. Fluid mixtures. *Phys. Rev.* **58**, 269–275.

Eckart, C. 1940c The thermodynamics of irreversible processes. III. Relativistic theory of the simple fluid. *Phys. Rev.* **58**, 919–924.

Eckmann, J.-P., Pillet, C.-A. and Rey-Bellet, L. 1999 Non-equilibrium statistical mechanics of anharmonic chains coupled to two heat baths at different temperatures. *Commun. Math. Phys.* **201**, 657–697.

Ehlers, W. 1989 Poröse Medien, ein kontinuumsmechanisches Modell auf der Basis der Mischungstheorie. Habilitation, Universität-Gesamthochschule Essen.

Einstein, A. 1970 Autobiographical notes. In *Albert Einstein: Philosopher-Scientist* (ed. P. A. Schilpp). Library of Living Philosophers, vol. VII, p. 33. Cambridge University Press.

Ellis, R. S. 1985 *Entropy, Large Deviations and Statistical Mechanics*. Springer.

Evans, D. J., Cohen, E. G. D. and Morriss, G. P. 1993 Probability of second law violations in steady flows. *Phys. Rev. Lett.* **71**, 2401–2404.

Falk, G. and Jung, H. 1959 Axiomatik der Thermodynamik. In *Handbuch der Physik* (ed. S. Flügge), Vol. III/2. Springer.

Fannes, M. and Tuyls, P. 2000 *Inf. Dim. Analysis, Quantum Prob. Rel. Top.* **2**, 511.

Fannes, M., Nachtergaele, B. and Werner, R. F. 1992 *Commun. Math. Phys.* **144**, 443.

Feller, W. 1957 *An Introduction to Probability Theory and Its Applications*, Vol. I, 2nd edn. Wiley, Chapman and Hall.

Fermi, E. 1956 *Thermodynamics*, p. 101. Dover, New York.

Fleming, W. H. 1985 A stochastic control approach to some large deviations problems. *Recent Mathematical Methods in Dynamic Programming* (Rome, 1984), pp. 52–66. Lecture Notes in Mathematics, vol. 1119. Springer.

Föllmer, H. 1973 On entropy and information gain in random fields. *Z. Wahrscheinlichkeitsth.* **26**, 207–217.

Föllmer, H. 1988 Random fields and diffusion processes. In *École d'Été de Probabilités de Saint Flour XV–XVII, 1985–87* (ed. P. L. Hennequin), Lecture Notes in Mathematics, Vol. 1362, pp. 101–203. Springer.

Ford, J., Mantica, G. and Ristow, G. H. 1991 *Physica D* **50**, 493.

Freidlin, M. I. and Wentzell, A. P. 1984 *Random Perturbations of Dynamical Systems*. Springer.

Friedrichs, K. O. and Lax, P. D. 1971 Systems of conservation equations with a convex extension. *Proc. Natl Acad. Sci. USA* **68**, 1686–1688.

Gallavotti, G. 1996 Chaotic hypothesis: Onsager reciprocity and fluctuation–dissipation theorem. *J. Statist. Phys.* **84**, 899–926.

Gallavotti, G. 1998 Chaotic dynamics, fluctuations, nonequilibrium ensembles. *Chaos* **8**, 384–392.

Gallavotti, G. 1999 A local fluctuation theorem. *Physica A* **263**, 39–50.

Gallavotti, G. 1999 *Statistical Mechanics; a Short Treatise*. Springer Texts and Monographs in Physics.

Gallavotti, G. and Cohen, E. G. D. 1995a Dynamical ensembles in nonequilibrium statistical mechanics. *Phys. Rev. Lett.* **74**, 2694–2697.

Gallavotti, G. and Cohen, E. G. D. 1995b Dynamical ensembles in stationary states. *J. Statist. Phys.* **80**, 931–970.

Georgii, H.-O. 1979 *Canonical Gibbs Measures*, Lecture Notes in Mathematics, Vol. 760. Springer.

Georgii, H.-O. 1988 *Gibbs Measures and Phase Transitions*. Walter de Gruyter, Berlin and New York.

Georgii, H.-O. 1993 Large deviations and maximum entropy principle for interacting random fields on $\mathbb{Z}^d$. *Ann. Prob.* **21**, 1845–1875.

Georgii, H.-O. 1995 The equivalence of ensembles for classical systems of particles. *J. Statist. Phys.* **80**, 1341–1378.

Gibbs, J. W. 1878 *On the Equilibrium of Heterogeneous Substances*. Connecticut Academy of Sciences. Reprinted in *The Scientific Papers of J. Willard Gibbs* (Dover, New York, 1961).

Gibbs, J. W. 1906 *The Scientific Papers of J. Willard Gibbs*, Vol. 1. *Thermodynamics*. Longmans, London.

Gibbs, J. W. 1960 *The Scientific Papers*, Vol. I. Dover, New York.
Giles, R. 1964 *Mathematical Foundations of Thermodynamics*. Pergamon, Oxford.
Glansdorff, P. and Prigogine, I. 1971 *Thermodynamic Theory of Structure, Stability and Fluctuations*. Wiley Interscience, London.
Glimm, J. 1965 Solutions in the large for nonlinear hyperbolic systems of equations. *Commun. Pure Appl. Math.* **18**, 697–715.
Goldstein, S. 1998 Quantum mechanics without observers. *Phys. Today* (March), p. 42, (April), p. 38.
Goldstein, S. 2003 Boltzmann's approach to statistical mechanics. In *Chance in Physics: Foundations and Perspectives* (ed. D. Dürr). Springer.
Goldstein, S. and Lebowitz, J. L. 1995 Quantum mechanics. In *The Physical Review: the First Hundred Years* (ed. H. Stroke). AIP Press, New York.
Goldstein, S. and Lebowitz, J. L. 2003 On the (Boltzmann) entropy of nonequilibrium systems. *Physica D*. (In the press and on archive at cond-mat/0304251.)
Golodets, V. Ya. and Størmer, E. 1998 *Ergod. Theory Dynam. Sys.* **18**, 859.
Goodman, M. A. and Cowin, S. C. 1972 A continuum theory for granular materials. *Arch. Ration. Mech. Analysis* **44**, 249–266.
Grad, H. 1949 *On the Kinetic Theory of Rarefied Gases*. Communications in Pure and Applied Mathematics, vol. 2. Wiley.
Grad, H. 1958 *Principles of the Kinetic Theory of Gases*. Handbuch der Physik XII. Springer.
Green, A. E. and Laws, N. 1972 On a global entropy production inequality. *Q. J. Mech. Appl. Math.* **25**, 1–11.
Green, A. E. and Naghdi, P. M. 1972 On continuum thermodynamics. *Arch. Ration. Mech. Analysis* **48**, 352–378.
Greven, A. and den Hollander, F. 1991a Population growth in random media. I. Variational formula and phase diagram. *J. Statist. Phys.* **65**, 1123–1146.
Greven, A. and den Hollander, F. 1991b Population growth in random media. II. Wave front propagation. *J. Statist. Phys.* **65**, 1147–1154.
Greven, A. and den Hollander, F. 1992 Branching random walk in random environment: phase transitions for local and global growth rates. *Prob. Theory Relat. Fields* **91**, 195–249.
Greven, A. and den Hollander, F. 1994 On a variational problem for an infinite particle system in a random medium. II. The local growth rate. *Probab. Theory Relat. Fields* **100**, 301–328.
Guo, M. Z., Papanicolaou, G. C. and Varadhan, S. R. S. 1988 Nonlinear diffusion limit for a system with nearest neighbor interactions. *Commun. Math. Phys.* **118**(1), 31–59.
Gurtin, M. E. and Pipkin, A. C. 1969 A general theory of heat conduction with finite wave speed. *Arch. Ration. Mech. Analysis* **31**, 113–126.
Haber, F. 1905 *Thermodynamik technischer Gasreaktionen*. R. Oldenbourg, München.
Herstein, I. N. and Milnor, J. 1953 An axiomatic approach to measurable utility. *Econometrica* **21**, 291–297.
Hoeffding, W. 1965 Asymptotically optimal tests for multinomial distributions. *Ann. Math. Statist.* **36**, 369–400.
Holden, H. and Risebro, N. H. 2002 *Front Tracking for Hyperbolic Conservation Laws*. Springer.
Holley, R. 1971a Pressure and Helmholtz free energy in a dynamic model of a lattice gas. *Proc. Sixth Berkeley Symp. Prob. Math. Statist.* **3**, 565–578.

Holley, R. 1971b Free energy in a Markovian model of a lattice spin system. *Commun. Math. Phys.* **23**, 87–99.

Hollinger, H. B. and Zenzen, M. J. 1985 *The Nature of Irreversibility*. Reidel, Dordrecht.

Huang, K. 1963 *Statistical Mechanics*. Wiley.

Hutter, K. 1975 On thermodynamics and thermostatics of viscous thermoelastic solids in the electromagnetic fields. A Lagrangian formulation. *Arch. Ration. Mech. Analysis* **54**, 339–366.

Hutter, K. 1977a A thermodynamic theory of fluids and solids in the electromagnetic fields. *Arch. Ration. Mech. Analysis* **64**, 269–289.

Hutter, K. 1977b The foundation of thermodynamics, its basic postulates and implications. A review of modern thermodynamics. *Acta Mech.* **27**, 1–54.

Hutter, K., Jöhnk, K. and Svendsen, B. 1994 On interfacial transition conditions in two phase gravity flow. *Z. Angew. Math. Phys.* **45**, 746–762.

Iguchi, T. and LeFloch, P. G. 2003 Existence theory for hyperbolic systems of conservation laws with general flux functions. *Arch. Ration. Mech. Analysis* **168**, 165–244.

Illner, R. and Neunzert, H. 1987 The concept of irreversibility in the kinetic theory of gases. *Transp. Th. Statist. Phys.* **16**, 89–112.

Isakov, S. N. 1984 Nonanalytic features of the first order phase transition in the Ising model. *Commun. Math. Phys.* **95**, 427–443.

Jaynes, E. T. 1957 Information theory and statistical mechanics. *Phys. Rev.* **106**, 620–630, **108**, 171–190.

Jaynes, E. T. 1982 On the rationale of maximum entropy methods. *Proc. IEEE* **70**, 939–952.

Jaynes, E. T. 1992 The Gibbs Paradox. In *Maximum-Entropy and Bayesian Methods* (ed. G. Erickson, P. Neudorfer and C. R. Smith). Kluwer, Dordrecht.

Jensen, L. 2000 Large deviations of the asymmetric simple exclusion process. PhD thesis, New York University.

John, F. 1974 Formation of singularities in one-dimensional nonlinear wave propagation. *Commun. Pure Appl. Math.* **27**, 377–405.

Johnson, O. 2000 Entropy inequalities and the central limit theorem. *Stoch. Proc. Appl.* **88**, 291–304.

Johnson, O. and Suhov, Y. 2000 Entropy and convergence on compact groups. *J. Theor. Prob.* **13**, 843–857.

Joseph, D. D. and Preziosi, L. 1989 Heat waves. *Rev. Mod. Phys.* **61**, 41–73.

Jou, D., Casas-Vázquez, J. and Lebon, G. 1988 Extended irreversible thermodynamics. *Rep. Prog. Phys.* **51**, 1105–1179.

Junk, M. 2000 Maximum entropy for reduced moment problems. *Math. Models Meth. Appl. Sci.* **10**, 1001–1026.

Junk, M. 2002 Maximum entropy moment systems and Galilean invariance. *Continuum Mech. Thermodyn.* **14**, 563–576.

Kac, M. 1959 *Probability and Related Topics in Physical Sciences*, Lectures in Applied Mathematics, Proceedings of the Summer Seminar, Boulder, CO, 1957, Vol. I. Interscience, London and New York.

Katok, A. 1980 Lyapunov exponents, entropy and periodic orbits for diffeomorphisms. *Publ. Math. IHES* **51**, 137–174.

Katok, A. and Hasselblatt, B. 1995 *Introduction to the Modern Theory of Dynamical Systems*. Cambridge University Press.

Kelvin, Lord 1852 On a universal tendency in nature to the dissipation of mechanical energy. *Trans. R. Soc. Edinb.* **20**, 139–142. (Also in Kestin (1976), pp. 194–197.)

Kestin, J. 1976 *The Second Law of Thermodynamics*. Dowden, Hutchinson and Ross, Stroudsburg, PA.

Khinchin, A. I. 1957 *Mathematical Foundations of Information Theory* (translated by R. A. Silverman and M. D. Friedman). Dover, New York.

Klippel, A. and Müller, I. 1997 Plant growth—a thermodynamicist's view. *Continuum Mech. Thermodyn.* **9**, 127–142.

Kogan, M. N. 1967 In *Proc. 5th Symp. on Rarefied Gas Dynamics*, 1, Suppl. 4. Academic.

Kruzkov, S. 1970 First-order quasilinear equations with several space variables. *Mat. USSR Sbornik* **10**, 217–273.

Künsch, H. 1984 Nonreversible stationary measures for infinite interacting particle systems. *Z. Wahrscheinlichkeitsth.* **66**, 407–424.

Kurchan, J. 1998 Fluctuation theorem for stochastic dynamics. *J. Phys.* A **31**, 3719–3729.

Landsberg, P. T. 1964 A deduction of Carathéodory's Principle from Kelvin's Principle. *Nature* **201**, 485–486.

Lanford III, O. E. 1973 Entropy and equilibrium states in classical statistical mechanics. In *Statistical Mechanics and Mathematical Problems* (ed. A. Lenard), pp. 1–113. Lecture Notes in Physics, 20. Springer.

Lanford III, O. E. 1975 *Time evolution of large classical systems* (ed. J. Moser). Springer Lecture Notes in Physics, vol. 38, pp. 1–111.

Lanford III, O. E. and Ruelle, D. 1969 Observables at infinity and states with short range correlations in statistical mechanics. *Commun. Math. Phys.* **13**, 194–215.

Lax, P. D. 1957 Hyperbolic systems of conservation laws. *Commun. Pure Appl. Math.* **10**, 537–566.

Lax, P. D. 1971 Shock waves and entropy. In *Contributions to Functional Analysis* (ed. E. A. Zarantonello), pp. 603–634. Academic Press.

Lebowitz, J. L. 1993a Macroscopic laws and microscopic dynamics, time's arrow and Boltzmann's entropy. *Physica* A **194**, 1–97.

Lebowitz, J. L. 1993b Boltzmann's entropy and time's arrow. *Phys. Today* **46**, 32–38.

Lebowitz, J. L. 1994a Microscopic reversibility and macroscopic behavior: physical explanations and mathematical derivations. In *25 Years of Non-Equilibrium Statistical Mechanics, Proc. Sitges Conf., Barcelona, Spain, 1994*. Lecture Notes in Physics (ed. J. J. Brey, J. Marro, J. M. Rubí and M. San Miguel). Springer.

Lebowitz, J. L. 1994b Time's arrow and Boltzmann's entropy. *Physical Origins of Time Asymmetry* (ed. J. Halliwell and W. H. Zurek), pp. 131–146. Cambridge University Press.

Lebowitz, J. L. 1999 A century of statistical mechanics: a selective review of two central issues. *Rev. Mod. Phys.* **71**, 346–357.

Lebowitz, J. L. 1999 Microscopic origins of irreversible macroscopic behavior. *Physica* A **263**, 516–527.

Lebowitz, J. L. and Spohn, H. 1999 A Gallavotti–Cohen type symmetry in the large deviation functional for stochastic dynamics. *J. Statist. Phys.* **95**, 333–365.

Lebowitz, J. L., Prigogine, I. and Ruelle, D. 1999 Round table on irreversibility. In *Statistical Physics*, vol. XX, pp. 516–527, 528–539, 540–544. North-Holland.

Ledrappier, F. 1984 Proprietes ergodiques des mesures de Sinai. *Publ. Math. IHES* **59**, 163–188.

Ledrappier, F. and Strelcyn, J.-M. 1982 A proof of the estimation from below in Pesin entropy formula. *Ergod. Theory Dynam. Sys.* **2**, 203–219.

Ledrappier, F. and Young, L.-S. 1985a The metric entropy of diffeomorphisms. I. *Ann. Math.* **122**, 509–539.

Ledrappier, F. and Young, L.-S. 1985b The metric entropy of diffeomorphisms. II. *Ann. Math.* **122**, 540–574.

Ledrappier, F. and Young, L.-S. 1988 Entropy formula for random transformations. *Prob. Theor. Rel. Fields* **80**, 217–240.

LeFloch, P. G. 2002 *Hyperbolic Systems of Conservation Laws*. Birkhäuser.

Leslie, F. M. 1968 Some constitutive equations for liquid crystals. *Arch. Ration. Mech. Analysis* **28**, 265–283.

Levermore, C. D. 1996 Moment closure hierarchies for kinetic theories. *J. Statist. Phys.* **83**.

Lewis, J. T., Pfister, C.-E. and Sullivan, W. G. 1995 Entropy, concentration of probability and conditional limit theorems. *Markov Process. Rel. Fields* **1**, 319–386.

Li, M. and Vitanyi, P. 1993 *An Introduction to Kolmogorov Complexity and Its Applications*. Springer.

Li, M. and Vitanyi, P. 1997 *An Introduction to Kolmogorov Complexity and Its Application*. Springer.

Lieb, E. H. 1999 Some problems in statistical mechanics that I would like to see solved. 1998 IUPAP Boltzmann Prize Lecture. *Physica A* **263**, 491–499.

Lieb, E. H. and Yngvason, J. 1998 A guide to entropy and the second law of thermodynamics. *Notices Am. Math. Soc.* **45**, 571–581; mp_arc 98–339; arXiv math-ph/9805005.

Lieb, E. H. and Yngvason, J. 1999 The physics and mathematics of the second law of thermodynamics. *Phys. Rep.* **310**, 1–96. Erratum, **314** (1999), 669. (Also at http://xxx.lanl.gov/abs/cond-mat/9708200.)

Lieb, E. H. and Yngvason, J. 1999 The physics and mathematics of the second law of thermodynamics. *Phys. Rep.* **310**, 1–96; Erratum **314**, 669; arXiv cond-mat/9708200; http://www.esi.ac.at/ESI-Preprints.html #469.

Lieb, E. H. and Yngvason, J. 2000a A fresh look at entropy and the second law of thermodynamics. *Phys. Today* **53**, 32–37; mp_arc 00-123; arXiv math-ph/0003028. (See also Letters to the Editor, *Phys. Today* **53** (2000), 11–14, 106.)

Lieb, E. H. and Yngvason, J. 2000b The mathematics of the second law of thermodynamics. In *Visions in Mathematics, Towards 2000* (ed. A. Alon, J. Bourgain, A. Connes, M. Gromov and V. Milman). GAFA, Geom. Funct. Anal. Special Volume—GAFA, pp. 334–358; mp_arc 00-332.

Lieb, E. H. and Yngvason, J. 2002 The mathematical structure of the second law of thermodynamics. In *Contemporary Developments in Mathematics 2001* (ed. A. J. de Jong et al.), pp. 89–129. International Press.

Lindblad, G. 1988 Dynamical entropy for quantum systems. In *Quantum Probability and Applications II*. Lecture Notes in Mathematics, vol. 1303, p. 183.

Linnik, Yu. V. 1959 An information-theoretic proof of the central limit theorem with the Lindeberg condition. *Theor. Prob. Appl.* **4**, 288–299.

Lipster, R. S. and Shiryayev, A. N. 1978 *Statistics of Random Processes. II. Applications*. Springer.

Liu, I.-S. 1972 Method of Lagrange multipliers for exploitation of the entropy principle. *Arch. Ration. Mech. Analysis* **46**, 131–148.

Liu, I.-S. and Müller, I. 1972 On the thermodynamics and thermostatics of fluids in electromagnetic fields. *Arch. Ration. Mech. Analysis* **46**, 149–176.

Liu, I.-S. and Müller, I. 1984 Thermodynamics of mixtures of fluids. In *Rational Thermodynamics* (ed. C. Truesdell), pp. 264–285. Springer.

Liu, T.-P. 1976 The entropy condition and the admissibility of shocks. *J. Math. Analysis Appl.* **53**, 78–88.

Liu, T.-P. 1981 Admissible solutions of hyperbolic conservation laws. *Mem. AMS* **30**, No. 240.

Maes, C. 1999 The fluctuation theorem as a Gibbs property. *J. Statist. Phys.* **95**, 367–392.

Maes, C. and Netocny, K. 2003 Time-reversal and entropy. *J. Statist. Phys.* **110**, 269–310.

Maes, C. and Redig, F. 2000 Positivity of entropy production. *J. Statist. Phys.* **101**, 3–16.

Maes, C., Redig, F. and van Moffaert, A. 2000 On the definition of entropy production via examples. *J. Math. Phys.* **41**, 1528–1554.

Maes, C., Redig, F. and Verschuere, M. 2000b No current without heat. *J. Statist. Phys.* **106**, 569–587.

Maes, C., Redig, F. and Verschuere, M. 2001a Entropy production for interacting particle systems. *Markov Process. Rel. Fields* **7**, 119–134.

Maes, C., Redig, F. and Verschuere, M. 2001b From global to local fluctuation theorems. *Moscow Math. J.* **1**, 421–438.

Mañé, R. 1987 *Ergodic Theory and Differentiable Dynamics*. Ergebnisse der Mathematik und ihrer Grenzgebiete, 3. Folge, Vol. 8. Springer.

Maxwell, J. C. 1866 On the dynamical theory of gases. *Phil. Trans. R. Soc. Lond.* **157**, 49–88.

Maxwell, J. C. 1990–95 *Scientific Letters and Papers* (ed. P. M Harman). Cambridge University Press.

Meixner, J. 1943 Zur Thermodynamik der irreversiblen Prozesse. *Z. Phys. Chem.* **538**, 235–263.

Meixner, J. 1969a Thermodynamik der Vorgänge in einfachen fluiden Medien und die Charakterisierung der Thermodynamik irreversibler Prozesse. *Z. Phys.* **219**, 79–104.

Meixner, J. 1969b Processes in simple thermodynamic materials. *Arch. Ration. Mech. Analysis* **33**, 33–53.

Mendoza, E. (ed.) 1960 *Reflections on the Motive Power of Fire by Sadi Carnot and Other Papers on the Second Law of Thermodynamics by E. Clapeyron and R. Clausius.* Dover, New York.

Minlos, R., Roelly, S. and Zessin, H. 2001 Gibbs states on space-time. *Potential Analysis* **13**(4).

Moulin-Ollagnier, J. and Pinchon, D. 1977 Free energy in spin-flip processes is non-increasing. *Commun. Math. Phys.* **55**, 29–35.

Müller, I. 1967a Zum Paradox der Wärmeleitungstheorie. *Z. Phys.* **198**.

Müller, I. 1967b On the entropy inequality. *Arch. Ration. Mech. Analysis* **26**, 118–141.

Müller, I. 1971 Die Kältefunktion, eine universelle Funktion in der Thermodynamik viskoser wärmeleitender Flüssigkeiten. *Arch. Ration. Mech. Analysis* **40**, 1–36.

Müller, I. 1985 *Thermodynamics*. Pitman.

Müller, I. 2001 *Grundzüge der Thermodynamik—mit historischen Anmerkungen*, 3rd edn. Springer.

Müller, I. and Ruggeri, T. 1993 *Extended Thermodynamics*. Springer Tracts in Natural Philosophy, Vol. 37. Springer.

Müller, I. and Ruggeri, T. 1998 *Rational Extended Thermodynamics*, 2nd edn. Springer.

Müller, I. and Ruggeri, T. 2003 Stationary heat conduction in radially symmetric situations—an application of extended thermodynamics. *J. Non-Newtonian Fluids*. (In the press.)

Müller, I., Reitebuch, D. and Weiss, W. 2002 Extended thermodynamic, consistent in order. *Continuum Mech. Thermodyn*. **15**, 113–146.

Myung, I. J., Balasubramanian, V. and Pitt, M. A. 2000 Counting probability distributions: differential geometry and model selection. *Proc. Natl Acad. Sci*. **97**, 11 170–11 175.

Nernst, W. 1924 *Die theoretischen und experimentellen Grundlagen des neuen Wärmesatzes*, 2nd edn. W. Knapp, Halle.

Neves, E. J. and Schonmann, R. H. 1991 Behavior of droplets for a class of Glauber dynamics at very low temperatures. *Commun. Math. Phys*. **137**, 209–230.

Newhouse, S. 1988 Entropy and volume. *Ergod. Theory Dynam. Sys*. **8**, 283–299.

Nietzsche, F. 1872 *Kritische Gesamtausgabe*, Vol. 3, A. Abt. Nachgelassene Fragmente. Walter de Gruyter, Berlin, New York.

Nohre, R. 1994 Some topics in descriptive complexity. PhD thesis, Linkoping University, Linkoping, Sweden.

Nunziato, J. W. and Passman, S. L. 1981 A multiphase mixture theory for fluid-saturated granular materials. In *Mechanics of Structured Media A* (ed. A. P. S. Selvadurai), pp. 243–254. Elsevier, Amsterdam.

Ohya, M., and Petz, D. 1993 *Quantum Entropy and Its Use*. Springer.

Olivieri, E. and Scoppola, E. 1995 Markov chains with exponentially small transition probabilities: first exit problem from a general domain. I. The reversible case. *J. Statist. Phys*. **79**, 613–647.

Olivieri, E. and Scoppola, E. 1996a Markov chains with exponentially small transition probabilities: first exit problem from a general domain. II. The general case. *J. Statist. Phys*. **84**, 987–1041.

Olivieri, E. and Scoppola, E. 1996b Metastability and typical exit paths in stochastic dynamics. In *Proc. 2nd European Congress of Mathematics, Budapest, 22–26 July*. Birkhäuser.

Park, Y. M. 1994 *Lett. Math. Phys*. **36**, 63.

Passman, S. L., Nunziato, J. W. and Walsh, E. K. 1984 A theory of multiphase mixtures. In *Rational Thermodynamics* (ed. C. Truesdell), pp. 286–325. Springer.

Pauli, W. 1973 *Thermodynamics and the Kinetic Theory of Gases* (ed. C. P. Enz). Pauli Lectures on Physics, Vol. 3. MIT Press, Cambridge, MA.

Pauling, L. 1935 The structure and entropy of ice and of other crystals with some randomness of atomic arrangement. *J. Am. Chem. Soc*. **57**, 2680–2684.

Penrose, O. and Lebowitz, J. L. 1971 Rigorous treatment of metastable states in van der Waals theory. *J. Statist. Phys*. **3**, 211–236.

Penrose, O. and Lebowitz, J. L. 1987 Towards a rigorous molecular theory of metastability. In *Fluctuation Phenomena* (ed. E. W. Montroll and J. L. Lebowitz), 2nd edn. North-Holland.

Penrose, R. 1989 *The Emperor's New Mind*. Oxford University Press.

Pesin, Ya. 1977 Characteristic Lyapunov exponents and smooth ergodic theory. *Russ. Math. Surv*. **32**, 55–114.

Petersen, K. 1989 *Ergodic Theory* (corrected reprint of the 1983 original). Cambridge Studies in Advanced Mathematics, vol. 2. Cambridge University Press.

Petersen, K. and Schmidt, K. 1997 Symmetric Gibbs measures. *Trans. Am. Math. Soc.* **349**, 2775–2811.

Petz, D. and Mosonyi, M. 2001 Stationary quantum source coding. *J. Math. Phys.* **42**, 4857.

Planck, M. 1897 *Vorlesungen über Thermodynamik.* Veit, Leipzig.

Planck, M. 1901 Über die Elementarquanta der Materie und der Elektrizität. *Annalen Phys.* **4**, 564–566.

Planck, M. 1912 *Über neuere thermodynamische Theorien* (Nernst'sches Wärmetheorem und Quantenhypothese, Read 16 December 1911 at the German Chemical Society, Berlin). Akad. Verlagsges., Leipzig.

Planck, M. 1926 Über die Begrundung des zweiten Hauptsatzes der Thermodynamik. *Sitzungsberichte der preußischen Akademie der Wissenschaften, Math. Phys. Klasse,* pp. 453–463.

Preston, C. J. 1976 *Random Fields.* Lecture Notes in Mathematics, Vol. 534. Springer.

Pugh, C. and Shub, M. 1989 Ergodic attractors. *Trans. AMS* **312**, 1–54.

Rao, C. R. 1973 *Linear Statistical Inference and its Applications.* Wiley.

Renyi, A. 1970 *Probability Theory.* North-Holland, Amsterdam.

Rezakhanlou, F. 1991 Hydrodynamic limit for attractive particle systems on Z^d. *Commun. Math. Phys.* **140**, 417–448.

Rissanen, J. 1978 Modeling by shortest data description. *Automatica* **14**, 465–471

Rissanen, J. 1986 Stochastic complexity and modeling. *Ann. Stat.* **14**, 1080–1100

Rissanen, J. 1989 *Stochastic Complexity in Statistical Inquiry.* World Scientific.

Rissanen, J. 1996 Fisher information and stochastic complexity. *IEEE Trans. Information Theory* **IT-42**, 40–47.

Rissanen, J. 2000 MDL denoising. *IEEE Trans. Information Theory* **IT-46**, 2537–2543.

Rissanen, J. 2001 Strong optimality of normalized ML models as universal codes and information in data. *IEEE Trans. Information Theory* **IT-47**, 1712–1717.

Robb, A. A. 1921 *The Absolute Relations of Time and Space.* Cambridge University Press.

Roberts, R. D. and Luce, F. S. 1968 Axiomatic thermodynamics and extensive measurement. *Synthese* **18**, 311–326.

Ruelle, D. 1967 A variational formulation of equilibrium statistical mechanics and the Gibbs phase rule. *Commun. Math. Phys.* **5**, 324–329.

Ruelle, D. 1978a *Thermodynamic Formalism.* Addison Wesley, Reading, MA.

Ruelle, D. 1978b An inequality of the entropy of differentiable maps. *Bol. Sc. Bra. Mat.* **98**, 83–87

Ruelle, D. 1999 Smooth dynamics and new theoretical ideas in nonequilibrium statistical mechanics. *J. Statist. Phys.* **95**, 393–468.

Sanov, I. N. 1957 On the probability of large deviations of random variables (in Russian). *Mat. Sbornik* **42**, 11–44. (English translation *Selected Translations in Mathematical Statistics and Probability I*, 1961, pp. 213–244.)

Schonmann, R. H. 1992 The pattern of escape from metastability of a stochastic Ising model. *Commun. Math. Phys.* **147**, 231–240.

Schonmann, R. H. and Shlosman, S. B. 1998 Wulff droplets and the metastable relaxation of kinetic Ising models. *Commun. Math. Phys.* **194**, 389–462.

Schrödinger, E. 1950 Irreversibility. *Proc. R. Irish Acad.* A **53**, 189–195.

Schrödinger, E. 1944 *What Is Life?* Cambridge University Press.

Schumacher, B. 1995 *Phys. Rev. A* **51**, 2738.
Schuster, H. G. 1995 *Deterministic Chaos*, 3rd edn. VCH, Weinheim.
Serre, D. 1999, 2000 *Systems of Conservation Laws*, Vols 1 and 2. Cambridge University Press.
Sewell, G. L. 1980 Stability, equilibrium and metastability in statistical mechanics. *Phys. Rep.* **57**, 307–342
Shannon, C. E. 1948 A mathematical theory of communication. *Bell Syst. Technol. J.* **27**, 379–423, 623–657. (Reprinted *Key Papers in the Development of Information Theory* (ed. D. Slepian), IEEE Press, New York (1974).)
Simon, B. 1993 *The Statistical Mechanics of Lattice Gases*, Vol. 1. Princeton University Press.
Sinai, Ya. G. 1972 Gibbs measures in ergodic theory. *Russ. Math. Surv.* **27**, 21–69.
Smale, S. 1967 Differentiable dynamical systems. *Bull. Am. Math. Soc.* **73**, 747–817.
Smith, C. and Wise, M. N. 1989 *Energy and Empire: a Biographical Study of Lord Kelvin*. Cambridge University Press.
Smoller, J. 1994 *Shock Waves and Reaction–Diffusion Equations*, 2nd edn. Springer.
Spengler, O. 1919 Die Entropie und der Mythos der Götterdämmerung. In *Der Untergang des Abendlandes. Umrisse einer Morphologie der Weltgeschichte*, Kapitel VI, Vol. 1, 3rd edn. C. H. Beck'sche Verlagsbuchhandlung, München.
Spitzer, F. 1971 Random fields and interacting particle systems. Notes on Lectures Given at the 1971 MAA Summer Seminar Williamstown, MA, Mathematical Association of America.
Struchtrup, H. 2003 Heat transfer in the transition regime: solution of boundary value problems for Grad's moment equations via kinetic schemes. *Phys. Fluids*. (In the press.)
Struchtrup, H. and Weiss, W. 1998 Maximum of the local entropy production becomes minimal in stationary processes. *Phys. Rev. Lett.* **80**, 5048–5051.
Svendsen, B. and Hutter, K. 1995 On the thermodymics of a mixture of isotropic materials with constraints. *Int. J. Engng Sci.* **33**, 2021–2054.
Sznitman, A.-S. 1998 *Brownian Motion, Obstacles and Random Media*. Springer.
Tartar, L. C. 1983 The compensated compactness method applied to systems of conservation laws. In *Systems of Nonlinear Partial Differential Equations* (ed. J. M. Ball), pp. 263–285. Reidel, Dordrecht.
Trouvé, A. 1996 Cycle decompositions and simulated annealing. *SIAM J. Control Optim.* **34**, 966–986.
Truesdell, C. 1957 Sulle basi della termomeccanica. *Rend. Accad. Lincei* **22**(8), 33–88, 158–166. (English translation in *Rational Mechanics of Materials*, Gordon & Breach, New York, 1965.)
Truesdell, C. 1969 *Rational Thermodynamics*. McGraw-Hill, New York.
Truesdell, C. 1986 What did Gibbs and Carathéodory leave us about thermodynamics? In *New Perspectives in Thermodynamics* (ed. J. Serrin), pp. 101–123. Springer.
Truesdell, C. and Noll, W. 1965 *The Non-Linear Field Theories of Mechanics*. Flügge's Handbuch der Physik, Vol. III/3. Springer.
Uffink, J. 2001 Bluff your way in the second law of thermodynamics. *Stud. Hist. Phil. Mod. Phys.* **32**, 305–394. (See also Uffink's contribution in this book, Chapter 7.)
Uffink, J. 2003 Bluff your way in the second law of thermodynamics. *Stud. Hist. Phil. Mod. Phys.* **32**, 305–394.
Vajda, I. 1989 *Theory of Statistical Inference and Information*. Kluwer, Dordrecht.

van der Waals, J. D. and Kohnstamm, Ph. 1927 *Lehrbuch der Thermostatik*. Barth, Leipzig.

van Enter, A. C. D., Fernandez, R. and Sokal, A. D. 1993 Regularity properties and pathologies of position-space renormalization-group transformations: scope and limitations of Gibbsian theory. *J. Statist. Phys.* **72**, 879–1167.

Ventcel, A. D. and Freidlin, M. I. 1970 Small random perturbations of dynamical systems. *Russ. Math. Surv.* (translation of *Usp. Mat. Nauk*) **25**, 3–55.

Vershik, A. M. 1974 A description of invariant measures for actions of certain infinite-dimensional groups (in Russian). *Dokl. Akad. Nauk SSSR* **218**, 749–752. (English translation *Sov. Math. Dokl.* **15**, 1396–1400 (1975).)

Voiculescu, D. 1995 *Commun. Math. Phys.* **170**, 249.

von Neumann, J. 1955 *Mathematical Foundations of Quantum Mechanics*, pp. 398–416. Princeton University Press. (Translated from German edition (1932, Springer) by R. T. Beyer.)

Walters, P. 1982 *An Introduction to Ergodic Theory*. Graduate Text in Mathematics, Vol. 79. Springer.

Wang, Y. and Hutter, K. 1999a Shearing flows in a Goodman–Cowin type granular material—theory and numerical results. *Particulate Sci. Technol.* **17**, 97–124.

Wang, Y. and Hutter, K. 1999b A constitutive model for multi-phase mixtures and its application in shearing flows of saturated soil–fluid mixtures. *Granular Matter* **1**(4), 163–181.

Wang, Y. and Hutter, K. 1999c Comparison of two entropy principles and their applications in granular flows with/without fluid. *Arch. Mech.* **51**, 605–632.

Wehrl, A. 1978 *Rev. Mod. Phys.* **50**, 221.

Weiss, W. 1990 Zur Hierarchie der erweiterten Thermodynamik. Dissertation, TU Berlin.

Weiss, W. and Müller, I. 1995 Light scattering and extended thermodynamics. *Continuum Mech. Thermodyn.* **7**, 123–178.

Wick, W. D. 1982 Monotonicity of the free energy in the stochastic Heisenberg model. *Commun. Math. Phys.* **83**, 107–122.

Woods, L. C. 1973 The bogus axioms of continuum mechanics. *Bull. Math. Applic.* **9**, 40.

Yomdin, Y. 1987 Volume growth and entropy. *Israel J. Math.* **57**, 285–300.

Young, L.-S. 1990 Some large deviation results for dynamical systems. *Trans. Am. Math. Soc.* **318**, 525–543.

Young, L.-S. 2003 What are SRB measures, and which dynamical systems have them? *J. Stat. Phys.* **108**, 733–754.

Index

additivity, 153
adiabatic
    accessibility, 134, 138, 142, 151
    processes, 128, 135
adiabats, 165
admissibility criteria, 112
affine equivalence, 155
Alicki–Fannes entropy, 292
alphabet, 301
Alt, H. W., 23
ammonia synthesis, 32
ARMA models, 306
arrow of time, 123, 132
atmospheres
    planetary, 28
Avogadro's number, 192

balance laws, 64
    source-free, 64
Bernoulli
    class, 310
    scheme, 329
Bernstein collection, 335
BGK
    equation, 85, 90
    model, 103
binary mixture, 64
binomial transformation, 332
bit-streams, 288
Boltzmann, L. E., 23, 37, 38
    constant, 24
    entropy, 253
    equation, 23, 84, 270
Born, M., 130
Bosch, K., 33
branching random walk in a random environment, 217
Burgers equation, 110, 213
BV space, 111

Callen, H. B., 133
cancellation law, 154

canonical relations
    thermodynamic, 70
Carathéodory, C., 22, 122, 133, 149
    Principle, 134, 137
Carnot, N. L. S., 20, 124
    cycles, 19, 125
        efficiency of, 19
    Principle, 125
    Theorem, 125–127
Cattaneo, C., 93
    equation, 61, 94
CDCN approach, 67, 68, 71
central limit theorem, 47
change of state
    quasistatic, 135
characteristic speeds
    finite, 98
chemical reactions, 65, 183
χ^2-test, 42
choice method
    random, 113
Clausius, R. J. E., 19, 122, 124, 149
Clausius–Duhem inequality, 57, 58, 60, 66, 74, 81, 109
Clausius–Kelvin Principle, 126
closed systems, 73
closure, 85
clumping, 223
CNT entropy, *see* Connes–Narnhofer–Thirring entropy
code, 301
    block, 40, 49
    prefix, 39, 301, 302
codeword, 301
coin tossing, 316
Coleman, B. D., 81
Coleman–Noll
    approach, 67, 69, 72, 74
    theory, 71
comparability, 152
Comparability Hypothesis, 140, 143

Comparison Hypothesis, 155
compensated compactness, 119
complexity, 300
   algorithmic, 279
   conditional, 305
   geometric, 309
   Kolmogorov, 300, 303–305
   mean, 302
compound system, 150
compression rate, 279
concavity, 97
condensation, 30
conductivity
   thermal, 81
Connes–Narnhofer–Thirring entropy, 290
conservation laws, 109, 110
consistency, 154
constitutive
   equations, 59, 60, 73
   functions, 66
   relations, 58, 61, 65, 67, 69, 70, 72–74, 109
   relations of Navier–Stokes and Fourier, 82, 90
   theory, 96
constraint
   conditions, 64
   relations, 67
contraction principle, 202
convex
   combination, 161
   conjugates, 44
convexity, 97
   requirement, 96
Cramér transform, 44
Csiszár's conditional limit theorem, 47, 220
Curie–Weiss model, 237
curve fitting, 306
cutting and stacking, 332
cycle
   reversible, 126
cyclic processes, 127

decompositions of states, 290
denoising, 311
density
   reference, 109

description length
   minimum, 310
diffusion
   paradox of, 95
dimension of a measure, 319
dimensions
   partial, 320
disorder
   measure, 28
dissipation, 251
   of energy, 127
distribution
   empirical, 38
   exponential, 45
   function, 23
   Poisson, 45
distribution of molecules, 25
Duhem, P. M. M., 81
dynamical system, 313
   random, 321

Eckart, C. H., 80
Eddington, A. S., 123
Einstein, A., 32, 136, 193
elasticity
   entropic, 26
empirical
   distribution, 38
   measure, 216
   temperature, 72
energy
   free, 29, 50, 204, 236, 237, 239
   Helmholtz free, 109
energy function, 244
ensembles
   equivalence, 51
entropy, 153, 203, 206, 251, 269, 303
   Alicki–Fannes, 292
   Bernoulli, 237
   Boltzmann, 253
   configurational, 26
   Connes–Narnhofer–Thirring, 290
   convex, 116
   density, 49
   differential, 45
   flux, 81, 113
      modified, 62
      nonconvective, 91
   functional, 199
   Gibbs, 274

growth, 26
inequality, 60, 72, 96, 114
  general, 72
Kolmogorov–Sinai, 52, 204, 279
maximization, 43, 89
maximization of nonequilibrium, 83
mean, 239
metric, 315
of reaction, 34
of the partition α, 315
principle of increasing, 45, 52
production density, 81, 91
quantum Boltzmann, 274
rate condition, 117
relative, 41, 200, 203, 205, 210, 216, 325
Shannon, 279
specific, 91
techniques, 215
theories, 63
time, 233
topological, 52, 314
von Neumann, 274, 279
entropy condition
  Lax, 112
  Liu, 112, 116, 117
Entropy Conjecture, 323
entropy principles, 57, 63, 73, 74
  generalized CDCN approach, 58
  Lieb and Yngvason, 141
  Müller, 58, 74
  Müller–Liu, 71
  weak version, 143
entropy production
  Markov chains, 45
  minimum, 89
  principle of minimal, 92
equation of state
  caloric, 21, 81
  thermal, 21, 81
equations of balance
  of entropy, 96
  of mechanics, 95
  of thermodynamics, 95
equilibrium
  principle of local, 80
  thermodynamic, 50, 70
  thermostatic, 61
equilibrium states, 150

equipartition
  asymptotic, 40, 49
equivalence
  affine, 155
equivalence of ensembles, 51
ergodic theory, 52
ergodicity, 330, 334
evaporation, 30
exclusion, 52
  process, 263
exponential
  distribution, 45
  families, 44
  growth rates, 218
exponents
  Lyapunov, 317
extended irreversible
  thermodynamics, 58, 60, 62, 70, 74
extensivity, 153
extent of reaction, 33

families
  exponential, 44
Fick's law, 95
field equations, 66
first law of thermodynamics, 20, 149
Fisher information matrix, 308
fluctuation theorem, 259
fluctuations, 263
fluid, viscous, heat-conducting, 82
flux, 110
  entropy, 81
  thermodynamic, 81
forces
  entropic, 26
  reaction, 68, 72
  thermodynamic, 81
formation energy, 243
Fourier's law, 61, 81, 93
14-moment system, 98
frames
  Galilei, 96
free energy, 29, 50, 204, 236, 237, 239

Galilei, G., 88
  frames, 96
gas, viscous, heat-conducting, 83
gases
  kinetic theory, 59
Gaussian models, 311

generator, 331
Gibbs, J. W., 21, 122
    distributions, 44, 258
        microcanonical, 51
    entropy, 274
    equation, 21, 61, 70, 74, 80, 82
        generalized, 62
    measure, 49, 203, 204
    Principle, 132
    relation, 70, 72, 73
Gibbs–Jaynes Principle, 44, 50, 220
Giles, R., 150
Glauber processes, 52
Grad's 13-moment approximation, 62
Grad, H., 84
growth rates, 219
    exponential, 218
    global, 219
    local, 219

H-Theorem, 190
Haber, F., 33
halting problem
    undecidability, 304
heat conduction
    paradox of, 93, 95
heat flux, 81
heat of reaction, 32
heating, 20
Helmholtz free energy, 109
Hoeffding Theorem, 42
Holley Theorem, 52
horseshoe, 324
Host, B., 335
hydrodynamic scaling, 211
hyperbolic
    strictly, 110
hyperbolic form
    symmetric, 96
hyperbolic system
    symmetric, 97
hyperbolicity
    symmetric, 98
hypothesis
    Onsager, 102

I-divergence, 41
I-projection, 43
i-shock, 112

ideal gas
    monatomic, 98
information, 39, 49, 309
    Kolmogorov, 306
    Kullback–Leibler, 41, 303
    Shannon–Wiener, 303
information gain, 41
information matrix
    Fisher, 308
integrability condition, 21
interactions
    electro-mechanical, 71
invariance
    scaling, 154
invariants
    Riemann, 114
irrecoverability, 123, 127
    second law, 129
irreversibility, 57–60, 121, 138, 141, 162, 252
    Planck's sense, 143
irreversible processes, 57, 60, 74, 141
irreversible thermodynamics, 57, 60, 63, 70
    extended, 58
    ordinary, 58
irrevocable changes, 123

Joule's constant, 126
Joule–Mayer principle, 126
jump conditions
    Rankine–Hugoniot, 112
jump set, 111

k-mixing, 330
Kawasaki processes, 52
Kelvin, Lord, 88, 122, 124, 149
    Principle, 130, 137
    temperature, 88
kinetic
    temperature, 82, 87, 92
    theory of gases, 59, 71
Kogan, M. N., 84
Kolmogorov, A. N., 329
    complexity, 300, 303–305
    information, 306
    minimal sufficient statistics, 306
    sufficient statistics, 303
Kolmogorov–Sinai entropy, 52, 204, 279

Kraft inequality, 301
Krieger, W., 331
Kullback–Leibler, 41, 303

Lagrange multipliers, 44, 69, 72, 97
Lagrangian formulation, 109
Landsberg, P. T., 137
large deviations, 200, 201, 203, 206
  theory, 43, 215
large-deviation
  principle, 201, 326
  rate functions, 200
lattice gas, 263
Legendre–Fenchel transform, 44, 97
Lieb, E. H., 122
Lieb–Yngvason approach, 144
light scattering, 101
  application of extended
    thermodynamics to, 100
limit
  quasistatic, 124
liquid crystals, 65, 73
localization versus delocalization, 224
Lord Kelvin, *see* Kelvin, Lord
Lyapunov exponents, 317

Markov chains, 45, 46
  quantum, 288
maximum
  entropy, 43, 89, 233
  likelihood estimate, 307
  nonequilibrium entropy, 83
maximum likelihood
  normalized, 308
Maxwell, J. C., 23
Maxwell–Boltzmann, 37
MDL, *see* minimum description
  length
mean
  complexity, 302
  entropy, 330
  free path, 91
  regression of a microscopic
    fluctuation, 101
measure
  empirical, 216
  Sinai–Ruelle–Bowen, 318
measure-preserving transformations, 329
Mendoza, E., 125

mesoscopic systems, 192
metaphysical 'principle', 68
metastability, 233–235, 237, 239,
  241–243, 248
microcanonical Gibbs distributions,
  51
micromorphic, 65
micropolar, 65
microscopic fluctuation
  mean regression, 101
microstate, 25
minimum
  description length, 310
  entropy production, 89
Minkowski spacetime, 136
minmax principle, 92
mixing, 183
  weakly, 335
mixtures, 71
Mizel, V., 81
model
  ARMA, 306
  BGK, 103
  cost, 309
  parametric, 306
  selection, 310
moment equations
  for 13 moments, 85
  for 14 moments, 89
moments
  balance equations, 84
moments of the distribution function,
  98
momentum, 109
monatomic ideal gas, 98
monotonicity, 153
Müller's entropy principle, 58, 74
Müller–Liu's entropy principle, 71

Navier–Stokes's law, 61, 81, 95
Navier–Stokes–Fourier
  equations, 86
  theory, 86
Nernst, H. W., 31
NML, *see* normalized maximum
  likelihood
Noll, W., 81
nonequilibrium
  states, 190
  thermodynamic, 73

nonlinearity
  genuine, 111
nonTRI, 124, 127

Onsager, L., 101
  hypothesis, 102
open systems, 73
optimal selection from possible
  growth strategies, 226
Ornstein, D., 331

paradox
  of diffusion, 95
  of heat conduction, 93, 95
particle density
  global, 217
  local, 217
partition
  generating, 330
partitions of unit, 290
Pascal transformation, *see* binomial
  transformation
Pauling, L., 192
Pearson's χ^2-test, 42
periodic points, 323
perpetuum mobile, 125
Pesin's Formula, 317
Pfeffer tube, 30
phase
  changes, 65
  transitions, 223, 236, 237
phenomenological equations, 81
  Fourier, 86
  Navier–Stokes, 86
photosynthesis, 33
Pinsker conjecture, *see* weak Pinsker
  conjecture
placement, 108
Planck, M. K. E., 25, 121, 149, 192
  constant, 25
Poincaré recurrence, 270
Poisson distribution, 45
porous mixture, 64
possible worlds, 132
potential
  thermodynamic, 73
prediction, 306
pressure, 50, 235
  dynamic, 81
  topological, 52, 326

Prigogine, I., 92
principle of
  increasing entropy, 45, 52
  local equilibrium, 80
  minimal entropy production, 92
  relativity, 96
processes
  adiabatic, 128, 135
  cyclic, 127
  exclusion, 52, 263
  irreversible, 57, 60, 74, 141
  quasistatic, 124
  reversible, 60, 74, 142
  spin-flip, 52
  thermodynamic, 66, 72, 96
propagation speed, 94
pulse speed, 98, 100
  lower bound, 100

q-bit, 282
quantum
  Boltzmann entropy, 274
  Markov chains, 288
  statistics, 273
quasistatic
  change of state, 135
  limit, 124
  processes, 124

random
  choice method, 113
  dynamical system, 321
  environment, 217
  SRB measures, 322
random walk, 217
Rankine–Hugoniot jump conditions,
  112
rate function, 202, 204, 208, 233
rates of escape, 326
recombination, 154
recurrence
  Poincaré, 270
reflexivity, 154
relativity
  principle of, 96
requirement of convexity, 96
residual inequality, 70
reversibility
  Planck's sense, 123, 142
  time, 191

reversible, 124
  cycle, 126
  processes, 60, 74, 142
Riemann
  invariants, 114
Rokhlin, V., 331
Ruelle's Inequality, 317
Ruelle–Föllmer Theorem, 50

Sanov's Theorem, 43, 216
saturation condition, 64
scaled copy, 151
scaling
  hydrodynamic, 211
  invariance, 154
scattered light
  spectral distribution, 101
scattering spectrum, 101
Schrödinger, E., 34
second law of thermodynamics, 21, 43, 45, 57, 58, 63, 66, 67, 72, 73, 109, 121, 149, 251, 270
  Clausius' version, 128
  irrecoverability, 129
  Müller–Liu version, 58
sectors
  forward, 161
semipermeable membrane, 30
sensitivity to initial conditions, 285
set
  regular, 111
  residual, 111
Shannon, C. E., 37, 39, 329
  entropy, 279
Shannon–McMillan–Breiman Theorem, 49, 315
Shannon–Wiener information, 303
shift of finite type, 323
shock
  structures, 103
  tube experiment, 103
Sinai, Ya. G., 330
Sinai–Ruelle–Bowen measure, 318
source-coding theorem, 39, 40
spacetime
  Minkowski, 136
specific entropy, 91
spectral distribution
  in moderately rarefied gas, 103
  scattered light, 101

spectrum
  simple Lebesgue, 330
speed of propagation, 94
splitting, 154
SRB measure, 318
  random, 322
stability, 154
stationary
  fields, 48
  processes, 48
statistical mechanics, 59, 190
statistics
  Kolmogorov minimal sufficient, 306
  Kolmogorov sufficient, 303
  quantum, 273
Stein's lemma, 41
strain, 108
stress, 109
  deviatoric, 81
summarizing property, 305
survival of the fittest, 219
survival versus extinction, 223
symmetric
  hyperbolic form, 96
  hyperbolic system, 97
  hyperbolicity, 98
symmetrizable systems, 114
systems
  13-moment, 98
  14-moment, 98
  closed, 73
  dynamical, 313
  mesoscopic, 192
  open, 73
  random dynamical, 321
  simple, 135, 136, 160
  thermodynamic, 150

telegraph equation, 94
temperature
  absolute, 88
  defined by entropy, 177
  empirical, 72
  Kelvin, 88
  kinetic, 92
  thermodynamic, 82, 88, 92
test, 41

thermal
  conductivity, 81
  contact, 166
  equation of state, 21
  join, 167
  motion, 25
  splitting, 167
thermodynamic
  canonical relations, 70
  equilibrium, 70
  fluxes, 81
  forces, 81
  nonequilibrium, 73
  potential, 73
  processes, 66, 72, 96
  system, 150
  temperature, 82, 88, 92
thermodynamics, 74
  classical irreversible, 62
  entropy free, 63
  extended, 61, 93, 95
  extended irreversible, 58, 60, 70, 74
  first law, 20, 149
  irreversible, 57, 60, 63, 70
  irreversible processes, 61, 74, 79, 80
  ordinary irreversible, 58, 60
  phenomenological, 57
  rational, 58, 63, 74, 81
  second law, 21, 43, 45, 57, 58, 63, 66, 67, 72, 73, 109, 121, 149, 251, 270
  third law, 31
  zeroth law, 82, 88, 149, 169
thermoelasticity
  isothermal, 110
thermomechanics
  isothermal, 109
thermostatic equilibrium, 61
thermostatics, 59–61, 70, 74
third law of thermodynamics, 31
13-moment
  distribution, 83
  system, 98
Thomson, W., *see* Kelvin, Lord
time
  arrow of, 123, 132
  entropy, 233
  ravages of, 123
  reversibility, 191
    time-reversal invariant, 258

time-reversal invariant, 122, 124
TIP, *see* thermodynamics of irreversible processes
topological
  entropy, 52, 314
  pressure, 52, 326
transitivity, 154
TRI, *see* time-reversal invariant
Truesdell, C., 132

undecidability of the halting problem, 304

van der Waals, J. D., 30
  theory, 235
vanishing viscosity method, 113, 115
variational
  calculus, 215
  formula, 44, 50, 52, 204
  principles, 50, 132, 316
velocity, 108
  diffusion, 64
Vershik, A. M., 332
viscosity
  bulk, 81
  shear, 81
viscosity method
  vanishing, 113, 115
volume
  fractions, 64
  growth, 322
von Neumann, J., 37
  entropy, 274, 279

wave
  acceleration, 98
  front tracking method, 113
  tracing method, 113
wavelet transforms, 311
weak Pinsker conjecture, 331
weakly mixing, 335
well-posedness, 98
Wiener measure, 45
Woods, L. C., 82
work coordinates, 160
working, 20

Yngvason, J., 122

zeroth law of thermodynamics, 82, 88, 149, 169